XDSL Architecture

Padmanand Warrier

Balaji Kumar

McGraw-Hill
New York   San Francisco   Washington, D.C.
Auckland   Bogotá   Caracas   Lisbon   London   Madrid
Mexico City   Milan   Montreal   New Delhi   San Juan
Singapore   Sydney   Tokyo   Toronto

McGraw-Hill

A Division of The McGraw-Hill Companies

Copyright © 2000 by The McGraw-Hill Companies, Inc. All rights reserved. Printed in the United States of America. Except as permitted under the United States Copyright Act of 1976, no part of this publication may be reproduced or distributed in any form or by any means, or stored in a data base or retrieval system, without the prior written permission of the publisher.

1 2 3 4 5 6 7 8 9 0   AGM/AGM   9 0 4 3 2 1 0 9

ISBN 0-07-135006-3

The sponsoring editor for this book was Steven Elliot and the production supervisor was Clare Stanley. It was set in Life by Patricia Wallenburg.

Printed and bound by Quebecor/Martinsburg.

Throughout this book, trademarked names are used. Rather than put a trademark symbol after every occurrence of a trademarked name, we use names in an editorial fashion only, and to the benefit of the trademark owner, with no intention of infringement of the trademark. Where such designations appear in this book, they have been printed with initial caps.

This book is printed on recycled, acid-free paper containing a minimum of 50% recycled, de-inked fiber.

To Swami who made it all come together

—Padmanand Warrier

This is dedicated to Anup and Durga

—Balaji Kumar

Acknowledgments

We would like to thank several people who have contributed to making this book possible. First, we would like to acknowledge the help of Jay Ranade and Steve Elliott at McGraw-Hill. Next, we would like to acknowledge the effort of the reviewers—Ashok Baragi, David Dorsey, Julia Marsh, Babak Nabili, and Deb Robinson. We would also like to acknowledge the efforts of several people who contributed to the success of the UAWG, and all the people at Texas Instruments, MCI and Telehub whom I have grown to know over several months. Specifically, we would like to acknowledge the assistance of the following individuals: David Allan, Todd Andreini, Jim Carlo, James Collinge, Terry Cole, Kevin Cone, Gene Edmond, Robert Ferguson, Mark Jeffrey, Tim Kwok, Chris Hansen, Peter Melsa, Farhad Mighani, Tim Murphy, Prasad Nimmagadda, Mike Polley, Paul Shieh, Peter Silverman, Bill Timm, Danny Van Bruyssel, Herman Verbueken, Ray Wang and Ben Wiseman. Finally, no work of this nature is possible without the support and understanding of our family—Latha, Vishnu, Durga, and Anup.

Contents

Contents

Introduction

The 1990s have inaugurated the second revolution of telecommunications—high-speed access. Changes have already occurred so rapidly in the telecommunications and computer environment that it is hard to believe more is to come during this decade. This book gives the reader a detailed look at DSL technologies and architectures that will enable the future of high-speed access.

The primary objective of this book is to present a comprehensive view of one aspect of DSL technologies and architectures, which encompasses multimedia applications where voice, video, and data are integrated. The reader will learn the various flavors of DSL technologies and its applicable services and architectures. Among the different DSL technologies mentioned, the most important, ADSL, is covered in detail. Other DSL flavors such as HDSL, HDSL2, VDSL, etc. are also covered.

Details are given on DSL network design aspects with respect to providing an integrated access network environment. This book provides additional information on the Internet resources that provide up-to-date information on the DSL offering in the market place.

Targeted Audience

The target audience for this book is professionals and advanced students. This is designed as a handbook addressing all the pertinent issues related to XDSL from technology capability to its limitation in real-world deployment covering all aspects (technology, architecture, and network design).

This book is a valuable asset for professionals in the telecommunication and computer industry who are involved in understanding the systems-level issues, it will facilitate their designing and implementing DSL-based networks.

Benefits for the Reader

1. Readers can gain a comprehensive, systematic understanding of XDSL technologies and architectures

2. Readers can gain practical knowledge on development of copper-based technologies and their deployment.
3. Readers can have an overall view of copper-based access network architecture from basic technologies to network architecture design.

Readers can use this as a reference book for copper-based networks.

Organization of the Book

This book is organized into three parts.

Part 1 provides the background to broadband communications and its evolution. We then introduce the various broadband access technology options available. Part 2 describes the different DSL technologies. Here, HDSL, HDSL2, ADSL, ADSL lite, and VDSL broadband access technologies are covered. Part 3 discusses the DSL-based access architecture with respect to the physical architecture, service architecture and the design of both physical and service architectures. Figure 1 illustrates the organization of the book.

Authors' Disclaimer

Every effort has been made to include the latest information available at the time of writing. Much of the information, which was at a draft stage at the time of writing, may have become standard by the time of publication. We have made every effort to write in a way that includes the reader who has little background knowledge. Also, please excuse any personal biases, which may have crept into the text because of our background or our work environment.

Figure 1
Organization of the Book

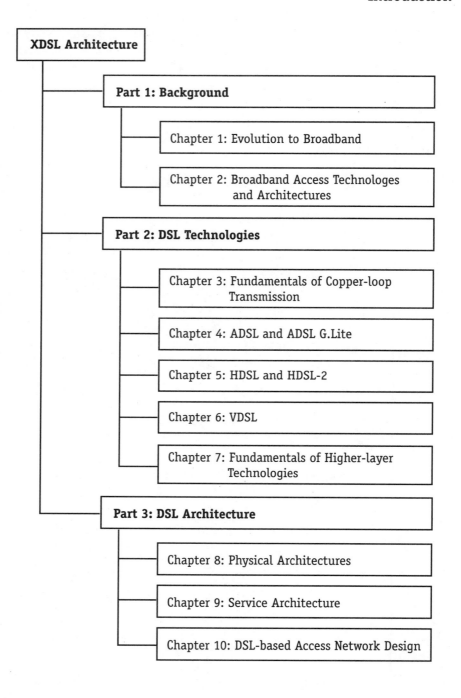

Acronyms and Abbreviations

2B1Q	2 Binary, 1 Quaternary
AAL	ATM Adaptation Layer
ADSL	Asymmetric Digital Subscriber Line
CDV	Cell Delay Variation
CLR	Cell Loss Ratio
IDLC	Integrated Digital Loop Carrier
AD-PCM	Adaptive Differential Pulse Code Modulation
AM	Amplitude Modulation
AMI	Alternate Mark Inversion
AN	Access Node
ANSI	American National Standards Institute
APS	Automatic Protection Switching
ARQ	Automatic Repeat reQuest
AT&T	American Telephone & Telegraph
ATM	Asynchronous Transfer Mode
ATU	ADSL Termination Unit
AWG	American Wire Gauge
BECN	Backward Explicit Congestion Notification
BER	Bit-Error Rate
BH	Busy Hour
BISDN	Broadband Integrated Services Digital Network
BOM	Beginning Of Message
BRI	Basic Rate Interface
CAC	Connection Admission Control
CAD	Computer-aided Design
CAE	Computer-aided Engineering
CAM	Computer-aided Manufacturing
CAP	Carrierless Amplitude Modulation
CATV	Cable Television or Community Antenna Television
CBDS	Constant Bit Rate Data Service
CBR	Continuous Bit Rate, or Constant Bit Rate
CCITT	Consultative Committee on International Telegraph and Telephone
CFM	Configuration Management
CIR	Committed Information Rate
CLEC	Competitive Local Exchange Carrier
CLLM	Consolidated Link-Layer Management
CLP	Cell Loss Priority

Acronyms and Abbreviations

CMT	Connection Management
CO	Central Office
COI	Community of Interest
COM	Continuation of Message
COMSAT	Communications Satellite Corporation
CPE	Customer Premise Equipment
CPN	Customer Premises Node
CRC	Cyclic Redundancy Check
CS	Convergence Sublayer
CSU	Channel Service Unit
DAS	Dual Attachment Stations
DBS	Direct Broadcast Satellite
DCC	Data Communications Channels
DCE	Data Communications Equipment
DE	Discard Eligibility
DLC	Digital Loop Carrier
DLCI	Data-Link Connection Identifier
DMT	Discrete Multitone
DOJ	Department Of Justice
DSLAM	DSL Access Multiplexer
DSP	Digital Signal Processor
DSU	Data Service Unit
DTE	Data Terminal Equipment
DTP	Data Transport Protocol
DTPM	Data Transport Protocol Machine
EA	Extended Address
ECM	Coordination Management
ECN	Explicit Congestion Notification
ECSA	Exchange Carriers Standards Association
EO	End Office
EOM	End of Message
ETSI	European Telecommunications Standards Institute
FCC	Federal Communications Commission
FCS	Frame Check Sequence
FDM	Frequency Division Multiplexing
FEC	Forward Error Control
FECN	Forward Explicit Congestion Notification
FEP	Front-end Processor

FM	Frequency Modulation
FR	Frame Relay
FRI	Frame Relay Interface
FSK	Frequency Shift Keying
FSN	Full Service Network
FTAM	File Transfer Access and Management
FTTC	Fiber to the Curb
FTTN	Fiber to the Node
FTTH	Fiber to the Home
GAN	Global Area Network
GEOS	Geo-Synchronous Satellites
GFC	Generic Flow Control
HDLC	High-Level Data Link Control
HDSL	High-Speed Digital Subscriber Line
HDT	Host Digital Terminal
HDTV	High-Definition Television
HE	Header Extension
HEC	Header Error Control
HFC	Hybrid Fiber/Coax
HIPPI	High-Performance Parallel Interface
HOB	Head of Bus
HPNA	Home Phoneline Networking Alliance
HRC	Hybrid Ring Control
HSSI	High-Speed Serial Interface
HTU-C	HDSL Termination Unit–Central
HTU-R	HDSL Termination Unit–Remote
I/O	Input/Output
IAO	Intraoffice Optical Interface
IBM	International Business Machines
IC	Integrated Circuit
ICA	International Copper Association
ICI	Intercarrier Interface
ICIP	Intercarrier Interface Protocol
IDSL	ISDN Basic Access DSLs
IEC	Interexchange Carriers
IN	Intelligent Network
INTUG	International Trade and User Groups
IP	Intelligent Peripheral/Internet Protocol

Acronyms and Abbreviations

ISDN	Integrate Services Digital Network
ISO	International Organization for Standardization
ISP	Internet Service Provider
ISSI	Inter-switching System Interface
ITFS	Instructional Television Fixed Service
ITU	International Telecommunications Union
IWU	Internetworking Unit
IXC	Interexchange Carrier
JPEG	Joint Photographic Experts Group
LAN	Local Area Network
LAP-B	Link Access Protocol–B
LATA	Local Access Transport Area
LEA	Line Extender Amplifier
LEC	Local Exchange Carrier
LED	Light-Emitting Diodes
LEOS	Low Earth Orbiting Satellite
LLC	Logical Link Control
LMDS	Local Multipoint Distribution Service
LME	Layer Management Entity
LMP	Layer Management Protocol
LOH	Line Overhead
LTE	Line Terminating Equipment
LTU	Line Termination Unit
MAC	Media Access Control
MAN	Metropolitan Area Network
MDF	Main Distribution Frame
MDS	Multipoint Distribution Service
MDSL	Medium-Speed Digital Subscriber Line
MEOS	Medium Earth-Orbiting Satellite
MFJ	Modified Final Judgment
MHS	Message-Handling System
MIB	Management Information Base
MMDS	Multichannel Multipoint Distribution Service
MMF	Multimode Fiber
MPEG	Motion Picture Experts Group
MSO	Multi-System Operator
NAP	Network Access Provider
NID	Network Interface Device

NIF	Neighborhood Information Frame
N-ISDN	Narrowband ISDN
NIUF	North American ISDN User's Forum
NME	Network Management Entity
NNI	Network-Network Interface
NSAP	Network Source Access Point
NTIA	National Telecommunications and Information Administration
NTP	Network Transport Provider
NTSC	National Television System Committee
NTU	Network Termination Unit
NVOD	Near Video On Demand
O/E	Optical to Electrical
OAM	Operations, Administration And Maintenance
OAM&P	Operations, Administration, Maintenance and Provisioning
OC	Optical Carrier
OCI	Optical Carrier Interface
ONI	Optical Network Interface
ONU	Optical Network Unit
OS	Operations System
OSI	Open Systems Interconnection
OSS	Operations Systems
OTA	Office of Technology Assessment
PA	Prearbitrated
PCS	Personal Communications Services
PDH	Plesiochronous Digital Hierarchy
PDU	Protocol Data Unit
PES	Packetized Elementary Stream
PFM	Parameter Frame Management
PHY	Physical Layer Protocol
PLPC	Physical Layer Convergence Protocol
PM	Phase Modulation
PMD	Physical Layer Medium Dependent
POH	Path Overhead
PON	Passive Optical Network
POP	Point of Presence
POTS	Plain Old Telephone Service

PPL	Phase Locked Loop
PPV	Pay Per View
PPP	Point-to-Point Protocol
PRI	Primary Rate Interface
PRM	Protocol Reference Model
PS	Program Stream
PSTN	Public Switched Telephone Network
PT	Payload Type
PTE	Path-Terminating Equipment
PTM	Packet Transfer Mode
PTT	Post, Telephone and Telegraph
PVC	Permanent Virtual Circuit
QA	Queued Arbitrated
QAM	Quadrature Amplitude Modulation
QoS	Quality of Service
RBOC	Regional Bell Operating Company
RME	Routing Management Entity
RMN	Remote Multiplexer Node
RMP	Routing Management Protocol
RMS	Root Mean Square
RMT	Ring Management
SAP	Service Access Point
SAR	Segmentation and Reassembly Sublayer
SAS	Single Attachment Stations
SCP	Service Control Point
SDLC	Synchronous Data Link Control
SDM	Space Division Multiplexing
SDMT	Synchronized DMT
SDSL	Symmetric Digital Subscriber Line
SDU	Service Data Unit
SIF	Status Information Frame
SMF	Single Mode Fiber
SMS	Service Management System
SMT	Station Management
SNA	System Network Architecture
SNI	Subscriber Network Interface
SRF	Status Report Frame
SS7	Signaling System Number 7

SSP	Service Switching Point
STB	Set-Top Box
STP	Shielded Twisted Pair
STV	Sprint Telecommunications Venture
SVC	Switched Virtual Circuit or Signaling Virtual Circuit
TA	Trunk Amplifier
TA 1996	Telecommunications Act of 1996
TC	Transmission Convergence
TDD	Time Division Duplexing
TDMA	Time Division Multiple Access
TP	Transaction Processing
TRT	Token Rotation Timer
TS	Transport Stream
TTRT	Target Token Rotation Time
TVX	Valid Transmission Timer
UAWG	Universal ADSL Working Group
UNI	User-Network Interface
UTOPIA	Universal Test and Operations Physical Interface for ATM
UTP	Unshielded Twisted Pair
VBR	Variable Bit Rate
VCI	Virtual Channel Identifier
VDSL	Very High-Bit Rate Digital Subscriber Line
VDT	Video Dial Tone
VIP	Video Information Provider
VoD	Video on Demand
VPI	Virtual Path Identifier
WAN	Wide Area Network
WCA	Wireless Cable Association
WDM	Wavelength Division Multiplexing
XC	Cross Connect

Evolution
to Broadband

Introduction

Today, we live in a *content*-driven society. Newsprint, television, and electronic media are engaged daily in what Andy Grove of Intel Corporation once described as the "war for the eyeballs." Not only are we bombarded every day with a combination of voice, text, and video content, we are also expected to react to it. Faxes, pagers, cellular phones, and e-mails have become necessities. Further, this constant interaction with dynamically changing content has an impact on much more than our work. We are also looking to be entertained and educated in new ways and for ways to stay connected no matter where we are physically located. This, then, is the age of networked multimedia—*rich, dynamic content that is available to everyone, everytime, everywhere.*

From the present, let us take a trip on an imaginary time machine and travel a few years into the future. Picture this scenario: You've just driven home from work. As you enter your home and walk through the living room, you have the following conversation with your home computer, also known as your personal "electronic assistant":

"Do I have any messages?"

Electronic assistant: "Yes, you have three new video messages. Do you want to play them?"

"Yes, I want them on my bedroom screen."

Your electronic assistant displays the video messages on an enormous flat screen on the bedroom wall. After dinner, you would like to help your teenage child with homework on ancient cultures. The assignment is a paper on the Mughal Emperors of India. Although you don't have a clue about the topic, you don't worry. First, you ask your electronic assistant to gather enough information to complete the paper. The electronic assistant searches all available libraries and video images to compile the data. In a few minutes the electronic assistant asks, "Are you ready to view what I found on Mughal Emperors?" Once you are ready, the electronic assistant displays the information on the screen in your child's bedroom.

Now that your child is busy writing the report, you decide to relax and watch a movie. You ask the electronic assistant for a list of the latest hits. It brings up a list of choices, you select one, and ask the electronic assistant to play it. The electronic assistant extracts the movie from the database at a video store nearby and plays it as if it were being played from your home VCR—the only difference being that the VCR is at a remote location.

Suddenly, the electronic assistant interrupts your viewing the movie to announce that a video call is coming in. You ask the electronic assistant to put the movie on pause and answer the video call. When you complete the call, the electronic assistant automatically switches the movie from pause to play so that you can continue to view it from the point at which you stopped before the video call.

This scenario is not several years away. Most of the technological building blocks that can deliver these services are in place *today*. Just as television is delivered to your home via a video network (in the form of cable, satellite, or over-the-air broadcasts), broadband networks will soon be able to deliver a plethora of integrated voice, data, and video services. Technological innovations and competitive factors will make it possible for consumers to enjoy these services at affordable prices.

Rapid technological innovations are occurring from the "last mile" to your home down to the "last inch" in your home. Consider, for example, the Home Phoneline Networking Alliance (HomePNA).[1] The objective of this association of technology companies is to work together to ensure the adoption of a single networking standard that does not require any rewiring in your home. It uses the existing phone wires installed in your residence to deliver a variety of integrated high-speed services, as shown in Figure 1.1. Other alliances such as HomeRF[2] and Bluetooth[3] are investigating ways to dispense with wires entirely.

Figure 1.1

Vision for the networked home (Source: Home Phoneline Networking Alliance)

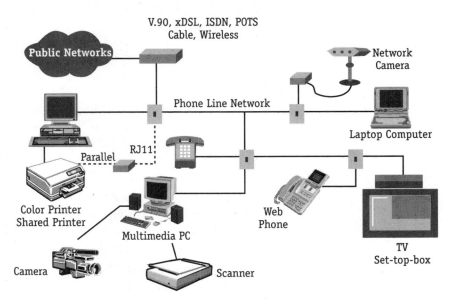

How is this possible? What are the factors that can enable the vision of "affordable, ubiquitous, high-speed" access to consumers all the way to their homes? To answer these questions, we need to get back in our imaginary time machine, and return to the past. An examination of the history of our communication networks can help us to better understand how we got to where we are today—and where we can go tomorrow.

What Is a Communications Network?

A communications network is a common resource shared by many customers who need to communicate with users at other locations. Not everyone uses the network all the time, so it is logical—often economical—to share this important resource. A communication network consists of the following elements (shown in Figure 1.2):

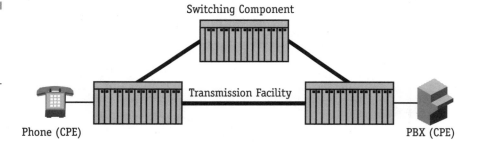

Figure 1.2

Elements of a communications network

- Station equipment or customer premises equipment (CPE) sends and receives user information and exchanges control information to gain access to the network and to place calls. Since station or customer premises equipment either begins or terminates the flow of information, i.e., it acts as either information sources or information sinks, it is often referred to an *endpoint*. Information flow within the network is commonly referred to as *traffic*.
- Transmission facilities provide the paths to carry the traffic and establish connectivity between endpoints. In general, transmission facilities consist of a medium such as air, copper wires, coaxial cables, or fiber optic cables. Therefore, transmission facilities are often referred to as *links*. Physical links are often shared among

multiple simultaneous connections between information sources and information sinks. The portion of a physical link dedicated to a single connection is often referred to as a *channel*. A channel either can be full duplex or half duplex. A full-duplex channel allows an endpoint to transmit and receive simultaneously; by contrast, in a *half-duplex channel* the endpoint can either be transmitting or receiving, but not both simultaneously.

■ Switching components interconnect transmission facilities at various locations and route traffic through the network. Switching makes it possible to share transmission facilities among economically multiple users. Switching components are often referred to as *network nodes*. The network node into which the source endpoint connects is called an *ingress node*. The network node from which the sink endpoint is connected is called an *egress node*.

Architecture of Communication Networks

There are several ways to categorize communication networks: by topology, traffic transport mechanism, functional area, or dominant application. In this section, we briefly discuss categories such as meshed, star, and bus topologies, switching and multiplexing mechanisms, and premises, access, backbone, and service domains. The reader who has knowledge of these topics may skip ahead to later sections in this chapter that discuss the history and architectures of voice, data, and video networks.

Categorization of Communication Networks Based on Topology

There are three basic types of topologies—mesh, star, and bus. In a mesh network, all endpoints are interconnected to each other. In a star network, the endpoints are not directly connected to each other; rather, they are connected directly to a network node. Endpoints can therefore only communicate to each other via the intermediate network node. Finally, in a bus network, all endpoints are connected to each other over a common bus; therefore, any endpoint may communicate to another over the bus.

The three basic topologies are illustrated in Figure 1.3.

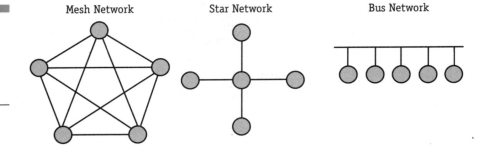

Figure 1.3
Three basic
topologies for
communication
networks

From these basic topologies, the following network topologies can be derived:

■ **Fully-meshed point-to-point networks.** In a fully-meshed point-to-point network, every endpoint is directly connected to every other endpoint at least once. If the number of endpoints is small, the number of links to support a fully-meshed network is small. However, as the number of endpoints increases, the number of links rapidly increases. For example, to connect endpoints A, B, and C in a fully-meshed network requires 2^3 or 8 links. To connect endpoints A to H in a fully-meshed network requires 2^8 or 256 links. Clearly, this becomes impractical as the number of endpoints grows. For this reason, fully-meshed network architectures are rarely implemented.

■ **Switched point-to-point networks.** Switched networks are designed to solve the problem of fully-meshed networks. In switched networks, endpoints are connected to switching components in the network, i.e., network nodes. The network nodes are then interconnected to each other in some arbitrary fashion; however, the links are designed so that it is possible to set up at least one unique path between any two network nodes (not endpoints).* Since the number of network nodes is much smaller than the number of endpoints, this architecture can be scaled up. This architecture supports sharing

---

* The network nodes themselves could be interconnected in a full mesh. Due to economic considerations, however, network designers typically include only the minimum number of links necessary to ensure a unique path between any two network nodes, with some redundancy of paths between critical nodes in the network.

of critical network resources, including the physical links between network nodes. Switched point-to-point networks can be further categorized by how the switching between the ingress node and the egress node takes place: it is circuit switched or packet switched.

The difference between fully-meshed point-to-point networks and switched point-to-point networks is illustrated in Figure 1.4.

Figure 1.4

Fully meshed versus switched point-to-point networks

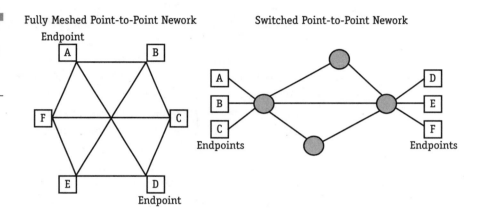

Broadcast networks. Broadcast networks are associated with endpoints that merely broadcast the information; it is up to the receiver to process or ignore the broadcast. Broadcast TV is an example of a star network with one-way communication and a single endpoint acting as the source of the information broadcast to other endpoints. Satellite communication is an example of a star network that supports two-way communication; the hub of the star is the satellite. The satellite may, in turn, relay the message to multiple satellite receivers as a broadcast. Finally, Ethernet is an example of a bus network where each endpoint broadcasts over the common bus. If another endpoint broadcasts simultaneously, there is a collision, so each endpoint backs off for a random time period and tries again.

Categorization of Communication Networks Based on Traffic Transport Mechanisms

Another way to categorize networks is based on the mechanism used within the network to transport traffic across a switched network. A switched network provides both switching and multiplexing mechanisms to transport the traffic.

Switching Mechanisms

There are two kinds of switching mechanisms:

■ **Circuit switching.** Circuit switching implies a dedicated communication path between two endpoints. The path is a connected segment of links between network nodes. On each physical link, a channel is dedicated to the connection. Circuit switching involves three phases: circuit establishment, signal transfer (i.e., information transfer or data transfer), and circuit termination. Note that the circuit path is established before data transmission begins. Thus, channel capacity must be reserved between each pair of nodes in the path, and each node must have sufficient internal switching capacity to handle the requested connection. The switches must be intelligent to make these allocations and route the call through the network.

Circuit switching is suited for communications that require sending the information in a continuous manner, and for which the information requires guaranteed bandwidth and low tolerance for delay once the traffic flow begins. Since voice telephony is an application that has these requirements, the telephone network is an example of a network that uses circuit switching.

■ **Packet switching.** In contrast to circuit switching, packet switching does not require a dedicated communication path between endpoints. It is sometimes appropriate to break the information up to be transmitted into smaller chunks called *packets*. The packets of information can now be routed over multiple parallel paths in the network to the endpoint destination where the receiver reassembles them. As a result, multiple parallel paths in the network can be shared among multiple senders and receivers. The sharing of network resources among multiple senders and receivers does have a drawback—since the actual load on the network is not known a priori, the capacity of the network could be overburdened at any given time if the number of senders far exceeds the capacity. Therefore, some buffering capability is necessary within the network.

Packet switching involves the following phases:

▪ Disassemble the data into packets with sequence numbers so that the receiver can reassemble them in the proper order.

▪ Add routing information so the network knows where to ship the packet.

▪ Route the packets across one or more paths to the receiver.

▪ Reassemble the packets when they reach the receiving end. Note that packet switching requires that each node in the network have enough intelligence to route the packet appropriately.

Given these characteristics, packet switching is best suited for communications that do not require sending the information in a continuous manner, but do require guaranteed delivery without corruption of the original information. Since file transfer is an application that has these requirements, the data network is an example of a network that uses packet switching.

Historically, information was divided into packets of varying length up to a maximum size. Since these variable length packets are often referred to as frames, this form of switching is called *frame relay*. Another important form of packet switching is called *cell relay*. In cell relay, the fixed, small size packets are called cells. Cell relay is more suited to today's network environment where transmission links are fast and more reliable than they used to be. Using a technique known as *fast packet switching*, cell relay makes it possible to switch cells through the network efficiently regardless of whether a particular cell is carrying voice, data, or video. Asynchronous transfer mode (ATM) is an example of both a network technology and an architecture based on the principles of cell relay.

Multiplexing Mechanisms

There are several multiplexing mechanisms, among them:

▪ **Space division multiplexing.** In this technique, each connection takes up one physical line between two switches and a path within each switch. If all the lines and/or all the paths are used up, a new connection is blocked until one of the existing connections is broken.

▪ **Frequency division multiplexing.** In this technique, the frequency spectrum of a physical line is divided into multiple bands, thereby allowing the single line to be multiplexed among several simultaneous connections with each connection occupying a different band in the frequency spectrum.

■ **Time division multiplexing.** In this technique, the bandwidth of the physical line is divided into 1 to N timeslots organized together as a "train" repeating every T seconds. There can be M such "trains" (numbered 1 to M), each containing N timeslots and repeating with time period T, as shown in Figure 1.5. The "trains" are often referred to as channels.

Figure 1.5
Time division
multiplexing

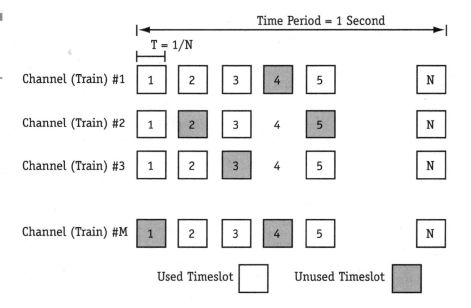

Time division multiplexing is commonly used to provide T1 services. Referring to Figure 1.5, for T1, N=8000, M= 24, and T=1/8000 (i.e., 125 μsecs). Each timeslot can carry one octet (8 bits). Therefore, the T1 capacity is ((24 (channels) × 8 (bits/timeslot)) + 1 (framing bit)) × 8000 (timeslots/second), or 1.544 bits/sec.

A connection between two endpoints on a particular link is preassigned a fixed timeslot (between 1 and N) on a fixed train number (between 1 and M). Information for that connection is always carried in that preassigned timeslot. If a connection has any information to send, it transmits that information during its preassigned timeslot. If there are more data to send, it has to wait until its timeslot arrives again. Conversely, if it has no data to send, the timeslot goes empty, regardless of whether there is another connection that could have utilized that empty timeslot at that time period. This technique is also called *synchronous transfer mode (STM)*.

■ **Statistical multiplexing.** Statistical multiplexing is a solution to the unused timeslot problem of STM. In statistical multiplexing, instead of always identifying a connection by a preassigned timeslot, data are packetized into small units with a small header that contains the connection ID. The packetized data are asynchronously carried in a bucket, the size of which is sufficiently small to multiplex connections efficiently. As long as there are data to send from any one of the multiplexed connections, a portion of the data that fits is sent in the first available bucket. Hence, the probability of empty buckets is minimized. Furthermore, the data rate from the multiplexed connections can be matched (whenever possible) to the line capacity so that, statistically, congestion is avoided. Even if there is network congestion resulting in packet loss, only a small portion of the original data is lost due to the small bucket size. Statistical multiplexing is illustrated in Figure 1.6.

Figure 1.6
Statistical
multiplexing

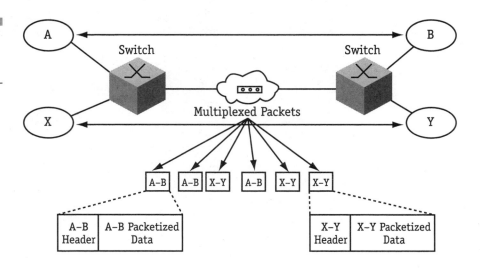

These concepts of statistical multiplexing using small-sized packets are embodied in *asynchronous transfer mode (ATM)*. ATM gets its name from the fact that packetized data are asynchronously carried in a bucket with no time designation for the bucket. ATM is also known as cell relay since the small buckets are often referred to as cells. Another name for ATM is fast packet switching, since some of its fundamental principles are rooted in packet switching. ATM is an efficient way to transport large amounts of data quickly. Therefore, ATM is being widely adopted today for high-speed backbone networks.

Categorization of Communication Networks Based on Functional Area

As communications networks have evolved in complexity, it has become necessary to develop a reference model to study the end-to-end architecture and to group network components logically by functional area. The end-to-end architecture reference model[4] is shown in Figure 1.7.

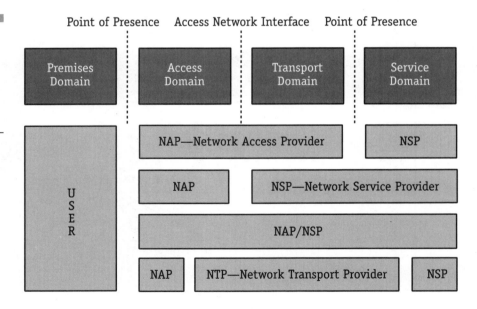

Figure 1.7
Reference model for end-to-end architecture (Source: ADSL Forum)

The reference model shows four separate domains—premises, access, transport, and service. These labels are somewhat arbitrary, but serve to categorize network components by the functional areas that make up an end-to-end architecture. This categorization is necessary to provide a level of abstraction and independence among the different domains. For example, an equipment manufacturer who develops products for the premises domain need not be concerned about what goes on in the transport domain, except for the standard interfaces used to communicate between domains. Each of the domains has a network associated with it:

■ **Customer premises network.** The customer premises include residences, home offices, and corporate offices. Actually, the customer premises network is multiple subnetworks; for example, one network for telephony and another for data. The telephony subnetwork includes analog single/multi-line telephones, key systems or PBXs, fax machines, and the like. The data subnetwork consists of PCs on some kind of premises distribution network (PDN). Typically, this is a local area network (LAN). In the future, a premises appliance network could also be considered a part of the customer premises network.

■ **Access network.** The access network is the focus of this book—specifically, copper loop access. It consists of the transmission medium and the equipment necessary to transport voice, data, or video from multiple customer premises to a common collection point.

■ **Transport network.** The transport network, also called the backbone network, regional network, or the core network, provides connectivity within a geographic region and access to points of presence (POPs) or service providers.

■ **Service provider network.** The service provider network includes various subnetworks such as the POPs, content provider networks, corporate networks, and regional operations centers.

In the end-to-end architecture, the access and transport domains must be "transparent" to the applications in the premises and service domains. In other words, end-user applications assume a generic transport layer should function without change and knowledge about intervening domains. This transparency allows the access and transport domains to be owned and operated independently by providers. It also give customers a choice of providers.

When we refer to Figure 1.7, there are four possible domain scenarios depending on provider ownership by network access providers

(NAPs), network service providers (NSPs), or network transport providers (NTPs). The four basic choices are:

- NAP owns the access and transport domains; NSP owns the service domain.
- NAP owns the access domain; NSP owns the transport and service domains.
- Combined NAP/NSP entity owns access, transport, and service domains.
- NAP owns the access domain, NTP owns the transport domain, and NSP owns the service domain.

An important requirement that comes out of these ownership scenarios is the necessity for an NAP to provide connectivity "without prejudice" to different NSPs, or, put another way, "equal access."

NAPs are typically local telephone companies, which provide access for telephony services to the public switched telephone network (PSTN). In addition to carrying voice, the access network often carries low-speed data with dial-up networking. NSPs provide connectivity across a backbone network to the content servers in the service domain. For Internet access, NSPs are referred to as Internet service providers (ISPs). NTPs are typically the long distance telephone companies.

Certain interface reference points are of particular interest when discussing DSL technologies and their architectures. These reference points are a legacy of ISDN-based architecture. They are named as R, S, T, U, and V according to the practice of the International Telecommunications Union (ITU)*. The following is a brief description of each of the interface reference points (see Figure 1.8):

- The *R interface* reference point is also known as the R(ate) reference point. It represents a non-DSL interface between user equipment
- The *S interface* reference point is also known as the S(ystem) reference point. It corresponds to an individual DSL-equipped terminal's interface.

---

* The ITU does not have a reference point for the interface between the transport domain network (also known as the regional broadband network) and the service domain network.

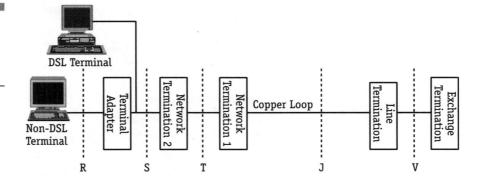

Figure 1.8
*Reference points
for end-to-end
architecture*

- The *T interface* reference point is also known as the T(erminal) reference point. It represents the DSL network termination point at the customer premises.
- The *U interface* reference point is also known as the U(ser) reference point. It delineates the interface at the subscriber line between the premises domain and access domain networks.
- The *V interface* reference point is also known as the R(ate) reference point. It delineates the interface between the access domain and transport domain networks.

For DSL, the U interface is further broken up into the interfaces at either end of the copper loop. The U interface at the end of the loop that connects to the central office side is labeled U-C, whereas the U interface at the end of the loop that connects to the premises (or remote) side is labeled U-R.

Categorization of Communication Networks Based on Dominant Applications

Yet another way to categorize a communications network is by its dominant application. A network used for voice communication is called a *telecommunications network*, and one used for communication among computers is called a *data communications network*. This separation does not mean that data cannot be carried over the telecommunications (voice) network or vice versa. It only means that the network was originally designed and optimized for a specific application and transporting any other type of information sometimes results in inefficient use of network resources.

The communications networks that have emerged are based on the dominant type of service they have historically provided:

■ **Telecommunications networks** (a.k.a., the public switched telephone network). Telecommunications networks have evolved from the original intent of providing ubiquitous voice services. The design was based on the circuit switched point-to-point architecture. Today, the core network of telecommunications networks is evolving to ATM architecture.

■ **Data communications networks** (a.k.a., computer communications networks). Data communications networks have evolved from their original purpose of providing communications between two computers. Data networks for wide area communications were designed using the packet switched point-to-point architecture. Today, the core of data communications networks is also evolving to ATM architecture.

■ **Video networks.** Video networks consist of cable TV networks, satellite networks, and off-the-air broadcast networks. Their original purpose was to provide video services over transmission facilities that could be shared among many users and their design was based on broadcast architecture. Today, the core of video networks (e.g., satellite networks[5]) is also moving toward an ATM architecture.

Now we should examine the history and architecture of each of these networks. This will help increase our understanding of how they converge into a common broadband access network that connects to customer premises.

Telecommunications Networks

This section discusses the history and architecture of telecommunications networks in detail.

History of Telecommunications Networks

Telecommunications networks are the most mature form of communications networks. Telecommunications first came into existence with the development of the telegraph in the 1830s and 1840s. For the first time, news and information could be transmitted great distances almost instantaneously. The invention of the telephone in 1875 by Alexander Graham Bell fundamentally transformed telecommunications. The telephone system assumed its modern form with the development of the dial-up telephone, and it spread in the middle of the twentieth century. A company known as American Telephone and Telegraph (AT&T) subsequently emerged to provide telephone services throughout the United States As a result of its significant market presence and the several thousand people it employed, it was often colloquially referred to as "Ma Bell."

Until the 1980s, the world telecommunications system had a relatively simple administrative structure. In the United States, telephone service was supplied by a regulated monopoly—AT&T. It was the dominant supplier, providing not only local and long distance voice services, but the switching and customer premises equipment as well. During this same time, Western Union Corporation provided telegraph services in the United States. In most other countries, both services were government owned and operated via agencies known as PTTs (for Post, Telephone, and Telegraph).

Beginning in 1983 in the United States, however, the situation became far more complex. As a result of an antitrust suit filed by the federal government, AT&T agreed in a court settlement to divest itself of the local operating companies that provided basic telephone service in various geographic regions of the United States. These local companies remained regulated entities and were known as regional Bell operating companies (RBOCs). In return, AT&T would continue to offer long distance service in competition with a half dozen major and many minor competitors. For a while AT&T also retained ownership of a subsidiary that produced telephone equipment, computers, and other electronic devices. Today, AT&T has chosen to focus itself on long distance and enhanced services and has further divested itself of its other businesses.

Another significant regulatory event in the history of telecommunications was the passage of the Telecommunications Act of 1996 by the U.S. Congress. One of the key motivations for this legislation was to accelerate competition by allowing long distance companies and local access companies to compete in each other's markets. According to this law, all local exchange carriers (LECs), including new entrants, must:

- Permit resale on a nondiscriminatory basis
- Provide number portability to the extent technically feasible
- Provide dialing parity and access to telephone numbers as well as operator and directory services
- Allow access to poles, ducts, conduits, and rights of way
- Establish reciprocal compensation agreements with other carriers for the transport and termination of traffic.

Additionally, the incumbent LECs must allow interconnection at any technically feasible point for the transmission and routing of telephone exchange and exchange access services. The interconnection must be of the same quality as that offered to an LEC subsidiary and must be made available at reasonable and nondiscriminatory rates, terms, and conditions. This interconnection requirement has opened the market to competition by alternative carriers, also known as CLECs (competitive local exchange carriers). Although industry observers have noted that, despite the 1996 legislation, competition has been stuck in the doldrums, recent announcements of significant mergers and acquisitions indicate that true competition to provide integrated access to consumers may finally be on its way.*

Although AT&T's monopoly ended in 1984, its dominance in the marketplace for many decades did have some positive influences in the development of voice technology. Several techniques and de facto standards in dialing, switching, and transmission techniques were originally developed in the deployment of AT&T's network. Some of these practices and standards, such as the North American Numbering Plan,† are still prevalent today. Overall, however, deregulation has been beneficial to the consumer in lowering the cost of voice services. It has also pro-

---

* In an interesting turn of events in the late 1990s, some of the RBOCs, prompted by increased competition, chose to merge their businesses. This caused industry analysts to observe that "Ma Bell is getting back together." Long distance companies and RBOCs also announced plans to merge or acquire cable operators.

† The North American Numbering Plan uses 10 digits to represent a phone number, i.e., three digits for area code, three digits for exchange, and four digits for the local number served by the exchange.

vided the opportunity for new companies to enter the market and to differentiate themselves through innovations and the creation of new standards.

In most countries around the world, the government-owned PTTs are not as segmented as in the United States. Most countries have, however, adopted a telecommunications network hierarchy and architecture similar to that of the United States. With the privatization of PTTs around the world, competition now thrives in every segment of telecommunications, and U.S. telecommunications stands as an example for other countries.* However, many local PTTs are still the dominant telecommunications players in their respective countries.

From the beginning, telecommunications networks were designed to provide universal voice services to the general public; the presence of these services can be seen almost everywhere within the United States and in other developed countries. These basic voice services are often referred to as "plain old telephone service" (POTS). Telephone companies are now beginning to differentiate themselves by providing enhanced services, such as "call waiting" and "caller ID". Another service includes data access using Integrated Services Digital Network (ISDN).

Architecture of Telecommunications Networks

The U.S. public telephone network architecture serves as a good example of a mature telecommunications architecture. The 1984 divestiture of AT&T resulted in the separation of local access services and long distance services. The local access providers or LECs divide their service regions into smaller geographical areas called local area transport areas (LATAs). Companies that offer long distance services do so by interconnecting the LECs, hence they are known as *interexchange carriers* or *IXCs*. Telephone companies are commonly referred to as *carriers* or *telcos*.

A telecommunications network architecture consists of multiple LECs interconnected by multiple IXCs. In the United States, LEC serv-

---

* For example, in Canada and the United Kingdom, other carriers have entered the market to compete against Telecom Canada and British Telecommunications, plc. respectively.

ices are provided by the RBOCs, GTE, Sprint, and others; companies such as AT&T, MCI, GTE, and Sprint provide long distance services.*

Although both the routing and the specific technologies of telecommunications networks have evolved since the AT&T divestiture, the overall architecture still uses the basic components of a communications network. In a telephone network, each subscriber is connected via the local loop to a switching center known as an end office (EO) or a central office (CO). Typically, a central office can support thousands of subscribers in a localized area. Clearly, it is impractical for each CO to be connected with a direct link. If that were to happen, we would probably need several million links. Therefore, intermediate switching nodes are used. These intermediate nodes, also called *access tandems*, provide traffic aggregation and reduce the number of links required to connect the central offices. Each of these nodes in the network has an average of 10 to 15 central office switches connected to it. Traffic routing in the LEC network is determined by the connections between the access tandem and central offices. The switching centers are connected by links called *trunks*. These trunks are designed to carry multiple voice frequency circuits using multiplexing techniques such as frequency division multiplexing (FDM), synchronous time division multiplexing (TDM), or wavelength division multiplexing (WDM).

Signaling System 7 (SS7) is the international standard for the signaling protocols used in telecommunications networks. It is SS7 that enables value-added features for subscribers such as caller ID, call forwarding, call waiting, and toll-free services. Within the network itself, SS7 provides mechanisms for call setup/teardown, call routing, call management, and authentication. SS7 signaling occurs out of band (i.e., on dedicated channels), rather than in band with voice channels. This provides several advantages, including the fact that use of the voice channels does not interfere with the operations, administration, maintenance, and provisioning (OAM&P) of the network.

Figure 1.9 shows an example of how a call is completed through the interconnection between IXCs and LECs. In the figure, a telephone call

---

* Note that the same company can be both an LEC and an IXC, as long as it does not provide IXC services in regional markets where it provides LEC services. With the Telecommunications Act of 1996 in the United States, this demarcation is disappearing since IXCs are not allowed to compete for LEC services, and vice versa. However, although the line between local and long distance access providers is blurring, litigation has stalled the process in several U.S. states in the late 1990s.

originates in LATA X. If the call terminates within the same LATA, the traffic goes no farther than the access tandem node. The traffic is routed to the appropriate central office and on to the destination phone. If the call has a destination address in a different LATA—for example, LATA Y—the call is handed to the appropriate IXC (depending on which carrier the call originator has subscribed), which switches the call to the appropriate LATA. At the LATA, the call is handed to the local exchange carrier serving the destination address. This LEC then routes the call appropriately to the destination. All signaling within the network is done through SS7 messaging.

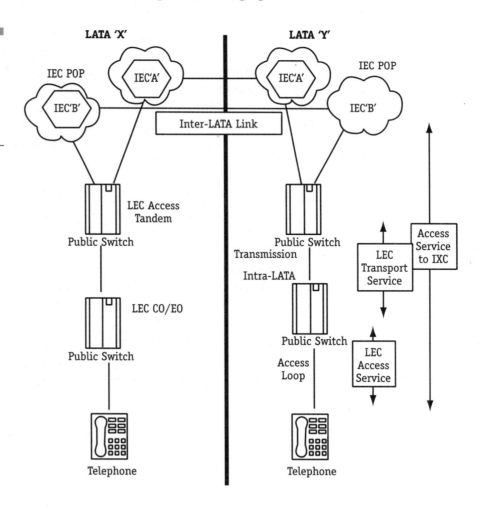

Figure 1.9

An example of interconnection between IXCs and LECs in the United States

Data Communications Networks

This section discusses the history and architecture of data communications networks in detail.

History of Data Communications Networks

The history of data communications is quite different from that of telecommunications. The industry is less mature; however, by comparison, its rate of growth has been phenomenal. The unprecedented technology revolution involving computers did not begin until the later part of twentieth century. It is now predicted that traffic generated from computers will dominate communications networks in the twenty-first century as did voice-dominated traffic in the twentieth century.

Early computer networks developed by companies such as Xerox, IBM, and DEC were not standardized. Rather, they were proprietary, which meant a network built by one company did not communicate with a network built by another company. These proprietary networks were also designed to meet 1950s communication requirements, which were batch-based and minimal. The processor communicated with its peripheral via input/output (I/O) devices over short distances at a very low speed. The 1960s brought the concept of *timesharing*, where users were connected to computers via a dumb terminal, as shown in Figure 1.10.

The 1970s saw the dominance of International Business Machines (IBM) in the area of mainframe communications. Its dominance resulted in several significant contributions to the advancement of data communications, including the development of protocols such as Synchronous Data Link Control (SDLC) and the Systems Network Architecture (SNA). Simultaneously, the 1970s also saw the development of integrated circuit (IC) technology and the microprocessor, which made it possible to bring personal computers to the desktop. This development drastically changed the way people viewed computers. The growth of local area networks (LANs) in the 1980s enabled personal computers to communicate with each other, thus facilitating the migration from centralized computing to distributed computing. Figures 1.11 and 1.12 show centralized and distributed computing, respectively.

Figure 1.10
Time-shared computer system

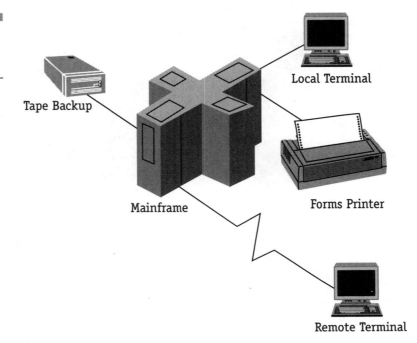

Tape Backup

Local Terminal

Forms Printer

Mainframe

Remote Terminal

Figure 1.11
Centralized computing

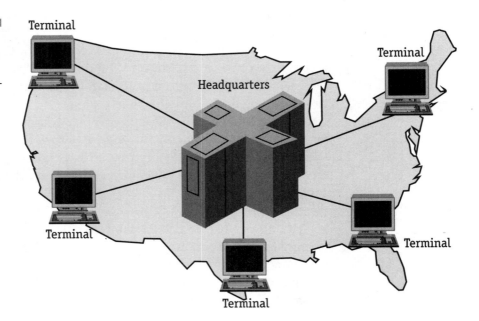

Terminal

Terminal

Headquarters

Terminal

Terminal

Terminal

Terminal

Figure 1.12
*Distributed
computing*

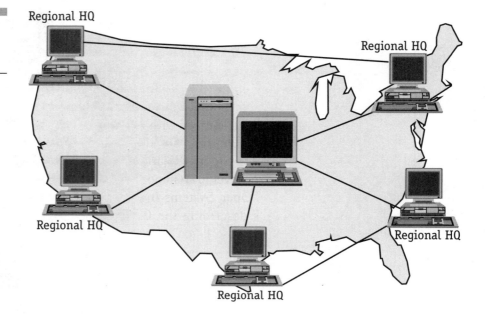

In a centralized computing environment, the main system or processor is centrally located, and all remote locations are connected via a direct link. All the information, i.e., the entire database, is located at the central system.

In a distributed computing environment, the processors, or main computers, are distributed at different locations, with each node having a complete copy or a portion of the database. The user accesses the information from the nearest processor, which keeps current by periodically updating the information in its databases.

Distributed computing became popular because of inherent advantages such as the elimination of single points of failure, better distribution of resources, and opportunities for parallel processing. Distributed processing has had a significant impact on the evolution of data communications.

■ The distributed processing concept was originally developed by a project underwritten by an agency of the Department of Defense known as the Advanced Research Projects Agency (ARPA). This agency was asked to find a solution to a difficult problem: how to get different computers to communicate with one another as if they were one computer. This led to a suite of protocols called TCP/IP (transmission control protocol/Internet protocol) that provided the building blocks for today's Internet, which connects millions of users around the world.

- A distributed processing network architecture requires a high degree of compatibility and interoperability among network elements, particularly with respect to its physical and logical interfaces and controls. To address this issue, the International Organization for Standardization (ISO) established a subcommittee in 1977 to develop a standard architecture to achieve the long-term goal of Open Systems Interconnection (OSI). There are numerous sources describing the OSI architecture in great detail in works such as those listed in reference number 7 for those who wish to know more. It is assumed that the reader is familiar with the seven-layer OSI architecture; this knowledge is necessary to understand the physical and higher layer issues of copper loop access networks discussed later in this book.

- Distributed processing gave rise to the client/server computing concept popularized in the 1980s. In client/server computing, the network provides connectivity between a group of clients and a group of servers, with each server providing specialized services to a shared group of clients such as file/print services, database services, and even communication gateway services (to connect to other networks outside the geographic area). As computers became more powerful and cost effective, client/server computing further evolved into peer-to-peer computing, where a single computer could act both as a client and as a server to other computers on the network.

Throughout the history of data communications, the networks were mostly designed to serve private interests. Therefore, while there are several large corporate networks in existence today, until the recent explosive growth of the Internet, there were only a handful of companies offering public data services.*

---

* In the 1980s, CompuServe became popular as one of the public data networks. Later, America Online (AOL) also grew in popularity. Today, most of these public data networks continue to offer differentiated services in addition to seamless access to Internet services.

Architecture of Data Communications Networks

Architecturally, data communications networks are often classified based on their geographic scope and capacity (measured by the speed at which the links operate). The four major categories are:

- **Local area networks (LANs).** LANs are typically used to interconnect computers and PCs within a relatively small area, such as within a building, office, or campus. A LAN typically operates at speeds ranging from 10 Mbps to 100 Mbps, connecting several hundred devices over a distance of up to approximately five miles. LANs became popular because they allowed many users to share scarce resources, such as mainframes, file servers, high-speed printers, and other expensive devices.

- **Metropolitan area networks (MANs).** A MAN, as the name implies, is a network covering a metropolitan area, connecting many LANs located at different office buildings. It has a larger geographical scope compared to a LAN, can range from five miles to a few hundred miles in length, and typically operates at a speed of 1.5 Mbps to 150 Mbps. One of the reasons MANs are connected via a much lower speed than LAN technology is the high cost of bandwidth; most of the traffic remains within the LAN environment. Therefore, traffic management (i.e. intelligently managing the efficient utilization of expensive bandwidth) becomes a critical factor in managing the public network cost.

- **Wide area networks (WANs).** A WAN is designed to interconnect computer systems over very large geographic areas, such as from one city to another within a country. A WAN can range from 100 miles to 1,000 miles, and the speed between the cities can vary from 1.5 Mbps to 2.4 Gbps. In a WAN, the cost of transmission is very high and the WAN is usually owned and operated by a public network. Businesses will lease a transmission system from the public network to connect their geographically diverse sites.

- **Global area networks (GANs).** As the name implies, GANs are network connections between countries around the globe. A good example of such a network is the Internet, which is really a network of networks connecting many countries together. A GAN's speed ranges from 1.5 Mbps to 100 Gbps and can cover thousands of miles.

Figure 1.13 illustrates an example of how a LAN, MAN, WAN, and GAN might fit together.

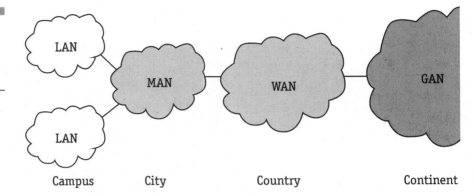

Figure 1.13

Interaction between LAN, MAN, WAN, and GAN

Video Networks

This section discusses the history and architecture of video networks in detail.

History of Video Networks

Video networks comprise both cable TV networks and satellite networks:

■ **Cable TV networks.** Cable TV was developed in the late 1940s as an alternative to broadcast TV, particularly in areas where reception was poor or unacceptable. It has also been called community antenna television (CATV), because the original design called for a common antenna that could be installed in a "community" (such as an apartment complex). The signals were distributed to individual homes via a coaxial cable tree-and-branch network. Cable TV got a boost in the 1970s from the satellite industry, when satellites made it possible for cable operators to receive additional channels and rebroadcast them on the cable network. These value-added channels enabled the market to grow rapidly beyond areas of poor reception. Today, cable systems carry a combination of local broadcast channels, special-interest channels (e.g., health, food, and travel), and premium channels (e.g., movies and sporting events). In the United States, about 60% of homes received video service via cable in 1998. An important observation about the history of the cable industry is that it developed as "islands of communication" within metro and rural areas. For example, the cable system (and operator) in Dallas is independent of the one in Houston.

■ **Satellite networks.** The origins of satellite communication can be traced back to Arthur C. Clarke's 1945 article in *Wireless World* entitled "Extra-Terrestrial Relays." It was not until the 1960s, however, that satellite technology got a boost with the prompting of the U.S. government. In 1961, NASA awarded contracts to AT&T, RCA, and Hughes Aircraft to begin developing satellite programs that were called TELSTAR, RELAY, and SYNCOM, respectively. The Communications Satellite Act of 1962 resulted in the formation of the Communications Satellite Corporation (COMSAT). In 1964, a similar international organization called INTELSAT was formed. Satellites played an important role in televising the 1964 Tokyo Olympics. In 1965, COMSAT's first satellite was launched based on geosynchronous technology for telecommunications. Global telecommunications are possible today because of the successful deployment of several geosynchronous satellites (GEOs) whose footprints cover most of the earth. In addition, several companies have committed themselves to satellites in low-earth and medium-earth orbits (LEOs and MEOs). Satellite companies have been investigating the use of satellite technology (in particular with LEOs and MEOs) to transport voice, data, and video. In the United States, one-way video broadcast has gained popularity due to the availability of digital technologies such as DBS and dish TV. An important observation about the satellite industry is that the satellite dish technology is often proprietary, which means that a dish receiver from one company is unlikely to work with one belonging to another company.

Architecture of Video Networks

Cable TV Networks

Although different cable TV network architectures have been implemented around the world, the majority of these architectures are variations of what is known as the "tree-and-branch architecture," as shown in Figure 1.14.

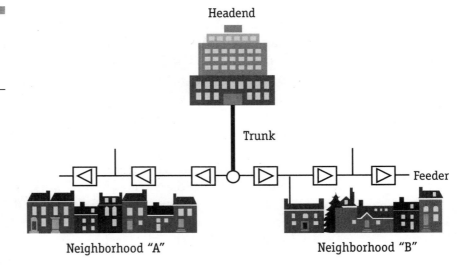

Figure 1.14

A tree-and-branch cable network architecture

This architecture is capable of delivering multiple channels of one-way distributed video service. Video signals from the different sources to be delivered to the subscriber are gathered at a central headend (equivalent to a central office switch in a telecommunications network). Typical sources of these signals are satellite earth stations, off-air antennas, videotape playback, and super trunking, which provides delivery of signals from studios at remote locations.

At the headend, multiple video sources at different frequencies are combined into a single broadband signal and transmitted over coaxial cables known as trunks (i.e., each trunk is a single coaxial cable). From the headend, various trunks radiate out to distribution plants in local neighborhoods. Extending from the distribution plants are feeder cables; each home is a drop off the feeder cable. Trunk and feeder amplifiers are required every 1,000 to 2,000 feet to overcome losses from cable branching and transmission. The electronic components in the network include the amplifiers and the converter boxes at the subscriber premises. The passive components in the network include the coaxial cable connectors, signal splitters, and signal taps for the drop. The network requires a power station for about every two miles of cable to drive the amplifiers. Hence, during a power failure, the cable system is inoperable, and cable services are lost.

Most cable TV networks currently use analog technology and coaxial cable in the backbone. One of the main reasons for using analog technology is that most of today's TVs receive only analog signals. It is not possible to easily change from analog TVs to digital ones because there are more than 200 million TVs in the United States alone. To save additional equipment costs for converting the signals, cable TV providers continue to use analog technology in their networks. However, analog technology can often cause quality-of-service problems in the network since analog signals are difficult to regenerate accurately.

The biggest concerns of the cable TV system are its limited capacity, one-way only services, and service quality that deteriorates as a result of noise and distortion. Modern cable TV network architecture calls for what is known as hybrid fiber/coax (HFC). In an HFC architecture, the coax cable for the trunks is replaced with fiber. HFC architecture offers several advantages including higher capacity, two-way communication, and better reliability. However, it also requires significant up-front capital investment to retrofit the legacy architecture.

Satellite Networks

Unlike cable TV networks, which rely on underground coax or fiber optic cables to carry the video signals, satellite networks utilize the airwaves as the medium to carry the signals.

As shown in Figure 1.15, a satellite network consists of earth stations that communicate with satellites in space. Satellites are placed into orbit around the earth so that they appear always to be in a fixed position over the earth. These are called *geosynchronous orbiting satellites*. Today, companies are also deploying low earth-orbit satellites and medium earth-orbit satellites, so called because of their relative position above the earth. The link from an earth station to a satellite is called an *uplink*; the reverse link (from the satellite to an earth station) is called a *downlink*. The part of the satellite that transmits the signal to the earth station is called a *transponder*. Satellites have amplifiers to boost the signals they receive from earth stations before retransmitting them on the downlinks. In addition, separate frequencies are used for uplinks and downlinks to avoid interference.

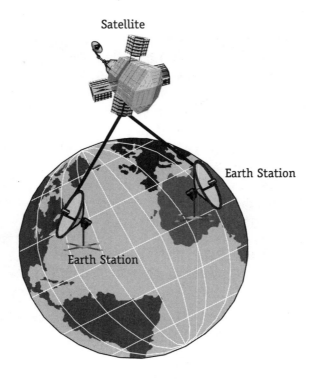

Figure 1.15
A satellite network

Figure 1.15
A satellite network

Satellite

Earth Station

Earth Station

A major concern of satellite networks is propagation delay. This is the time it takes for a signal to travel from the sending earth station up to the satellite, plus the time it takes for the satellite to amplify and convert the signal to a different frequency, plus the time it takes for the signal to travel to the receiving earth station. Propagation delays can range from 500 milliseconds to 3,000 milliseconds (3 seconds). Modern satellite technology, however, has improved upon transmission methods to reduce propagation delay.

Broadband Networks—The Convergence of Networks for High-Speed Communications

In the previous sections, we reviewed the history and architectures of networks based on their dominant application (voice, data, and video). We saw that the history and purpose for which these networks were developed heavily influenced their architectures. In recent times, the idea of a unified modern network architecture that can simultaneously deliver all three services has captured the imagination of technologists, business people, and politicians alike. The media has popularized the notion of an "Information Superhighway" that will seamlessly deliver these services to consumers.

Actually, this vision of an integrated communications network is not new. In 1984, when the ITU developed standards for the Integrated Services Digital Network (ISDN), it was hoped that this would provide a ubiquitous multimedia services network, which digitally integrated voice, data, and video. Affordable, ubiquitous, high-speed access to networked multimedia services is as much of a goal today as it was more than a decade ago.

How can this be possible? Have we not made technological advances in all these years? To answer these questions, let us consider the technology spirals illustrated in Figure 1.16.

Figure 1.16

Technology spirals of evolution of the computer and communication industries

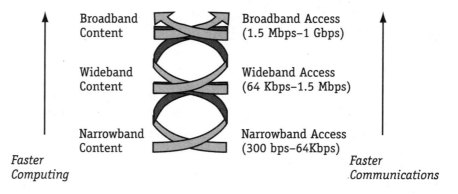

Broadband Content

Wideband Content

Narrowband Content

Broadband Access (1.5 Mbps–1 Gbps)

Wideband Access (64 Kbps–1.5 Mbps)

Narrowband Access (300 bps–64Kbps)

Faster Computing

Faster Communications

As Figure 1.16 illustrates, we have experienced spirals of evolution in both computing and communication technologies in the latter half of the twentieth century. Faster computers have resulted in increasingly enriched multimedia content. The communications infrastructure has evolved to meet the needs of delivering this content. As the communications infrastructure provided more bandwidth at affordable prices, computers have evolved in turn to take advantage of available bandwidth—thus completing the feedback cycle. Therefore, it is appropriate to think of the convergence of the computer and communication industries not as two straight lines that eventually meet in the future, but rather as two intertwining lines that are catalysts for each other's growth.

We should remember, though, the computer and communication industries have not uniformly grown apace. In particular, the communication industry has generally lagged behind in meeting the needs of the computer industry for access, as shown in Figure 1.17.

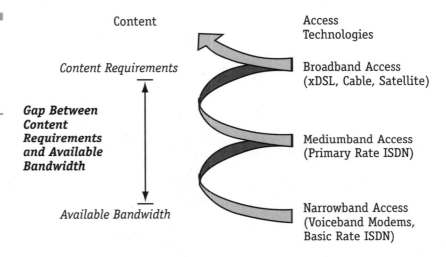

Figure 1.17
The gap between content require-ments and available access bandwidth

The reasons for this lag were:

- **Technological.** Until the 1990s, the cost of computing power to perform the complex signal processing algorithms necessary for high-speed communications was still relatively high.
- **Regulatory.** Until the advent of deregulation in the telecommunications industries, telcos were generally restricted to certain businesses. They were prevented from entering into certain markets, and therefore, were only motivated to invest in their core business of providing phone services (not, for example, in data delivery).
- **Economic.** As we have seen, networks were primarily designed to carry a specific dominant application. A significant change in the type of content to be delivered required a network redesign and sometimes, a totally new design. However, unless this would generate a profit, service providers had no incentive to upgrade their networks.

All of this began to change in the mid to late 1990s. As we shall find in the next section, several technological, economic, and regulatory factors have made the latter part of the twentieth century experience a dramatic acceleration in the need to upgrade existing voice, data, and video networks. The new upgraded network represents a convergence of the parts of the existing communications network to enable high-speed transport of several gigabits of information. This converged communications network, shown in Figure 1.18, is known as a broadband communications network, or simply a broadband network.

Figure 1.18
The convergence of voice, data, and video networks to broadband

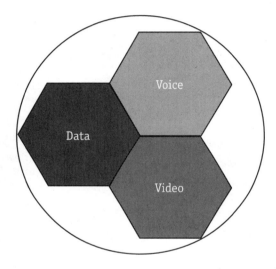

Technological, Economic, and Regulatory Drivers for Convergence

Technological Drivers

Some of the technology drivers pushing broadband convergence include:

- Availability of low-cost digital signal processors (DSPs) capable of executing complex algorithms in a cost-effective way
- Emergence of ATM as the unifying technology and architecture to carry voice, data, and video simultaneously over different kinds of transmission media
- Advances in fiber-optic transport technology leading to widespread deployment of the Synchronous Optical Network (SONET) in backbone networks

- Advances in "plug-and-play" technologies such as JINI[6] and Universal Plug and Play[7]
- Standardization efforts at various layers of the protocol stack
- An "always connected" model enabling new classes of applications such as automatic weather updates and stock quotes based on a predefined consumer profile.

Regulatory Drivers

Some of the regulatory drivers behind broadband convergence include:

- Environmental laws encouraging businesses to adopt telecommuting
- The Telecommunications Act of 1996 in the United States[8]
- The worldwide trend among countries to deregulate monopolies, for example in the United Kingdom and Japan[9].

Economic Drivers

Economic drivers include:

- Growth of the Internet
- The capability of the existing infrastructure to carry several megabits of data, i.e., copper loops, coaxial cable, and certain forms of wireless media
- Competitive pressures driving down the cost of computers, which has fueled the use of computers among consumers
- Demand for seamlessly integrated multimedia services over a common access network; customers want mobility, bandwidth on demand, security, flexibility, and end-to-end connectivity, among other options
- Revenue opportunities derived from market expansion to new classes of users and applications for the computer industry
- Revenue opportunities from services and equipment for the communications industry.

Chapter One

Evolutionary Process Toward Convergence

Although broadband promises to deliver multiple services on a unified network, the reality is difficult to achieve. The reason is that today's networks are historically designed to provide specific services in a cost-effective manner. These networks represent billions of dollars of investment over many years. It is cumbersome for a new broadband network to provide integrated services immediately and economically compete with existing specialized networks. Hence, to achieve the goal of a unified network, an evolutionary process must occur, and must include:

■ Provision of limited new services on an existing network
■ Enhancement of the network (or replacement of certain legacy portions of the network) to expand to new services while maintaining backward compatibility with existing services and revenue streams
■ Addition of new features to work with other networking services and systems to fill service gaps either temporarily or permanently.

In other words, new capabilities that have the best chance for success must first be implemented in the broadband network. Then other existing services must be slowly migrated as they can be economically justified.

Figure 1.19 illustrates the deficiencies of the current voice, data, and video networks. For example, the telephone network lacks broadband capability; the data network (Internet) lacks the necessary billing, security, and capacity; and the video network lacks switching, addressing, reverse channel capability for two-way communication, and intelligence to provide value-added services.

The broadband network must offer clear economic value to the customer by addressing the deficiencies highlighted in Figure 1.19 in a cost-effective manner. Fortunately, significant technological advances, such as ATM and SONET, allow service providers to migrate their existing networks so that they can offer broadband services.

Figure 1.20 illustrates an evolutionary path toward the broadband vision using ATM as the switching platform and SONET/SDH as the transmission platform. It also illustrates how the different existing networks can lead to the target broadband network by taking different paths suitable for the evolution from their current installed base. Figure 1.20 does not specify the timeline of the deployment of the different technologies; rather, it shows how some different technologies enable the evolution from an infrastructure point of view.

Figure 1.19
Deficiencies of current voice, data, and video networks

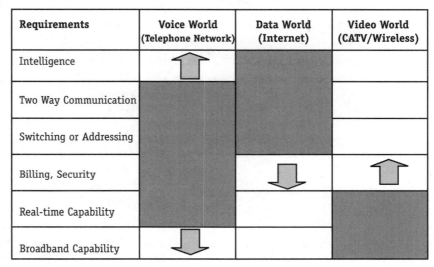

Figure 1.20
Evolutionary path toward broadband

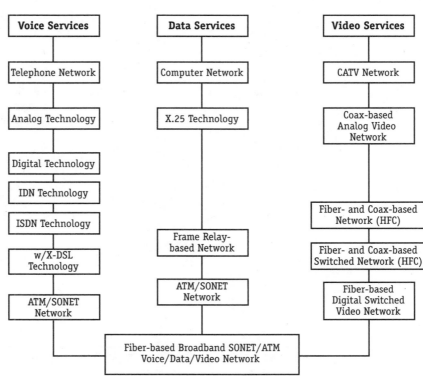

Table 1.1 shows a possible phased evolution from discrete narrowband/broadband networks to a converged broadband network that carries voice, data, and video.

Table 1.1

Possible phased evolution to a converged broadband network

	Phase 1	Phase 2	Phase 3	Phase 4
Description	Discrete voice, data, and video networks over narrowband/ broadband	High-speed data access over broadband	Integrated voice and high-speed data over broadband	Integrated voice, high-speed data, and video over broadband
Drivers		Consumers— pent-up demand for speed	Businesses— Need to share infrastructure	Service Providers— Need to maximize revenue
Market Segments*		• Consumers • Small business	• Consumers • Small business • Multi-dwelling units • Industrial	• Consumers • Small business • Multi-dwelling units • Industrial parks • Greenfield developments • Hospitality
Applications	Internet access	• High-speed Internet access	• Broadband access • Telephony	• Broadband access • Telephony • Local/net-work gaming • Video distribution • Video-conferencing • Content filtering • Utility metering • Appliance control • Industry-specific applications

* Individual market segments could evolve at different rates, and leapfrog over phases

Applications of Broadband Networks

The ultimate drivers for any networking technology are applications. In other words, unless the new technology can significantly add value to justify upgrading the existing infrastructure, no progress can take place. The evolution to broadband networks will succeed because of the plethora of both narrowband and broadband applications that such a network enhances and enables, as shown in Table 1.2.

Table 1.2

Possible applications

Narrowband Services Supported (Phase 1)	Narrowband Services Enhanced	Broadband Services Enabled
• E-mail • News groups • Chat rooms • Discussion lists	• Web surfing • Software distribution • Integrated voice and data access • Telecommuting • "Safe" Web browsing • Live multicast audio on demand (AoD) • E-commerce • Push services • Internet gaming • Remote home automation	• Live multicast video on demand (VoD) • High-quality Internet gaming • Videoconferencing/ videophone • Imaging

Let us look at a couple of applications to highlight the driving need for high-speed bandwidth. The first application, Internet access, is enhanced by broadband, while the second application, imaging, is enabled by it.

Internet Access

Few technological advances have had the impact on our society of the Internet. Today, access to the World Wide Web is an integral part of our society, touching businesses, educators, law enforcement, and families alike. The Internet is unique because it is a feedback loop that continually fuels its own explosive growth—a growth that seems far from abating. According to recent surveys, over 40% of U.S. households will be capable of high-speed Internet access by the dawn of the 21st century. Consumers are hungry for more than simple applications such as e-mail; they are looking for a rich, networked multimedia experience.

The Internet is fundamentally different from other communications technologies. It is unique because of its technical architecture, its lack of geographic boundaries, and its intrinsic fractal nature.[10] The Internet is driven by several self-reinforcing factors, which form a spiral of rapid evolution and explosive growth. At the center of this spiral is the key factor: access bandwidth.

Prior to 1994, traffic sent over the Internet was largely text-based information with file transfer and e-mail among the most popular services. The users were from universities and research institutions around the world. The surge in growth of the Internet during 1995 was due in part to the commercialization of the Internet for business use and the increasing graphical nature of the content. A significant aspect of this shift is that graphical images generally consist of a large number of bits. To transfer large graphical image files quickly with satisfactory performance meant that higher-speed access technologies were needed than those used to deliver relatively small text files. Each of these capabilities has been pushing the need for increasingly higher-speed access.

Internet access itself is not an application; rather, it is an enabler to a variety of applications including some narrowband services. Based on a BRG* survey, the following have been identified as potential applications that will drive Internet access demand. They are:

- **E-mail and messaging.** E-mail, file transfer, Web browsing, custom applications
- **Collaboration.** Decision-making applications, document management, project management, videoconferencing, and workgroup applications
- **Customer support.** Customer requests for information, customer service for product inquiries, order status, technical specifications, and troubleshooting
- **E-commerce.** Online customer orders and processing

---

* Market research firm specializing in information technology.

- **Information access/research.** Database access, database searches, and data warehousing
- **Internal/intranet information.** Access to employee information (e.g., human resources), technical information (e.g., reference manuals), and marketing information (e.g. competitive analysis).

Imaging

Imaging is defined as the process that digitizes and stores or retrieves documents, drawings, photographs, and other information in bitmapped format. Systems that digitize, compress, and store this information are called imaging systems. Imaging applications can be categorized as follows:

- **Capture and store intensive.** Items such as checks, drafts, and credit card slips are characterized by the need to process a large number of items quickly. These items are stored and retrieved only when the need arises. American Express, for example, uses imaging to capture all credit card receipts.
- **Retrieval and distribution intensive.** Items such as claim forms and loan documents are characterized by low volume and small storage needs, but a high frequency and need for nationwide retrieval. For example, in the insurance industry, a claim could be processed electronically by making an image of the claim document, thus enabling quick transfer for proper authorization.
- **Resolution and distribution intensive.** Items such as x-rays, scans, seismic data, and satellite photos are characterized by a unique, near one-to-one capture-to-access ratio with little need for online storage. They are usually stored in a tape backup system that can be retrieved in the future if the need arises. For example, a patient's x-rays could be digitized and electronically transferred from the x-ray lab to the doctor's office. This has tremendous benefit for rural communities where it is possible to have a specialist remotely "consult" on patient records. This application is known as *telemedicine*.

The better to understand why imaging is a broadband application, we should look at the size of the imaging files. Let us say that we are scanning a page at 300 dots per inch (dpi), using approximately 700K bytes of memory. If compressed, the page is between 80K and 250K bytes (depending on page content and compression ratio), which is about 0.25M to 2M bits (each byte is 8 bits). Using a 16-level gray scale, the size of the file is four times larger; with true color, it would be 24 times larger. In other words, a true-color image can involve a file of 40M (compressed). This application is currently LAN-based, but once it appears in MANs and WANs, it will become one of the predominant applications from a user perspective because it uses most of the available bandwidth. Clearly, the benefits of telemedicine would be impossible without a broadband network.

Summary

In this chapter, we reviewed several ways to categorize communications networks, particularly by their dominant applications. We studied the history and architecture of three major networks—voice, data, and video networks. Due to technological, regulatory, and economic drivers, each of these networks is evolving into a unified broadband network based on ATM as the switching platform and SONET/SDH as the transmission platform. This evolution allows service providers to migrate to new services in a cost-effective manner. Broadband enables several applications including Internet access, telecommuting, video-conferencing, video telephony, and imaging.

References

1. Information is available at the official Website at **www.homepna.org**.
2. Information is available at the official Website at **www.bluetooth.com**.
3. Information is available at the official Website at **www.homerf.org**.
4. ADSL Forum, "TR-010: Requirements and Reference Models for ADSL Access Networks: The 'SNAG' Document."
5. Chitre, P., and Yegenoglu, F., "Next-Generation Satellite Networks: Architectures and Implementations," *IEEE Communications Magazine*, Vol. 37, No. 3 (March 1999).

6. Information is available at the Website at **www.sun.com/jini**.

7. Information is available at the official Website at **www.upnp.org** or at **www.microsoft.com/homenet/default.htm.**

8. Telecommunications Act of 1996, Pub. L. No.104-104, 110 Stat. 56.

9. Ishikawa, H., and Nishimura, K. "Impact and Preliminary Results of Telecommunications Deregulation in Japan." *IEEE Communications Magazine*, Vol. 36, No. 7 (July 1998).

10. Rutkowski, A.M., "Internet as Fractal: Technology, Architecture, and Evolution," *The Internet as Paradigm* (Aspen Institute [Seminar], 1997).

Broadband Access Technologies and Architectures

Overview

Imagine a highway construction project where the on-ramps are either full of stoplights every few feet, under construction, or yet to be built. This analogy highlights the current situation with the "information superhighway." While politicians have rallied around the many benefits of bringing Internet access to the mass market, as a practical goal, this is difficult to achieve without a catalyst. As noted in Chapter 1, until the mid 1990s, the technological, regulatory, and economic drivers were not strong enough to provide any incentive for service providers to upgrade their access infrastructure.

The explosive Internet growth in the mid to late 1990s also exacerbated the problems caused by the lack of a readily available and affordable high-speed access infrastructure. Telcos began to feel the impact of public use of the voice infrastructure to transport data. Voice switches began to experience increased hold times counter to the engineering models on which they were initially deployed.*

In response to solving this "electronic traffic jam," the telecommunications, data networking, cable, and satellite industries began to accelerate efforts to pave the way for high-speed on-ramps. In this chapter, we shall review some of the important broadband technologies and architectures that have been proposed by these industries as solutions to the problem of providing "affordable, ubiquitous, high-speed" access. Let's begin by reviewing the key requirements. We will use them later as metrics to evaluate each solution.

Needs of Customers, Service Providers, and Equipment Manufacturers

The need for affordable, ubiquitous, high-speed access is driven by performance, choice, convenience, and reduced cost of ownership. The access network[1] is influenced by the needs of three distinct groups of people:

---

* Typical voice calls (on which voice switch deployments were patterned) are on the order of a few minutes. By contrast, data calls that use the voice network are several minutes to an hour in length.

■ Customers (end-users)
■ Service providers
■ Equipment manufacturers.

Needs of Customers

The following are the typical needs of customers:

■ **Utilization of existing access infrastructure.** Residential and small office/home office (SOHO) users need to migrate to multimedia services without replacing their existing wiring and investment in devices such as PCs, analog telephone sets, and fax machines. Businesses need to integrate voice and data services in order to achieve economies of scale. Most businesses today maintain separate infrastructures for voice and data, which is inefficient and expensive. Businesses also have telecommuters who need to access the same infrastructure as those working on-site, including not only the LAN, but also the PBX.

■ **Low cost of service and equipment.** Customers need services with low maintenance costs. The cost of installation and equipment must also not be a barrier.

■ **Upgradable services.** Residential and SOHO users need the ability to incrementally and economically add components of multimedia services such as supplementary voice and data services and pay per view. This is because they share a broadband link for business and entertainment, even though their needs may be quite different.

■ **Availability.** Residential/SOHO users need to be able to access and choose simultaneously between multiple service providers transparently and seamlessly.

Needs of Service Providers

The typical needs of service providers include:

■ **Utilization of existing access infrastructure.** Service providers need to be able to utilize their infrastructure efficiently to maximize their investment and reduce the cost of widespread deployment. For example, telcos are motivated to preserve their existing copper loops and maximize the number of services they can deliver over a single copper pair.

- **Low cost of equipment.** Service providers need cost-effective ways to deploy and manage emerging network services.
- **Upgradable services.** Service providers need the ability to offer widely deployed services to a variety of customers from residences to small- and medium-sized businesses to large corporations.

Needs of Equipment Manufacturers

Equipment manufacturers typically need:

- **Market acceptance.** Manufacturers need to judge market acceptance of enhanced features and services before they invest in technology innovations.
- **Universal standards.** Manufacturers need standard interfaces between the customer premises network and the access network. This allows them to lower their total delivered costs.

Key Requirements

From the needs discussed in the previous section, we can identify key requirements based on the three attributes *affordable, ubiquitous, high-speed* access.

Affordable Access

Utilization of Existing Access Infrastructure

Infrastructure covers several aspects. New services should ideally make use of existing wiring, equipment, standards, and billing systems. Furthermore, providers should not have to incur large capital expenses to upgrade their networks for these new services. Ideally, it should be possible to upgrade the network "as you grow," i.e., upgrade only those portions of the network required for a paying customer.

Cost of Equipment

The cost of equipment must not be an inhibiting factor. Consumers are sensitive to price differences of as little as $50; therefore, new services must not require significant upgrade costs at the premises. Typically, customer premises equipment (CPE) costs should be $250 or less to make transition to the new service attractive for consumers. Installation costs are also an important consideration. In a competitive market, installation costs are typically waived or factored into the first few months of service.

Cost of Service

The cost of the service should also not be a deterrent. Savvy consumers understand that ongoing costs far outweigh the one-time cost of equipment. In the United States, consumers have gotten accustomed to monthly dial-up Internet access fees of $19.95 (or less) plus monthly charges for a phone line. Therefore, access fees significantly greater (for example, $100 a month) are likely to be met with some resistance. Businesses, however, may be willing to pay higher access fees in return for guaranteed service quality. Even so, competitive pressures continue to force businesses to keep their ongoing costs down.

Universal Standards

Standards are important in driving volume and reducing costs to enable mass-market adoption of new services. New services must be based on technology that is deployable worldwide with strong support from both the computer and communication industries, including service providers and equipment manufacturers. Proprietary standards promoted by a single vendor are unlikely to result in reduced costs.

Scalability

Scalability implies coverage and resilience to rapid growth. Coverage means that the service can be deployed to reach as many potential customers as possible. Resilience to rapid growth means that the service can continue to be deployed at the initial levels of high quality without violating customer expectations or service-level agreements (e.g., service degradation or availability issues).

Ubiquitous Access

Market Acceptance

In the initial months of deployment, market acceptance of new services by several users is critical to providing the feedback loop that generates the fuel for further expansion. Equipment manufacturers need to be assured of widespread market acceptance before they will make the necessary investment in technology to develop products.

Availability

Affordable technology, in itself, is not sufficient for mass deployment. Regardless of a technology's affordability or performance, if users are unable to purchase the equipment and/or access new services due to scarcity, they will be turned off and the market won't get an opportunity to grow beyond initial trials.

Ease of Deployment

The service must be easy to provision and order. ISDN is often cited as an example of a service perceived to be too complex to provision, order, and install. Hence, it has met with limited acceptance in the worldwide marketplace. (It has, however, been successfully mass deployed in certain countries where the sole service providers have actively promoted the service and deployed a limited set of options to make the installation less complex.)

Ease of Use

Consumers today are familiar with the paradigm of hooking up the built-in modem in their computers to the phone jack in the wall, configuring the modem using their computer's operating system, and connecting to the Internet. Any new service must be just as easy to set up and use. Automatic service provisioning and configuration is a valuable mechanism to reduce complexity.

High-speed Access

Speed Beyond Voiceband Modems

The Internet has often been referred to as the "World Wide Wait," as a result of users frustrated with slow access. Even 56K modems are barely sufficient to meet the bandwidth requirements of rich, dynamic, multimedia content. High-speed bandwidth to match broadband content is, therefore, a key requirement.

Broadband Access Technologies

Broadband access technologies have been spawned by the efforts of the telecommunications and video networking industries to move toward convergence at the physical layer. By contrast, the efforts of the data networking industry toward convergence have generally focused on the networking and transport layers (e.g., IP telephony). The reason for this difference in approach is that, once content has been digitized, it is treated as data. In other words, from a networking perspective, the network merely transports data in digital format. It makes no difference whether the bits actually represent voice, data, or video streams at the application level. Consequently, the data networking industry did not have to do anything different to adapt a data stream for transport across a network. On the other hand, the telecommunications and video networking industries had to figure out ways to transport their content as packetized data cost-effectively by working with existing access infrastructure. This was an important catalyst in the development of the following classes of broadband access technologies:

- **Copper-loop access technologies**, also known as digital subscriber line (DSL) technologies. DSL technologies are collectively referred to as XDSL. XDSL was developed by the telecommunications industry to make use of the several million miles of existing copper loop infrastructure around the world.
- **Cable access technologies over fiber/coaxial cable.** Cable access was developed by the cable TV portion of the video networking industry to take advantage of the cable infrastructure that feeds

video channels to several million (mostly residential) subscribers around the world.

■ **Satellite access technologies over wireless medium.** Satellite access was developed by the wireless portion of the video networking industry to make use of the satellite infrastructure that feeds video channels to cable TV headend offices and to several hundred (mostly residential) subscribers around the world.

Copper-Loop Access Technologies (XDSL)

The roots of copper-loop access technologies can be traced back to ISDN. In the early 1980s, the idea of a digital subscriber line to provide access to an integrated services digital network was initiated. The transmission throughput requirement was 160 Kbps (2B+1D channels for ISDN BRI, plus overhead). In the late 1980s, with advances in digital signal processing, there was motivation to investigate even higher transmission throughput that would approach T1 speeds. The project was named high bit-rate digital subscriber line (HDSL). HDSL was a way of providing T1 service without the need for intermediate devices called *repeaters*. Later, in the early 1990s, Bellcore* engineers realized that not only symmetric services, but also asymmetric services, could be supported. The asymmetry would give higher bandwidth in one direction over the other (typically, higher downstream than upstream bandwidth). This asymmetry was well suited to the video on demand (VoD) trials being proposed at that time, and the development of asymmetric digital subscriber line (ADSL) resulted. Further advances in technology allowed engineers to consider even higher speeds than ADSL, but over shorter loop lengths. This resulted in the development of very high bit-rate digital subscriber line (VDSL).

There are now a plethora of DSL technologies, collectively known as XDSL. XDSL enables telephone operating companies with existing twisted-pair copper loops to offer single/multiple video channels or high-speed data (symmetrical and asymmetrical bandwidth) along with telephone service. Since XDSL uses existing loops, it is economically attractive for service providers to implement new services, such as high-speed Internet access. In 1999, for example, there were approximately

---

* In 1999, Bellcore renamed itself Telcordia Technologies.

150 million telephone lines in the United States, and approximately 870 million lines worldwide. Of these, about 70% of the copper loops providing the telephone service are within 18,000 feet of the customer premises and, therefore, are potential XDSL candidates. In general, all XDSL technologies share the following characteristics:

■ They are copper-loop access technologies, also known as "last mile" technologies since they operate over the last mile to the customer premises. In other words, XDSL technologies represent a radical departure for a mass deployed telecommunications infrastructure. They do not provide end-to-end connection through the switched voice network, as with POTS or ISDN; rather, they provide each customer with a dedicated copper pair (or pairs) to the central office.
■ All XDSL technologies are designed for point-to-point connections between the premises and the telcos' central offices. Therefore, they usually require equipment based on identical technologies at both ends of the copper loop.

The following XDSL technologies have either been standardized or are undergoing standardization:*

■ ADSL
■ ADSL "lite"
■ HDSL
■ HDSL2
■ VDSL

Each of the above technologies is discussed in greater detail in Chapters 4, 5, and 6. For now, we will do a quick overview.

ADSL

ADSL was first developed in the 1980s as the telecommunications industry's answer to the cable industry' request to support video on

---

* Not all XDSL technologies have been standardized. Proprietary variants of DSL technology have contributed to the "alphabet soup" of DSL technologies. Proprietary implementations require the same vendor's implementation at both ends of the wire, which "locks in" a service provider to that particular vendor. In large part due to a push from the providers, the industry is moving toward convergence on XDSL technologies that are standardized.

demand. In the mid 1990s, however, it was quickly recognized as a viable technology to enable access to high-speed services such as the Internet. ADSL delivers *asymmetric* transmission rates typically up to 9 Mbps downstream (from the CO to the premises) and 16 Kbps to 640 Kbps upstream (from the premises to the CO) as shown in Figure 2.1. Like all copper transmission systems, the higher the bit rate, the shorter the range.

On the other hand, loop reach is of paramount importance to a telco providing new service. It depends on the routing factor, which is a function of the radius of loop length, wire gauge, and the type/location of the central office and surrounding area. The longer the loop reach, the bigger the circle of coverage; hence, the higher probability of serving new customers from a single CO. Therefore, most operators are currently providing service at rates below the highest achievable rates to ensure a minimum rate for customers within a serving area. This also enables operators to study the technological and economic impact of the new service on their existing backbones before ratcheting up access bandwidth. ADSL is specifically designed to operate on the line at the same time as POTS by operating at frequencies above the voice band. ADSL has been widely tested around the world, and the initial service offerings began to appear in late 1996. ADSL deployments, with limited availability, began in 1998; large-scale deployments started in 1999.

There are several standards bodies associated with ADSL. In the United States, the American National Standards Institute (ANSI) T1E1.4 committee[2] developed the original physical-layer standard for ADSL known as T1.413. In Europe, the European Technical Standards Institute (ETSI)[3] subsequently built upon the work of the ANSI group. On a worldwide basis, the International Telecommunications Union (ITU)[4] determined the G.992.1 standard building upon the recommendations of ANSI, ETSI, and the work of several other countries. For the higher layers, the ADSL Forum[5] has developed recommendations to address ADSL access architectures. The ADSL Forum has also worked in concert with the ATM Forum[6] to address specific issues of ADSL with ATM transport.

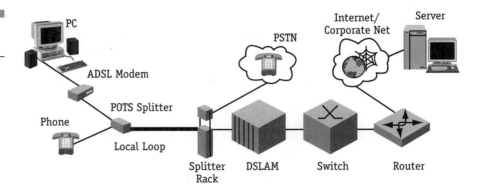

Figure 2.1

ADSL architecture

ADSL "Lite"

A new form of ADSL, known as ADSL "lite," "splitterless ADSL," or "consumer ADSL" emerged in early 1998. These terms are used to connote a form of ADSL optimized for mass deployable applications such as high-speed Internet access. As mentioned before, ADSL was originally developed to satisfy VoD. With the shift in focus to the Internet, the industry discovered that a simpler form of ADSL, i.e., a "lite" version, could be more economically deployed for consumer, rather than commercial, deployment. The downstream bandwidth requirements for consumer use are lower than full-rate ADSL, i.e., in the 1.5 Mbps range rather than the 9 Mbps range. Another change was the motivation to eliminate (or at least make optional) the premises splitter required by full-rate ADSL, as shown in Figure 2.2. If we compare the "lite" architecture with full-rate ADSL, we can see that only the premises network configuration changes; the rest of the end-to-end architecture remains the same as in full-rate ADSL. This is an important benefit to service providers.

As in the case of full-rate ADSL, there are several standards bodies associated with the lite version. The most significant event in the history of Lite ADSL was the formation of the Universal ADSL Working Group (UAWG)[7]. The UAWG, while not strictly a standards body, is an ad hoc body expressly formed by industry leaders to accelerate the standards process for lite ADSL. The UAWG is composed of representatives from the computer, telecommunications, and networking industries and represents significant technological and marketing clout. The efforts of the UAWG resulted in several contributions to the ITU, most of which were incorporated into G.992.2 (formerly known as "G.lite"), the international physical-layer standard for lite ADSL. At the higher layers, bodies such as the ADSL Forum and ATM Forum have embraced the evolution of G.992.2 to ensure its fit into the overall access network architecture.

Figure 2.2
ADSL "lite"
architecture

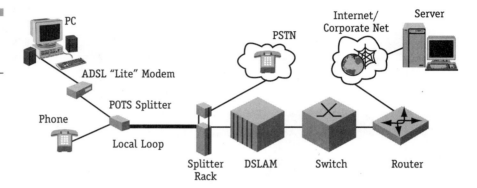

HDSL

In contrast to ADSL, HDSL is a *symmetric* baseband transmission system; that is, the data rates for the upstream and downstream directions are the same. It requires two copper pairs for T1 bit rates or two or three copper pairs for E1 bit rates, as shown in Figure 2.3. HDSL technology using two or three pairs is mature now and has been commercially deployed by many telcos in Europe and the United States over the past several years to carry T1/E1 services. The typical range of such systems is around 9,000 to 12,000 feet. HDSL was initially used by telcos for T1 services, mostly for PBX connectivity. Later, with the growth of the Internet, it was the choice for cost-effective Internet connectivity.

Figure 2.3
HDSL architecture

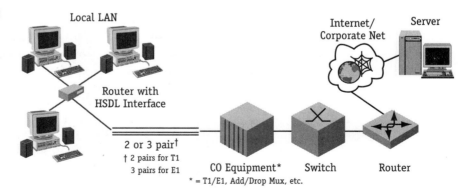

HDSL2

HDSL has enjoyed success as a T1/E1 replacement for many years. However, service providers have recently become motivated to find ways to maximize the use of their existing infrastructure to achieve economies of scale and reduce costs in a competitive environment. For example, copper pairs to businesses are becoming increasingly valuable as local telcos are under pressure to unbundle these pairs and lease them to their competitors. Additionally, telcos are facing increasing demands from customers to add additional lines. They must therefore find ways to conserve copper by maximizing the number of services over a single pair. This has provided the impetus for the next generation of HDSL, aptly called HDSL2, which is currently under development in standards committees. HDSL2 will offer the same deployment advantages as its predecessor, but will require only a single copper pair. As with the development process for the ADSL standard, the ANSI involvement in HDSL2 has been taken up by the ITU, which is working toward developing an international version of the standard. The finalized specification is expected to meet three critical requirements proposed by the ANSI T1E1.4 subcommittee:

▪ Full CSA (carrier serving area) loop reach to 12,000 feet
▪ Spectral compatibility with other services in the same cable
▪ Interoperability of equipment from different vendors.

In other words, HDSL2 will deliver full-duplex (symmetric) T1/E1 payload (2.3 Mbps) over one copper loop.

VDSL

VDSL technology is well suited for access architectures such as fiber-to-the-curb (FTTC) that use fiber. In simple terms, VDSL transmits high-speed data over short reaches of twisted-pair copper telephone lines, with a range of speeds dependent upon actual line length. The maximum downstream rate under consideration is 52 Mbps over lines up to 1,000 feet in length. Downstream speeds as low as 1.5 Mbps over lengths beyond 12,000 feet are also in the picture. Upstream rates in early models will be asymmetric, just as with ADSL, at speeds from 1.6 Mbps to 2.3 Mbps. In VDSL, both data channels will be separated in frequency from bands used for POTS and ISDN, enabling service providers to overlay VDSL on existing services. At present, the two high-speed channels will also be separated in frequency. As needs arise for higher-speed upstream channels or symmetric rates, VDSL systems may have to use echo cancellation.

While VDSL has not achieved the same degree of maturity as ADSL, it has advanced far enough to make it possible to begin discussions concerning realizable goals and standardizing data rates and range. Downstream rates derive from sub-multiples of the SONET and SDH canonical speed of 155.52 Mbps, namely 51.84 Mbps, 25.92 Mbps, and 12.96 Mbps. Each rate has a corresponding target range as shown in Table 2.1.

Table 2.1
VDSL bandwidth
vs. range

Bandwidth	Range
12.96–13.8 Mbps	4,500 ft (1,500 m)
25.92–27.6 Mbps	3,000 ft (1,000 m)
51.84–55.2 Mbps	1,000 ft (300 m)

Early versions of VDSL will certainly incorporate the slower asymmetric rate, but higher upstream and symmetric configurations may only be possible for very short lines.

Like ADSL, VDSL must transmit compressed video, which is a real-time signal unsuited to the error retransmission schemes used in data communications. To achieve error rates compatible with compressed video, VDSL will have to incorporate forward error correction (FEC) with sufficient interleaving to correct all errors created by impulsive noise events of some specified duration. However, interleaving introduces delay on the order of 40 times the maximum length of correctable impulse.

Summary of DSL Technologies

XDSL technologies can be categorized as:

■ **Asymmetric or symmetric.** Examples of asymmetric DSL technologies are ADSL and ADSL lite. Examples of symmetric DSL technologies are HDSL and SDSL.

■ **Single or multiple-pair.** As an example, SDSL uses a single copper pair; HDSL operates over two or three copper pairs.

■ **POTS band frequency overlap or non-overlap.** HDSL overlaps POTS frequencies, whereas ADSL does not overlap POTS in the 0–4 kHz range (so POTS can be concurrently supported with ADSL).

■ **Splitterless or splitter.** ADSL lite does not require a splitter (to separate the POTS band frequencies from ADSL frequencies); ADSL requires a splitter.

■ **Fixed or variable rate.** HDSL and SDSL provide fixed data rates up to a specific distance, whereas with ADSL, ADSL Lite, and VDSL, the data rates depend upon the loop distance.*

Although there are a variety of DSL technologies, each satisfies a particular market segment and requirements as summarized in Table 2.2.

Table 2.2

Summary of DSL technologies

	Asymmetric/ Symmetric	Features	Simultaneous POTS Support
ADSL	Asymmetric	Up to 8 Mbps downstream/ 640 Kbps upstream; single pair	Yes
ADSL Lite	Asymmetric	Up to 1.5 Mbps downstream/ 500 Kbps upstream; single pair	Yes, does not require a POTS splitter
HDSL	Symmetric	1.5 or 2.0 Mbps in either direction; 2 pair	No
SDSL	Symmetric	1.5 or 2.0 Mbps in either direction; single pair	Possibly
VDSL	Asymmetric; symmetric operation under discussion	Up to 52 Mbps downstream/ 1.5 Mbps upstream; single pair	Yes

---

* In some cases, the data rate can be dynamically adjusted to varying loop conditions caused by changes in the environment (e.g., temperature). This is known as rate adaptive DSL (RADSL), and it can operate in either symmetric or asymmetric mode.

Cable Access Technologies

Cable access technologies are primarily designed for transport of video signals. As of 1998, there were three basic technologies that leveraged cable infrastructure:

- AM (amplitude modulation)
- FM (frequency modulation)
- Digital.

Since the cable TV system based on coaxial analog technology cannot provide any of the additional channels or bandwidth necessary to provide residential broadband services, cable TV providers are required to adapt to new technology. AM and FM systems are familiar modulation schemes to most cable TV network providers, but digital technology is not. Therefore, a general trend among cable TV network designers is to continue to maintain or improve AM and FM technology for network upgrades. Another reason for delaying the deployment of digital technology in the cable network is that most TVs are still analog. Digital-to-analog conversion requires additional equipment, which is expensive.

In the long term, however, digital technology offers the most promise. A digital system does not require expensive telemetering and monitoring; hence, it can be built at a much lower cost. The advantages of using digital technology in the cable TV system are performance (far superior to the AM/FM system), ease of migration from a coaxial to a fiber-based system, quality of service (QoS), and long-term cost effectiveness. Digital technology also enables easy provisioning of additional services and channels. When the digital signal is compressed, more services can be provided since the bandwidth per channel is decreased.

Table 2.3 compares the bandwidth, cable length, distortion, video signal-to-noise ratio (SNR), and audio dynamic range of the three systems.

Fiber optic technology in cable TV networks is also important. Fiber's inherent characteristics eliminate most of the bottlenecks that currently exist in many cable TV networks. It can increase the channel capacity and reduce the number of amplifiers in the network, which in turn reduces distortion. In a coaxial system, there are repeaters every 1,000 to 2,000 feet to regenerate the signal, causing the deployment of numerous amplifiers in the network at additional capital, maintenance, and operations costs. By contrast, with fiber deployment, a typical cable

TV network operator can reduce the number of amplifiers to a handful, and still get substantial gain in bandwidth and reductions in cost. Fiber can be used to carry both analog and digital signals, so it does not need to be replaced if the technology evolves.

Table 2.3

Cable TV technologies

AM System	FM System	Digital System
Cable length: 18.75 miles	Cable length: 25 miles	Cable length: 25 miles
Bandwidth: 40 channels (420 MHz)	Bandwidth/fiber: 16 channels (700 MHz)	Bandwidth/fiber: 8–12 channels (600 MHz–1.2 GHz)
Intermodulation and reflection distortions	Intermodulation noise	No intermodulation noise; however, there is noise from analog-to-digital and digital-to-analog conversions
Channel spacing design	Channel spacing design	Channel spacing design not required
Regeneration on amplifier generates distortions	Regeneration on amplifier generates distortions	Regeneration without distortions
Video SNR: 55 dB	Video SNR: 65 dB	Video SNR: 57–67 dB
Audio dynamic range: 65 dB	Audio dynamic range: 65 dB	Audio dynamic range: 65–85 dB

In the mid 1990s, cable operators became excited about the possibility of using the cable infrastructure to provide broadband access, especially to residences. This resulted in the development of devices called cable modems.* A cable modem allows high-speed data access via a hybrid fiber/coax (HFC) network (described later in this chapter). Unlike traditional dial-up modems, cable modems are part of a point-to-multipoint system. In other words, multiple cable modems (from different subscribers) connect to a controller device at the headend owned by the cable operator. Like DSL modems, cable modems are "always on," meaning that they continuously communicate with the headend.

The downstream transmissions from the headend to the cable modem are carried on channels that are shared by multiple cable

---

* The term "cable modem" is somewhat misleading. A cable modem works more like a local area network interface card than a traditional modem.

modems. Each channel occupies a 6-Hz slice of downstream bandwidth, which translates to a raw data rate of 30 Mbps. The downstream channel can be tuned to any center frequency in the range from 65 to 850 MHz. Each cable modem is capable of receiving on one downstream channel at a time, but can be commanded by the headend to shift to an alternate downstream channel. Similarly, in the upstream direction, each channel occupies a 600-kHz slice of upstream bandwidth, which translates to a raw data rate of 768 Kbps. The downstream channel can be tuned to any center frequency in the range from 5 to 65 MHz.

As with the telecommunications industry, the cable industry got organized to develop standards for cable modems. There have been both standards organizations activities and industry ad hoc group activities:

- The 802.14 Working Group is the official group chartered by the Institute of Electrical and Electronics Engineers (IEEE) to create standards for data transport over traditional cable TV networks. This group was formed in the early 1990s with the intent of developing a specification. However, the group's efforts got bogged down in developing a technically elegant solution. They were further undermined when several North American cable operators chose to go their own way with an alternative specification called DOCSIS (see below). At this time, the future of the IEEE 802.14 specification is unknown. Although it was once believed that 802.14 would become a standard outside North America and/or would be used for business services, neither of these prospects seems likely today.

- CableLabs[8] was established in 1988 as a research and development consortium of cable television system operators. It represents about 85% of the cable subscribers in the United States, about 70% of the subscribers in Canada, and about 10% of the subscribers in Mexico. CableLabs administers an initiative aimed at the development and deployment of cable modem technologies called the Data Over Cable Service Interface Specification (DOCSIS) Project. This project is headed by a consortium of large cable operators called MCNS Holdings, LP (MCNS). MCNS, other cable companies, and CableLabs have cooperated to develop a common interface specification to be shared by all cable modems and their headend equipment. In part, the reason for the DOCSIS specification was that cable operators felt that the IEEE 802.14 specification would take too long to evolve to meet their deployment schedules. There was also a considerable difference of opinion between the two groups over the upper layer implementation—the MCNS cable operators preferred the IP-over-Ethernet approach; the IEEE 802.14 committee favored the ATM approach.

Satellite Technology

The electromagnetic spectrum is divided into several bands for multiple uses, as shown in Table 2.4.

Table 2.4

Allocation of bands in the electromagnetic spectrum

Bands	Frequency Range	Use
Very Low Frequency (VLF)	3 kHz–30 kHz	
Low Frequency (LF)	30 kHz–300 kHz	
Medium Frequency (MF)	300 kHz–3 MHz	AM Radio (500 kHz–1.6 MHz)
Short Waves	3 MHz–30 MHz	Citizen's Band Radio (26 MHz)
Very High Frequency (VHF)	30 MHz–300 MHz	FM Radio (88 MHz–106 MHz)
		VHF TV Channels (174–210 MHz)
Ultra High Frequency (UHF)	300 MHz–3,000 MHz	UHF TV Channels (470–884 MHz)
Super High Frequency (SHF)	3,000 MHz–30,000 MHz	Satellite C Band (3.4–4.8 GHz)
		Satellite Ku Band (10.7–12.75 GHz)
		Satellite Ka Band (19–22 GHz)

For satellite technology, it is the super high frequency (SHF) bands that are of interest. The ITU has allocated the SHF bands between 2.5 and 22 GHz for satellite transmissions. Since these are very high frequencies, the corresponding wavelengths are very short. That is why they are called *microwaves*. Microwaves travel directly along the line of sight (LOS) between a satellite and earth stations (that transmit or receive the signals). An advantage of microwaves at these frequencies is that they are not impeded by the earth's ionosphere, which means the signals can be beamed up to the satellite and back down again to the receiving earth station without significant attenuation loss.

Satellites are categorized by their distance from the earth. Initially, satellites operated in a geosynchronous orbit, which meant the speed of the satellite was synchronized so that it appeared to be stationary over a fixed geographical location above the equator. As the number of available "slots" in space got crowded, medium-earth orbit satellites and low-earth orbit satellites were introduced, both of which orbit around the earth at the poles at a lower altitude.

As shown in Table 2.5, satellite frequency bands are often referred to by their letter designations—a legacy from the days of microwave radar systems.

Table 2.5

Designation of satellite frequencies

Range	Letter Designation
2–3 GHz	S band
3–6 GHz	C band
7–9 GHz	X band
10–17 GHz	Ku band
18–22 GHz	Ka band

In the 1960s, the first satellites operated in the C band; however, since the frequency range in which they operated overlapped with many terrestrial microwave links, the power of C-band satellites had to be limited to avoid interference with earth-based links. Later, in the 1970s, satellites in the Ku band began to appear. Since the frequency range of Ku-band satellites did not overlap with terrestrial microwave, they could operate at much higher signal strengths than their C-band counterparts. For example, while C-band satellites often transmit at 33–38 dBW (decibels per watt), Ku-band satellites transmit at 47–52 dBW.

Ku-band satellites have another advantage over C-band satellites. Since they operate at higher frequencies, the wavelengths are much shorter, so the receiving antenna's beam width is much narrower. The combination of higher transmit power and narrower beam width results in a cost-effective satellite dish design that can be widely deployed because the signal can be accurately received in more locations.

Unlike copper-loop and cable-access technologies, satellite-access technologies have not been standardized as of 1999. Further, it is unlikely that standardization will occur any time soon since the number of satellite companies is naturally quite small, and each will choose to differentiate its services based on proprietary technologies.

Broadband Access Architecture

The following are examples of prevalent broadband access architectures:

∎ Integrated digital loop carrier (IDLC)
∎ Hybrid fiber/coax (HFC)
∎ Fiber-to-the-curb (FTTC) and other fiber architectures
∎ Multi-channel multipoint distribution service (MMDS)
∎ Local multipoint distribution service (LMDS)
∎ Direct broadcast satellite (DBS).

Some of the above architectures are discussed in greater detail in Chapter 8 of this book. For now, a quick overview will suffice.

IDLC (Integrated Digital Loop Carrier)

Currently, IDLC is the most widely deployed equipment in the public telephone network used to bring digital technology closer to subscribers. IDLC equipment uses fiber/digital technology to support various services: The access network rates are DS1, fractional DS1, 64 Kbps, and BRI. The transport network rates are usually DS1, DS3, OC-3, or OC-12. Most of the deployed equipment is SONET compatible and based on Bellcore specifications (as contained in TR-303 or GR-303). It provides a migration path from current analog to SONET-based transport and is designed to support both basic and enhanced telephony services. However, the hardware, software, and architectural features vary from vendor to vendor. In general, most equipment providers support the following system features:

- Remote software provisioning capability
- Compact and rugged shelf design to sustain outside plant operation
- Operations, administration, maintenance, and provisioning (OAM&P) features
- Flexible hardware architecture to accommodate future software upgrades
- Digital service to the end-users, if needed.

This architecture can be complemented with DSL technologies as the next phase of migration toward an end-to-end broadband network designed to handle the convergence of services. An IDLC-based system architecture is depicted in Figure 2.4.

Figure 2.4
IDLC architecture

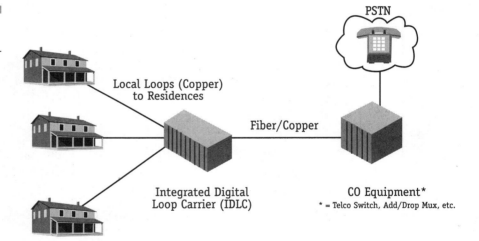

Hybrid Fiber/coax (HFC)

Hybrid fiber/coax (HFC) architecture was once considered the easiest and most logical way to provide multiple services to the end-user via the existing cable TV infrastructure because of the high bandwidth drop needed for each customer. Theoretically, HFC is capable of supporting all existing and emerging narrowband and broadband services including telephony, TV broadcasting, video dialtone, video on demand, and distance learning. It can support both analog and digital services and offers a migration path to an all-digital network in the last mile.

Figure 2.5

HFC architecture

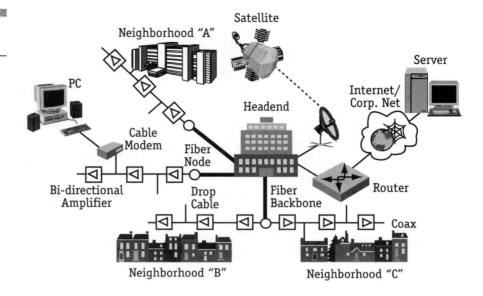

Figure 2.5 shows the HFC architecture from the headend down to the interface node between the fiber and coaxial cable. In HFC-based cable TV systems, the broadcast video and switched video signals are transported via fiber to the optical network interface (ONI). The ONI connects the fiber backbone to the coaxial cable distribution plant. At the ONI, the signals are frequency-shifted to the appropriate channel and fed to an amplifier for transmission over coaxial cable. The conventional analog, video, data, and voice signals can be carried simultaneously in different frequencies. The final segment of coaxial cable requires two-way amplifiers for the bidirectional signals on the cable. The optical network unit (ONU) performs additional functions such as separation of upstream and downstream signals.

Some of the current issues with HFC architectures for providing telephony service are grade of service, network-powering requirement, and ingress noise generated by non-cable subscribers. Also, the reverse channel currently uses the low frequency range that is bandwidth limited. For data services, the issue is still reverse bandwidth and ingress noise. Alternative proposals for reverse bandwidth, such as using telephone lines, are being considered.

HFC architectures may fall into one of three categories based on the signal type used:

■ **All-analog HFC.** In this case, the downstream bandwidth is utilized to provide broadcast analog video channels, premium pay-per-view (PPV) channels, and/or subscription channels. The upstream bandwidth in not typically used in all-analog HFC implementations.

■ **Mixture of analog and digital HFC.** This architecture permits transmission of both analog and digital signals and makes use of the back channels to provide interactive television, video on demand, and administrative features. A set-top box is required at the subscriber's premises to decompress and convert digital signals to analog format.

■ **All-digital HFC.** In this case, the digital signals travel via fiber and coax to a set-top box located on the subscriber premises. The back channel is used for control over content received through the forward channels and for administrative features.

Capacity

Hybrid fiber/coax utilizes the 0–50 MHz band for upstream traffic and the 50–750 MHz band, which translates to approximately a hundred 6-MHz analog video channels, for downstream traffic. In the case of digital signaling, each 6 MHz of the carrier channel can carry 27–38 Mbps of digital payload. A compressed digital video channel of VHS quality translates to 1.544 Mbps. By comparison, a video channel of NTSC (National Television System Committee) quality requires 7 Mbps. Typically, the lower end of the 50–450 MHz frequency is used for analog video. The upper end, from 450 to 750 MHz, is reserved for digital video. Some HFC implementations may use bandwidth above 850 MHz for upstream signaling.

One proposal for an HFC network claims a capacity of 70 analog video channels and 240 compressed digital video channels for broadcast purposes and 80 compressed digital video channels for VoD services (quite a leap from 40 cable television channels). This network utilizes the bandwidth from 0 to 30 MHz for upstream and from 30 to 750 MHz for downstream.

Services Supported

Hybrid fiber/coax networks offer a flexible suite of analog and digital services. Current analog television equipment is adequate to receive basic analog video through HFC. A digital television or a digital-to-analog converter within a set-top box and an analog television are required to receive digital video using HFC. It is also possible to send telephone signals through the HFC network by interfacing local central office switches with the headend. HFC, however, does not provide good telephone connections due to the use of a shared reverse channel, which can cause interference problems.* The following services are possible over HFC networks: analog video and sometimes voice, and digital video (compressed or uncompressed) and data.

Types of applications using HFC include:

- Regular analog TV
- High-speed, LAN-type (shared) data transport
- Telephone services
- Video on demand
- Home shopping
- Video games
- Pay per view
- Electronic news delivery.

Fiber-to-the-Curb (FTTC) and Other Fiber Architectures

These architectures are similar to the IDLC architecture, with the exception that fiber is used as the transmission medium in the distribution plant. If fiber is used all the way to the home, it is called fiber-to-the-home (FTTH). If fiber is used up to the curb of the home, it is called fiber-to-the-curb (FTTC). If fiber is used to the basement of the office building or complex, it is called fiber-to-the-building (FTTB). Finally, if fiber is deployed all the way to a node (which is equivalent to the pedestal in a neighborhood), it is called fiber-to-the-node (FTTN). All fiber architectures are actually variations of FTTC.

In an FTTC architecture, digital signals travel from the service provider to the central office via backbone links. The signals from the CO are routed through to an optical network unit. At the ONU, the

---

* Cable operators are looking toward voice over IP (VoIP). VoIP transports the voice traffic in data packets, which avoids the interference problem. However, VoIP has a potential latency issue due to unpredictable delays in the network; this is currently being addressed by standards organizations.

optical signal from the fiber is converted into an electrical signal and transported over copper or coax cable—even wireless in some cases—to the customer.

Figure 2.6
FTTC architecture

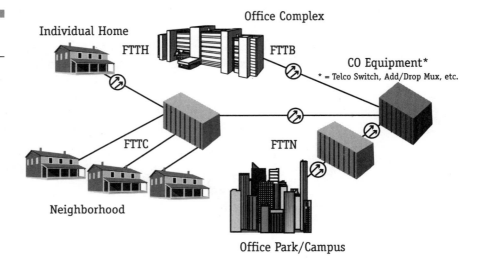

For telephone operating companies providing services, twisted-pair wires from the curb are used to connect to the customer premises. In the case of the cable TV provider, coaxial cables may be used. FTTC is usually implemented as a switched network with multiple fibers for bidirectional signal flow. It can also be implemented with an HFC architecture. Figure 2.6 illustrates an FTTC architecture in a typical telephone environment using a switched-star architecture.

Capacity

FTTC networks allow for dedicated downstream information transport from an ONU to each subscriber at 51-Mbps and upstream traffic of 1.62 Mbps per subscriber. The 51 Mbps switched information stream is capable of concurrently carrying six to seven unique, high-quality data programs to each subscriber. Along with this data traffic, FTTC can also support other high-speed data services and telephone services to each subscriber.

As mentioned earlier, if fiber goes all the way to the home, the architecture is called FTTH. In this type of network architecture, considerable bandwidth can be achieved for each customer. However, there is currently no business case to justify that amount of bandwidth for each customer. This is especially true if the current twisted-pair architecture using VDSL can provide the needed bandwidth in a cost-effective way for all the services needed by the customer.

Services Supported

FTTC and FTTH are optimized for the transport of digital signals end-to-end. The transport of analog video channels (while technically possible) is not practical or cost effective when compared with today's cable technology and its architecture. This fact limits these architectures to providing only digital services. Some of the services that can be provided in digital form are:

- Voice
- Digital video and data
- Telephone service (including video telephone services)
- High-speed LAN-type data transport
- Digital video (high definition)
- Video on demand, near video on demand, and interactive video on demand
- Highly interactive, real-time video games.

LMDS

Of the different wireless technologies, local multipoint distribution system (LMDS) is one of the few that addresses broadband multimedia. The FCC has allocated a 1-GHz frequency on the 28 to 32 GHz frequency band for a wide range of wireless broadband services. This wireless system is capable of providing advanced two-way multimedia services including telephony and high-speed data. This enables service providers such as long distance telephone operators—who don't have the infrastructure—to provide local access in a cost-effective way to both residential and business customers in the least amount of time.

The basic architecture of this system is illustrated in Figure 2.7. In this architecture, the last mile of the local loop is a wireless interface. The antenna at the customer's location must be in line of sight (LOS) with the cell site connected to the network, which provides all the customer-required services.

Figure 2.7

LMDS architecture

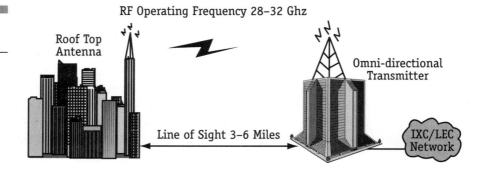

RF Operating Frequency 28–32 Ghz

Roof Top Antenna

Omni-directional Transmitter

Line of Sight 3–6 Miles

IXC/LEC Network

MMDS

Multi-channel multipoint distribution service or microwave multipoint distribution system is a type of broadcast network similar to LMDS but operating at the 2.4 GHz frequency. However, the operating bandwidth range in this frequency is limited when compared with LMDS. Currently, MMDS frequency is used by cable TV providers to connect multiple headend locations to broadcast analog video signals. With telecom deregulation, the use of this frequency is open to other services such as telephony and interactive services.

Unlike LMDS, MMDS is less susceptible to interference from external environmental factors such as rain and thunderstorms. Thus, the requirements for distance from the cell site are less stringent. Typically, MMDS covers a radius of 50 miles, whereas LMDS has a radius of 3–6 miles.

Capacity

The 2.2–2.7 GHz band is used to transmit video signals from towers to receiving antennae located at the subscribers' premises. Up to 33 channels can be used for broadcast purposes through this system. Subscribers within a 25–30-mile radius of the transmitting tower can pick up these signals. If the video signals are digitized and compressed, 100 to 150 channels are possible.

The FCC guarantees an MMDS license owner four of the 11 MMDS channels allocated in the 2.2–2.7 GHz range. Any additional channels must be acquired from institutions or other individuals who own the remaining spectrum allocated to wireless cable.

Services Supported

MMDS can be used to provide either analog or digital video. Receiving analog video requires a relatively simple antenna on the rooftop and a set-top box consisting of a down-converter and a de-scrambler. In the case of digital MMDS, the converter is more intricate and expensive. Vendors have recently developed equipment capable of providing high-speed data and voice service, in addition to traditional analog and digital video services.

DBS Architecture

DBS is the next generation of satellite-based video broadcast services. With the advent of digital technology and smaller dish antennae at each customer's location, this service has become very attractive. This, in turn, enables better quality video and audio. DBS provides a type of service similar to that currently provided by conventional analog satellite systems. The attractiveness of DBS is that the signal is transmitted in a digital format, which is decoded by the set-top box at each customer's location. This set-top box, in addition to converting signals from digital to analog, has built-in intelligence to provide many new advanced services such as interactive TV and information on demand. The basic DBS architecture is shown in Figure 2.8.

Capacity

True direct broadcast satellites operate in the broadcast satellite services (BSS) portion of the Ku-band spectrum. This is equivalent to the spectrum range of 12.2–12.7 GHz (DirecTV and USSB broadcast operate at these frequencies). Subscribers to DBS can receive 150–200 digital video channels using MPEG-2 compression. In addition to the video, some DBS service providers are also planning data broadcasts in the Ku band.

Services Supported

DBS systems provide digital video broadcast services to subscribers. Although they are capable of providing analog video, DBS systems typically utilize digital video using MPEG-2 compression to make the most efficient use of bandwidth. In the United States, commercial services are offered by DirecTV, which combined with USSB, Primestar, and Echostar. DirecTV and USSB provide a combined total of 200 digital channels, Primestar offers about 160 channels, and Echostar offers about 300 channels. Channels include choices such as pay-per-view movies, sporting events, educational programming, and special interest channels. Recent developments in DBS systems support one-way broadcast data, which allows for high-speed Internet access (up to 400 Kbps downstream*) and the use of a traditional telephone line as the control channel. An example of commercial service in the United States is DirectPC (offered by the same network as DirecTV).

Figure 2.8
DBS architecture

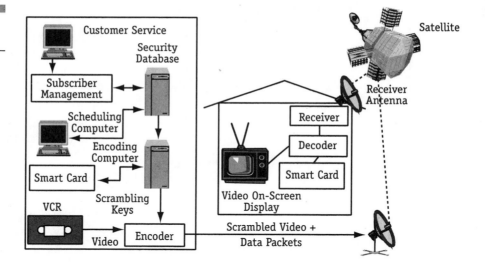

Analysis of Broadband Technologies and Architectures

In the previous sections, we discussed various broadband technologies and their respective architectures. Let us now examine our choices for

---

* Upstream speed is limited to modem speeds since a traditional modem dial-up over telephone lines is used for the upstream.

affordable, ubiquitous, high-speed access based on the criteria specified. Table 2.6 summarizes the choices:

Table 2.6

Summary of choices for affordable, ubiquitous, high-speed access

Attribute	Requirement	Copper Loop	Cable	Satellite
Affordable	Utilization of existing access infrastructure	Loops must be qualified	Two-way requires plant upgrade	Possible reuse of existing dishes
	Low cost of equipment	Yes	Yes	No
	Low cost of service	Possible	Yes	Possible
	Universal standards	Emerging	Emerging	No
	Scalability	Yes	May require cable plant upgrade	Depends on number of satellites
	Upgradable services	Yes	Yes	Yes
Ubiquitous	Market acceptance	New, trials have been positive	New, trials have been positive	New
	Availability	Limited, but widespread possible	Mainly residences	Limited, but widespread possible
	Ease of deployment	Yes	Requires plant upgrade	No, requires satellite launch
	Ease of use	Yes	Yes	Depends on satellite position
High-speed	Speed beyond voice-band modems	~1.5Mbps to 8 Mbps downstream dedicated per user	~30 Mbps (shared among multiple users)	400 Kbps downstream dedicated per user

From the table, it can be seen that satellite technology does not meet some of the key criteria for ubiquitous high-speed access:

■ Two-way communication is a problem since requests have to be made using standard telephone lines and modems, which tie up telephone network resources.

■ The number of satellites required in space limits the availability of service because of coverage. To increase coverage, it is necessary to launch more satellites. There is a limit to both the number of available "slots" in space for these satellites and the launch vehicles necessary to put them there. The latter is a severe gating factor today. The industry is developing methods to increase launch capacity; however, this requires substantial capital investment.

■ Another issue with satellite technology is the lack of standards. It is nearly impossible to purchase a satellite dish from one vendor, for example, and use it with another's system. In fact, even with a single vendor, it is sometimes necessary to purchase separate satellite dishes for broadcast video and data services. This is clearly unacceptable for consumers, particularly in neighborhoods where deed restrictions limit the number and location of dishes for aesthetic reasons. Vendors are developing mechanisms to support broadcast video and data services off the same system, but they are likely to remain proprietary.

This leaves copper-loop and cable-access technologies as the potential choices. Both these classes of technologies are viable; however, each of them suffers from certain barriers that may limit mass deployment. Fortunately, both industries have recognized these barriers, and increased competition among the industries has encouraged vendors to accelerate their efforts to solve these issues. We list below certain issues that continue to be difficult to surmount, particularly for the cable industry:

■ **Shared bandwidth.** Since everyone in the neighborhood is likely to share the same cable, bandwidth hogs can cause problems. For example, 10 Mbps shared among 200 users gives about 50 Kbps per user, which is equivalent to voiceband modems. Of course, it is not likely that everyone will "burst" at the same time, but as multimedia content puts increasing demands on bandwidth, a shared network is likely to get bogged down.

■ **Security concerns.** As with any shared-medium technology, security is a major consideration for cable. The industry press has been quick to raise the specter of subscribers inadvertently exposing data on their hard drives to snoops and malicious elements. Security is an even bigger concern for businesses, so they are more likely to shy away, even if the cable industry can extend its presence to business locations.

- **Regulatory issues.** For cable operators to offer a complete package, they must be in a position to provide video and voice services in addition to data services. However, to offer voice services in the United States, cable operators become subject to some of the regulatory restrictions imposed by the Telecommunications Act of 1996. Cable operators can enter the market as CLECs, but to do so, they must sign interconnect agreements with local telco operators. Interconnect agreements must be signed in every state in which the cable operator wishes to offer services, and they can be pretty complex to execute. This is a major drawback for organizations that are not experienced in dealing with these issues; therefore, most cable operators are concentrating on data-only services today.

- **Lifeline POTS.** One of the implications of the regulatory environment is the need to offer lifeline POTS to customers so that they always have a phone line available for emergencies (e.g., 911 calls). While lifeline POTS can be easily accommodated with DSL, it is not so straightforward with cable. To meet this requirement, the cable box has to have an optional forward battery backup in case of loss of power. Of course, the battery backup adds to the cost and consumer inconvenience.

- **Capital investment before two-way communication is possible.** Two-way communication is a key requirement because it avoids parallel infrastructures for requests and responses, as is the case with access via satellite. Cable access requires significant capital investment to upgrade the backbone networks to HFC before this is possible, and this investment must be made prior to signing on any subscribers. An analytical study done by Telenor Research and Development concluded that an FTTB architecture for the cable operator would be both high risk and uneconomical.[9] The study also showed that even FTTN architecture for the cable operator would result in significantly lower NPV than comparable FTTN architecture for the telephone network operator. An important factor is that DSL-based service does not require significant front-loaded expenditure; it can be rolled out in conjunction with demand from potential subscribers.

■ **Lack of presence at business locations.** Cable access can be a viable alternative for consumers and SOHOs; however, the lack of cable operator presence at business locations is a definite limitation on the revenue base. Moreover, cable installation is labor intensive and expensive, so it is unlikely that businesses will choose this over DSL, which can use the existing RJ-11 jacks.

This points to copper-loop access technologies as the class of technologies with the long-term promise of meeting key requirements, but only if the industry can demonstrate an appropriate standards-based, interoperable, cost-effective solution suitable for both commercial and consumer use. If the telecommunications industry should falter, the cable industry will continue to offer an appealing alternative for certain classes of customers.

Summary

There are three major categories of broadband technologies—copper loop, cable, and satellite. Each of these technologies is capable of providing a variety of multimedia services to customers. Architectures based on these technologies can provide a migration path to a broadband network from the current infrastructure. However, none of these classes of technologies is a perfect choice. Each has technological, economic, and regulatory issues that must be overcome before full-scale deployment can begin. In balance, however, copper-loop access technologies (XDSL) hold the best promise of meeting the broadest set of requirements with the fewest near- and long-term issues. The cable industry can offer a viable alternative to copper-loop access for certain classes of customers, especially consumers. Satellite technology is not yet mature enough for mass deployment for commercial or consumer use.

References

1. ADSL Forum, "TR-010: Requirements and Reference Models for ADSL Access Networks: The 'SNAG' Docement."
2. Information is available at the official Webside at **www.tl.org/tlel/el4home.htm**.
3. Information is available at the official Website at **www.etsi.org**.
4. Information is available at the official Website at **www.itu.int**.
5. Information is available at the official Website at **www.adsl.com**.
6. Information is available at the official Website at **www.atmforum.com**.
7. Information is available at the official Website at **www.uawg.org**.
8. Information is available at the official Website at **www.cablelabs.com**.
9. Ims, L. A., Stordahl, K., and Olsen, B. T. "Risk Analysis of Residential Broadband Upgrade in a Competitive and Changing Market," *IEEE Communications Magazine*, Vol. 35, No. 6 (June 1997).

Fundamentals of Copper-Loop Transmission

Overview

Digital subscriber line (DSL) technologies have evolved from the basics of copper-loop transmission. Like other media used in data communications, copper loops support both analog and digital transmission. DSL technologies have deep roots in the advanced analog transmission techniques developed over the years to improve the data rate over ordinary phone lines using modems. Furthermore, DSL technologies have built upon these techniques to take advantage of low-cost digital signal processors to further improve the available bit rates and transmission reliability across an ordinary pair of twisted copper wires. Over the years, these advances have made it possible to support data rates of not just Kbps, but Mbps. Although data rates are always limited, for a given distance, the maximum data rates have recently improved by orders of magnitude. In this chapter, we will review the challenges and limitations of data transmission over the copper loop, how various techniques were developed to overcome these issues, and how these techniques form the building blocks for DSL technologies.

Background of PSTN Technology

Information, Data, and Signals

Humans communicate by exchanging information, whereas computers communicate by exchanging data. One way to understand the difference is to think of useful data as information. The usefulness is determined by a human—information acts as a stimulus and produces a reaction; therefore, information is always associated with humans. On the other hand, computers are factual, i.e., devoid of human feelings and emotions. They deal with data and are not concerned with the "usefulness" of anything. Computers use signals to convey data.

For example, a traffic signal can be red, yellow, or green. As electric current flows through the traffic signal, the appropriate color—red, yellow, or green—lights up, conveying the state of the signal. The state of the signal is data. Humans associate the state of the signal with expected actions to perform, i.e., "red means stop," "yellow means caution," and "green means go." This association between the state of the signal and the expected human reaction is information.

Analog Signals

An analog signal is one that is continuously changing over time. Voice and video are examples of analog data that can be conveyed across media such as copper or air as analog signals. Based on Fourier analysis, any analog signal of practical interest can be constructed by summing an infinite number of sine and cosine waves. As an approximation, a finite sum of the first few *harmonics** can be used to represent an analog signal.

As shown in Figure 3.1, a sine wave has three attributes—amplitude (M), period (T), and phase ($\emptyset$). However, it is common to refer to the frequency (f) of a wave, rather than its period. The relationship between frequency and period is $f=1/T$. It can be mathematically expressed as $S(t)= M \sin (2\pi ft+\emptyset)$, where M is the peak amplitude, π is a constant, f is the signal frequency, and $\emptyset$ is the phase. Figure 3.1 shows two sine waves that are 90° apart from each other in phase.

Figure 3.1

Attributes of a wave

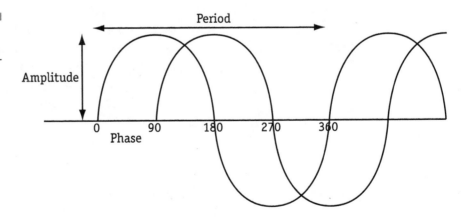

Period

Amplitude

0   90   180   270   360

Phase

---

* Harmonics are waves whose frequencies are integer multiples of the base frequency. For example, if the base frequency is f, the first harmonic has a frequency of $1f$, the second harmonic has a frequency of $2f$, the third harmonic a frequency of $3f$, and so on. Note that since the period of the wave is inversely proportional to the frequency, as the frequencies increase, the periods of successive harmonics get smaller and smaller.

Human speech is an example of analog data. When we speak, sound waves are generated in the air—each with its own amplitude, frequency, and phase. These sound waves travel through the air and reach the eardrums, causing them to vibrate in response. The vibrations of the eardrums translate into signals that are transmitted to the brain via the nervous system—the brain, in turn, deciphers these signals as sound. If the sound patterns are recognizable by our memory, they are deciphered as intelligible speech. Otherwise, they are interpreted as unintelligible—our typical reaction when we hear someone speak an unfamiliar foreign language.

There is one problem: our voices can travel only a finite distance within which we can be accurately heard. In other words, as the sound waves travel through the air, they are attenuated and distorted, just as any other medium is. As a result of the travel through the air, the attributes of a sound wave that leaves the speaker's mouth are different from the attributes of the same wave when it reaches the listener's ear. The human brain can compensate for these differences to some extent; however, as the distance between the speaker and the listener increases, there occurs a point when the listener can no longer accurately decipher what was said. We are familiar with examples in our daily lives where we often use megaphones or loudspeakers to amplify the voice of a speaker so that by the time the sound waves reach the listeners, they are strong enough to be intelligible. However, amplifying works well only for one-way broadcasts over a limited distance; it does not help in situations where a two-way, private conversation between two geographically separated people is required. The telephone solves this problem.

Conversion of Human Voice into Analog Signals

The purpose of the telephone is to convert human voice into an *electrical* analog signal. When a human speaks into a telephone handset, a thin membrane called a *diaphragm* within the mouthpiece (part of the handset) converts the vibrations of the speech into equivalent electrical signals. To accomplish this conversion, the diaphragm is connected to a carbon chamber. As the diaphragm vibrates in response to the sound waves, it compresses carbon molecules in the chamber. When the carbon molecules get compressed, they vary the flow of electricity relative to the applied pressure. Thus, electrical signals are generated in response to the input sound waves. These electrical signals are then transmitted over the wire to the subscriber at the far end.

At the receiver's side, the process is reversed. Much as there is the diaphragm in the transmitter, there is a diaphragm in the receiver. The receiver's diaphragm is connected to a magnet and an opposing electro-magnet on the outside. As the electromagnet receives the electrical signals over the wire, it creates a magnetic field that interacts with the magnet connected to the diaphragm. This interaction between the magnets causes the diaphragm to vibrate. The vibrations of the diaphragm produce sound waves in the air that reproduce human voice (i.e., the vibrations of the diaphragm closely mimic the vibrations of human vocal cords). The listener can pick up the sounds from the receiving handset. Figure 3.2 illustrates communication via two telephones.

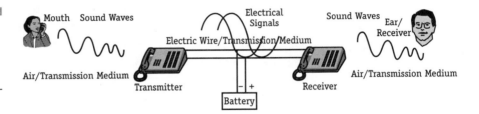

Figure 3.2
Communication between two telephones

The early telephone systems allowed people to communicate over copper wires but still suffered from a multiplicity of problems. One issue was the human intervention required to connect the pairs of wires between the sender and the receiver through a manual switchboard. Another issue had to do with distance. While telephones allowed people to communicate within a community, they still did not support calling across town, much less global communication. The development of automatic telephone switches solved these issues. This led to the next major advances in telecommunications and to the growth of the public switched telephone network (PSTN).

The PSTN and Analog Transmission

As described in Chapter 1, every subscriber in a modern PSTN is connected via a local copper loop to a switch in a switching center known as an *end office* (EO) or a *central office* (CO). In turn, CO switches are connected via intermediate switching nodes called *access tandems*. However, the details of the internal architecture of the PSTN are hidden from the subscriber. All the subscriber sees is the telephone connected to the end of the copper loop that terminates in the premises. This fact is both a benefit and a problem. On the one hand, it is still possible to operate successfully a telephone set purchased a few decades ago on the modern day PSTN. On the other hand, since the loop has been optimized primarily to carry voice and the fundamental loop characteristics have not changed in over 100 years, this imposes limitations on using the PSTN to transport data. We shall expand upon these limitations in the next section.

Before discussing the issues of the PSTN to transport data, however, we need first to describe the details of how the PSTN is used to transport voice. The human range of audible frequencies is roughly 0–20 kHz. Nonetheless, everyday speech can be adequately transmitted using only the 300 Hz–3.3 kHz range. The human brain on the receiver's side can compensate for the lack of the transmitted information at the higher frequencies. Designing the PSTN to support only 0–4 kHz greatly reduces the cost and complexity of the PSTN. The middle of the voice passband (around 2 kHz) is called the center frequency. Figure 3.3 illustrates a *voice passband*.

Figure 3.3

Voice passband

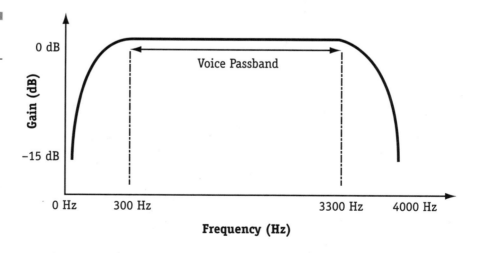

The extra frequency ranges at either end, i.e., from 0 to 300 Hz or from 3.3 to 4 kHz, provide *guardbands*. The need for guardbands will be described later. A voice channel, therefore, only occupies 4 kHz of bandwidth—about a fifth of the complete range of audible frequencies. This also explains why CD audio played over a telephone set never sounds the same as listening to it directly; the higher frequencies are automatically chopped off by the PSTN.

On the CO side, the switch has to deal with multiple voice channels coming in over individual copper loops, and then aggregate them onto a few outgoing trunk lines. The need to aggregate is based on economics. Trunks are expensive; therefore, it would be cost prohibitive to have a one-to-one correspondence between the number of physical lines that enter into the CO switch (representing several hundreds of subscribers) and the physical lines that exit from it. Aggregation of multiple voice channels represents a multiplexing function, and that is exactly what happens at the CO switch.

As discussed in Chapter 1, one technique of multiplexing is frequency division multiplexing (FDM). As shown in Figure 3.4, multiple voice channels can be carried over a single line by shifting the center frequency of each voice channel. The multiple voice channels are separated by the guardbands at either end, thereby improving the reliability of the receiver in demultiplexing the voice channels. Without guardbands, it would be difficult accurately to distinguish whether a particular frequency belonged to one voice channel or the other when transmitting signals over an error-prone medium.

Figure 3.4
Frequency division multiplexing of multiple voice channels

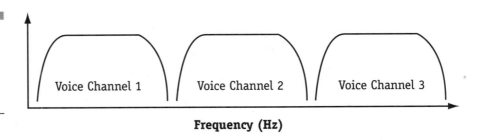

Analog transmission is ill suited for long-haul transmission for several reasons:

■ **Difficulty of accurate detection.** Accurately detecting a signal is always a problem. For example, is it 3V or 3.01V? The difference may or may not matter, depending on what is being represented. Nevertheless, accurate determination of the difference requires a complex implementation of detection circuitry that adds to cost.

■ **Difficulty of signal amplification/regeneration.** Another problem is signal amplification and/or regeneration of the signal to carry it across distances. This is required to counter the effect of signal attenuation due to distance. Unfortunately, amplifying the signal also amplifies the associated noise, which further exacerbates the problem of detection. Precisely regenerating an input analog signal also requires complex circuitry.

■ **Switching latency.** The problems of detection and regeneration slow the process of switching; accurate detection of the signal on the input side causes delay; there is still more delay to regenerate it accurately on the output side. These delays can accumulate and cause further distortion. For example, if speech were to be slowed down, it would sound very different to the listener.

To overcome these difficulties, digital transmission techniques were implemented within the core of the PSTN.

Evolution of the PSTN Core to Digital Transmission

Digital transmission solves the problems of analog transmission as follows:

■ **Simplified signal detection.** With digital transmission, the problem of detection is simpler. The mere presence of a positive voltage (regardless of actual value) can be interpreted as a one, and the absence of voltage can be interpreted as a zero. Therefore, even if the actual value of the transmitted signal changes as the result of an error-prone medium, it does not matter to the detection circuitry. As long as voltage is present, it will be accurately interpreted as a one, greatly simplifying the complexity of the circuitry.

■ **Simplified signal regeneration.** Signal regeneration merely requires repetition of the input pattern of 0s and 1s a far simpler problem than accurately replicating an analog signal with complex circuitry.

■ **Low-latency switching.** Switching digital signals can be done efficiently and cost effectively in silicon using gates and transistors.

■ **Simplified error detection and correction.** It is easier to detect a bit that has been incorrectly received (as with parity checking) than to detect whether an analog signal is accurately within the allowable range. Bit correction involves merely flipping the bit in error (i.e., a 0 back to a 1 or a 1 back to a 0). Furthermore, error detection and correction techniques can be implemented without a lot of overhead bits.

Digital transmission, however, does have its drawbacks. There is a need for circuitry that will convert the original analog signal into a digital form (after all, human voice is still analog data), and back again into analog form at the other end. Fortunately, advances in technology have made it possible to implement A/D (analog-to-digital) and D/A (digital-to-analog) converters very cost effectively in silicon.

Another problem of digital transmission is accurate representation of the original analog signal. This is done by sampling. By sampling the original signal often, it is possible to approximate the behavior of the signal. For example, consider an analog signal that changes linearly from 1V to 5V over five seconds. It is possible to represent this signal digitally with five samples, each taken one second apart—1V, 2V, 3V, 4V, and 5V. In real life, analog signals do not always change linearly, so the digital samples taken every second may actually look as follows: 1V, 2.8V, 4V, 4.5V, and 5V. With sufficiently frequent samples, it is possible to represent the change in value of the analog signal over a period of time. Sampling can represent the change in value, but not the value itself (i.e., the signal level at the point in time it is sampled). Associating a number of bits with each signal level can do this. As a simple example, suppose three bits represent each signal level, each of which can be either a 0 or a 1. We have 2^3 possible values, represented as 000, 001, 010,...,111 in binary notation. These values work out to decimal 0 through 7, which is sufficient to represent up to eight different signal levels. Note that these discrete values are only approximations of the true (analog) value of the signal at the point in time when the sample was taken.

Sampling and associating bits per sample provide some answers. However, one problem remains: What should be the *sample rate*, i.e. number of samples/second? The sample rate should be high enough to result in digital samples that accurately represent the behavior of the analog signal. However, it should not be so high as to put a needless burden on the process of sampling and converting to a digital value, then reversing the process at the other end. In other words, there is a point of diminishing return beyond which over-sampling adds very little in representing the original analog signal.*

Harry Nyquist researched this problem; his findings are now commonly known as Nyquist's theorem.[1] Nyquist's theorem is expressed by the equation $C=2B_w \log_2 L$ where C is the channel capacity (data rate), B_w is the available channel bandwidth, and L is the number of signaling levels. In the digital world, L levels can be represented by 2^n, where n is the number of bits (for example, eight signal levels can be represented by 2^3 bits). In other words, $L=2n$, and therefore, $n=\log_2 L$. Substituting for $\log_2 L$ in Nyquist's theorem, we can rewrite it as $C=2B_w n$, or $C/n=2B_w$ where C/n is nothing but the sample rate. Therefore, for a voice channel of 4 kHz, we will need to sample at 2 × 4,000, or 8,000 samples/second. Further, if each sample is represented by 8 bits, we have 8,000 samples/second × 8 bits/sample or 64,000 bits/sec or 64 Kbps.

For reasons discussed earlier, digital transmission offers several advantages for long-haul transmission. Therefore, the core of the PSTN has evolved to support digital transmission. However, the local loop transmission from the subscriber's telephone set to the CO switch is still analog, which explains why a 40-year-old telephone still works today. One of the main reasons the access technology has remained analog has to do with the success of the PSTN itself. Since the PSTN is so prevalent in many parts of the world, there is an established base of several million telephone sets based on analog transmission with access to the PSTN. It would be impossible to expect all these consumers to convert overnight to a purely digital network from end to end. Therefore, the conversion of the analog electrical signals, which represent the human voice, into digital samples takes place at the ingress CO switch, not at the subscriber's telephone. Modern digital telephone sets are exceptions to this scenario, but they depend on technologies such as ISDN. At the egress CO switch, the process is reversed so that the receiving subscriber can pick up the telephone and hear what was originally spoken.

---

* As mentioned earlier, an analog signal can be reasonably represented by the sum of the first few harmonics. Over-sampling only results in frequencies that will get cut off anyway either because of severe attenuation by travel over the copper loop or due to filters.

Digital transmission in the core gave rise to another form of multiplexing—time division multiplexing (TDM). CO switches had to deal with multiple 64-Kbps digitized voice channels, which necessitated the adoption of TDM. With TDM, it is possible to design an efficient switch to take bits from multiple input streams (in a round-robin fashion) and transmit the bits out on a single output. This is the principle behind the T-carrier network in the United States (E-carrier in Europe and J-carrier in Japan) and makes it possible to increase the backbone bandwidth significantly. For example, a T3 circuit provides 44.736 Mbps or support for 672 digitized voice channels (44.736 Mbps/64 Kbps = 672). Today, the PSTN core network is going through another phase of evolution—statistical multiplexing using asynchronous transfer mode (ATM). This allows for even higher backbone bandwidth. ATM fundamentals are covered in Chapter 7 of this book.

From the discussion so far, we understand that the PSTN has evolved into two distinct networks: the PSTN core network based on digital transmission and TDM (and is further evolving into an ATM core). In contrast, the PSTN access network is based on analog transmission to support only 4 kHz of bandwidth.* While this is sufficient to support voice, it imposes significant barriers for data transmission.

Challenges and Limitations of the PSTN Access Network for Data Transmission

The PSTN access network is not ideally suited for data transmission for the reasons discussed in the following sections.

Signal Attenuation Due to Distance

A signal traveling over copper is attenuated (reduced in strength) just as it is with any other media. As shown in Figure 3.5, the attenuation is distance related—the longer the signal travels, the more the attenua-

---

* ISDN is an end-to-end, full digital network, which includes the access network. However, given the limited deployment of ISDN lines worldwide in comparison to analog telephone lines, it remains the exception, rather than the rule. Therefore, the predominant access network transmission is still analog.

tion. Attenuation loss can be as much as 60–70 dB* over the longer loops.

Figure 3.5

Signal attenuation due to distance and frequency

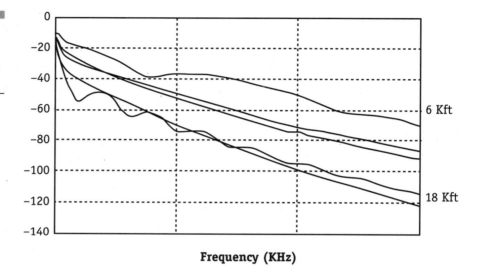

Frequency (KHz)

Accurate signal detection by the receiver depends on the signal-to-noise ratio (S/N). It follows that if the signal strength reduces in relation to the noise, it becomes more difficult for the receiver accurately to distinguish the signal from the surrounding noise. The obvious example from our daily lives is when we lean forward across the table to hear better in a crowded restaurant with lots of external noise; we are reducing the distance from the speaker.

Wire gauge and signal frequency are two of the other factors that have an impact on attenuation. Figure 3.5 shows that higher frequency signals are attenuated more than lower frequency signals.

Noisy Environment

Not only does a signal get attenuated as it travels down a copper wire, it also has to contend with the inherently noisy environment of the copper

---

* Signal power and signal loss are measured in logarithmic units called *decibels* (dB). A 3-dB increase in power equals a doubling of power, 6 dB is four times the power, and so on. Conversely, a 3-dB decrease in power is half the power, and so on. Therefore, a loss of 60–70 dB is quite significant.

loop. This further reduces the S/N ratio, worsening the problem of accurate detection at the receiving end. A copper loop experiences several sources of noise:

■ **White noise.** White noise is always present on the line and is caused by the movement of electrons in the line. In general, many noise sources add together, and the law of large numbers makes the cumulative interference look Gaussian.

■ **Crosstalk.** Crosstalk occurs when signals from two lines interfere with each other.* It can either be near-end crosstalk (NEXT) or far-end crosstalk (FEXT). Crosstalk is illustrated in Figure 3.6. A particular form of NEXT, called self-NEXT, is the interference in an adjacent wire pair when transmitting a like signal (e.g., ADSL to ADSL, HDSL to HDSL, etc.).

Figure 3.6
Near- and far-end crosstalk

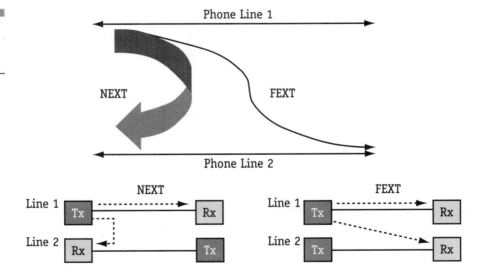

■ **Radio frequency interference (RFI).** Radio frequency signals intrude on phone lines which, being made of copper, act as antennae. At lower frequencies, the effects of RF signals are minimal because the wire pair is twisted and, therefore, balanced with

---

* Electrical signals in a wire pair generate an electomagnetic field that surrounds the wire pair. The field can induce an electrical signal onto adjacent wire pairs, thus causing interference. Crosstalk can be minimized, but not eliminated, by twisting the wire pairs.

respect to earth. However, balance decreases with increasing frequency, so DSL systems will experience RF noise called *RF ingress*. The extent of the interference depends on the proximity of the source (interferer) to the loop. RFI is illustrated in Figure 3.7.

The primary sources of such interference are AM radio broadcasts and ham radio. AM radio stations broadcast in the range from 560 to 1,600 kHz. However, their frequency is fixed; so their interference is predictable. In contrast, ham radio interference is unpredictable, since there can be frequency hopping and multiple power levels. Fortunately, ham radio interference is largely an issue only for VDSL (because ham radio frequency bands mainly overlap with the VDSL transmission band).

Figure 3.7

Radio Frequency Interference (RFI)

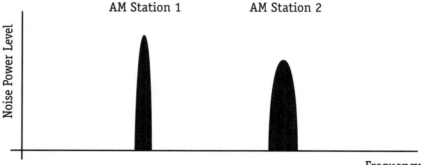

- **Impulse noise.** This is caused by temporary electromagnetic interference. Examples include the effect of lightning storms and home appliances switching on and off. Each of these effects results in the temporary injection of noise into the phone lines much like RFI. Impulse noise can last from a few microseconds to as long as a couple of milliseconds.

The problem of data transmission across a noisy channel was studied extensively by Claude Shannon. In 1948, Shannon's research led to the famous Shannon's law[2]: $C = B_w \times Log_2(1+S/N)$, where C is the channel capacity (data rate), B_w is the available channel bandwidth, and S/N is the signal-to-noise ratio. It follows that for a given channel bandwidth (for example, the 4-kHz voice passband), the way to increase channel capacity is to improve the S/N ratio. Consequently, improving the S/N ratio has become the focus of modem engineering.

Legacy of the Voice Access Network Architecture

■ **Analog transmission.** Despite the development of ISDN, the PSTN access network is predominantly based on analog transmission, with all its inherent problems of accurate detection of signal strength and accurate regeneration of the signal.

■ **Narrow bandwidth.** The PSTN access network was designed to carry analog voice channels within a narrow range of 4 kHz, so the CO has filters to condition the lines for voice traffic and eliminate frequencies above 4 kHz. Unfortunately, the 4 kHz range severely limits the bandwidth available to carry data.

■ **Bridged taps.** Telephone lines often have bridged taps that give a network provider greater flexibility in serving multiple customers, as shown in Figure 3.8. The serving area interface (SAI) crossconnects feeder cable from the central office to the distribution cable that serves the customer.* Bridged taps maximize the use of a distribution cable by allowing access to more than one customer's network interface (NI). Another reason for bridged taps is that they may be remnant branches that have been left over from previous use of the wire pairs. According to a Bellcore study, the average local loop in the United States has 22 splices, including bridged taps.[†]

Bridged taps are unterminated stubs of twisted pair cable, and their effect[††] shows up as notches in the frequency spectrum, as shown in Figure 3.9.

---

* Feeder cables are usually large cables that are placed along (or below) major streets. Distribution cables are smaller cables that are placed in easements in the subscribers' backyards.
† Bridged taps are common in the United States, but are seldom used in Europe and Asia.
†† When a pulse transmitted down a wire encounters a bridged tap, its energy is divided among the two paths. The pulse that takes the path of the unterminated end gets reflected back to the tap point. This reflected pulse also gets divided into two paths, causing echos back to the sender.

Figure 3.8
Bridged taps

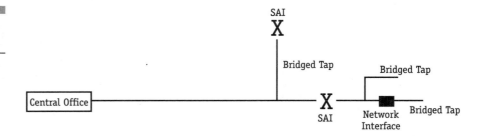

Figure 3.9
Impact of bridged taps

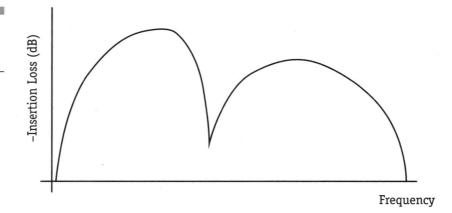

Bridged taps were originally designed into the public switched telephone network when it was commonly believed that the network would carry only voice, and not high-speed data. Bridged taps are an example of legacy network design that provides POTS flexibility, but has a negatively impact on data service.

■ **Load coils.** Long loops sometimes have load coils on them to flatten out the voiceband frequency response. Unfortunately, load coils act as low-pass filters, thereby preventing use of the higher frequencies for data transmission. For this reason, load coils are incompatible with DSL technologies. Since loops in the United States tend to be longer than in Europe, loaded loops are common here,* but rare in European countries.

---

* In the 1970s, about 20% of U.S. loops were loaded. In the 1990s, with the deployment of digital loop carriers that shorten the loop lengths, this percentage has dropped down to 10%–15%, depending upon the region.

■ **Loop conditions.** The loops are not always of uniform wire gauge. For example, in the United States, the first 10 Kft of wire from the CO is usually 26 American Wire Gauge (AWG),* then follows heavier gauge wire (typically 24 AWG) to keep the loop resistance to less than 1,300Ω. As in the case of bridged taps, signal reflections can result from the impedance change cause by splicing wires of two different gauges. Even with a uniform wire gauge, a loop has several wire splices since wire pairs are typically manufactured in 500-foot spools. A wire splice can potentially be the source of a problem because if wires are not sealed securely, then oxidation can occur. Oxidation causes the wire to develop high resistance, which defeats the design practice of keeping the total loop resistance less than 1,300Ω.

■ **Multiple analog/digital conversions.** Human voice is natively analog data. By contrast, a computer sends information natively as digital data. Consider what happens when digital data must travel across the public access network and the core network before emerging on the other side onto another public access network. The data must undergo the following manipulations, as shown in Figure 3.10:

1. Conversion from digital to analog form (a 4-kHz analog voice channel) goes from the computer across the local loop. This is done by the near-end modem.

2. Conversion from analog to digital form (a 64-Kbps digitized voice channel) is multiplexed and sent across the digital PSTN core. This is done by the ingress CO switch.

3. Conversion from digital to analog form (a 4-kHz "voice" sample) is done by the egress CO switch for the far-end local loop.

4. Conversion from analog to digital form is done by the far-end modem so as to be meaningful to the far-end computer.

---

* In the United States, wire diameter is measured as an AWG number, where the number N represents 1/Nth of an inch. For example, 26 AWG means 1/26th of an inch, 24 AWG means 1/24th of an inch, and so on. In other countries, the diameter of the wire is measured in millimeters.

Figure 3.10
*Multiple conversions
necessary for data
to transit the PSTN*

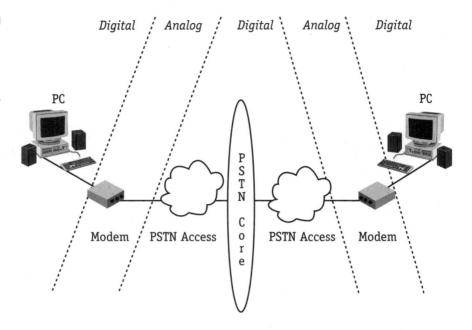

Digital / *Analog* / *Digital* \ *Analog* \ *Digital*

Obviously, the number of conversions necessary to accomplish end-to-end connectivity between two computers across the access network is inefficient* and reduces the data rate.

Techniques to Overcome the Challenges and Limitations of the PSTN Access Network

The motivation for transmitting data across the PSTN is strong—the PSTN is ubiquitous and cost-effective. Therefore, engineers have studied the problems of PSTN infrastructure and have come up with ingenious ways to combat the challenges of data transmission across the PSTN by improving the S/N ratio. The device that makes this possible is the modem. The word modem stands for MOdulator/DEModulator. What is being *modulated* is an analog carrier signal residing in the voiceband (0–4 kHz). The modulated analog carrier signal is sent over the PSTN and represents the analog equivalent of the digital input that originates from a computer.

---

* Today's 56K voiceband modems are able to achieve higher speeds by eliminating one of these conversion steps.

A modem has the following major functions:

- Line coding and modulation/demodulation
- Error control
- Data compression.

Modulation, error control, and data compression protocols are necessary to implement these functions in a uniform manner, so that a modem from one manufacturer can successfully interoperate with another.

Line Coding and Modulation/Demodulation

Line Coding

Line coding is the technique of putting out electrical pulses on a medium, such as a copper wire, so that the receiver can accurately interpret them. For example, if a sender transmits the digital pattern "0110," line coding involves sending a series of pulses on the wire that would accurately represent the pattern. The receiver interprets these pulses to recover the original pattern.

Various line-coding techniques have evolved over the years to suit transmission requirements over different media. In general, a line-coding technique should:

- Use as little bandwidth as possible to maximize the number of signals that can be transmitted
- Have a small or no DC component, since high DC-level signals can increase the cost of the circuitry
- Maintain synchronization between the sender and the receiver
- Result in a non-polarized signal so that sending the signals over a pair of wires will not be affected if the wires are reversed.

Modulation

Modulation is the process of encoding a digital signal into an analog signal that represents and conveys the information over an analog channel. In the simplest case, a single signal element, or symbol (a combination of the amplitude, frequency, and phase of the signal at particular point in time), represents one bit. With complex techniques, a single symbol represents multiple bits, thereby increasing the effective data rate. Therefore, the terms "bit rate" and "baud" do not mean the same thing. Bit rate refers to the raw data rate and is expressed as bits/second or b/s. Baud refers to the number of times the signal changes in one second, i.e., the number of symbols/second.

In general, a modulation technique should:

- Maximize the number of bits/symbols that can be supported
- Minimize complexity of receiver circuitry, since complexity can increase latency, and hence limit the data rate
- Optimize the root mean square (RMS) value of the power of the modulated carrier on the line, so that the impedance values of the components stay within the linear regions (non-linearity can cause signal distortion).

Modulation involves manipulating one (or more) of the three attributes of a signal—amplitude, frequency, and phase. An outline of the common modulation techniques follows.

Amplitude Modulation (AM)

Amplitude modulation, also known as amplitude shift keying (ASK), manipulates only the carrier signal *amplitude*. As shown in Figure 3.11, in amplitude modulation, a binary 1 is represented with higher carrier signal amplitude, and a binary 0 is represented with zero amplitude. A disadvantage of this method is that small changes in amplitude are difficult to detect reliably.

Frequency Modulation (FM)

Frequency modulation, also known as frequency shift keying (FSK), manipulates only the carrier signal *frequency*. As shown in Figure 3.12, in frequency modulation, a binary 1 is represented by a higher carrier signal frequency, and a binary 0 is represented by a lower carrier signal frequency. However, distortion caused by the medium makes frequency changes extremely difficult to detect.

Figure 3.11
Amplitude
modulation

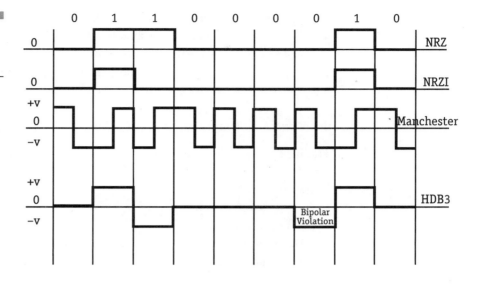

Figure 3.12
Frequency
modulation

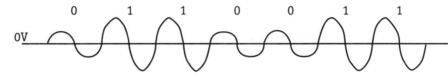

Phase Modulation (PM)

Phase modulation, or phase shift keying (PSK), manipulates only the carrier signal *phase*. As shown in Figure 3.13, in phase modulation, a binary 1 is represented by a phase shift in the carrier signal frequency, and a binary 0 is represented by the absence of a phase shift in the carrier signal frequency. A disadvantage of PM is that it requires the sender and receiver to keep in phase synchronization—this adds to the complexity of the receiver circuitry.

Figure 3.13
Phase modulation

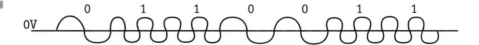

A variation of phase modulation that does not require perfect phase synchronization between the sender and the receiver is differential phase modulation (DPM) or differential phase shift keying (DPSK).

With this technique, the receiver only has to detect relative phase shifts, not the absolute phase, which makes the circuitry less complex. A popular form of DPSK is quadrature phase shift keying (QPSK). With QPSK, there are four possible phase shifts, with each symbol representing two bits.

Baseband Line Codes

The fundamental ideas discussed above led to the development and evolution of line codes to suit different requirements. Line codes can be characterized as either baseband (that can transmit energy at DC, i.e., $f=0$) or passband (that transmit energy at a frequency spectrum shifted above DC). In this book, we shall take an overview of some line codes. For additional details on encoding and baseband and passband codes, the reader is encouraged to refer to references 3 and 4.

Some of the commonly used baseband line-coding techniques are outlined below.

NRZ—Non-return to Zero

NRZ is the simplest form of line coding. With NRZ, a "0" bit is represented by the absence of voltage (i.e., a voltage level of 0 volts); a "1" bit is represented by the presence of voltage (i.e., a voltage level of +V volts). NRZ has the following characteristics:

- The bandwidth can range up to half the data rate (for a string of alternating 1s and 0s).
- The average DC component is $\frac{1}{2}$ of +V volts (for a string of alternating 1s and 0s).
- There is difficulty maintaining synchronization between the sender and receiver when there is a sequence of only 1s or 0s.
- The signal is polarized, so reversing the wire pair will cause the receiver to interpret the pulses differently.

NRZI–Non-return to Zero Inverted

As seen before, a problem with NRZ is that a sequence of consecutive 1 bits means the signal stays high (at +V) for an extended period of time. A protracted high signal may cause clock drift, since there are no inherent changes in the line to keep synchronization.* The reverse situation is also a problem, i.e., a sequence of consecutive 0 bits means the signal stays low (at 0V) for an extended period. An extended period of low signal is also caused by a dead link. So, in addition to the difficulty of maintaining synchronization, there is no way for the receiver to distinguish between a dead link and a string of 0 bits in the data stream.

NRZI solves the clock synchronization problem of consecutive 1 bits as follows. NRZI works by manipulating the relative signal level rather than the absolute signal level. A "0" bit is represented by no change in signal level from the previous bit time; a "1" bit is represented by a change in signal level from the previous bit time. NRZI has the following characteristics:

- The bandwidth can range up to half the data rate (for a string of alternating 1s and 0s).
- The average DC component is $\frac{1}{2}$ of +V volts (for a string of alternating 1s and 0s).
- There is better synchronization between the sender and receiver since a sequence of 1s will cause a transition. However, the problem of consecutive 0s still remains.
- The signal is non-polarized (since the receiver is looking at differential voltage levels, rather than absolute values). Therefore, reversing the wire pair will not cause the receiver to interpret the pulses incorrectly.

Manchester

Manchester coding, also called phase coding (PE), solves the problem of both consecutive 1s and 0s. In Manchester coding, there is always a transition at the middle of each bit time (clock cycle). A "0" bit is represented by a high-to-low transition; a "1" bit is represented by a low-to-high transition.

Manchester coding has the following characteristics:

---

* The process of encoding bits (by the sender) and decoding bits (by the receiver) is driven by a clock. Every bit time (clock cycle) the sender transmits a bit down the line, the receiver must decode the bit in the same bit time while accounting for the transmission delay for the bit to travel down the wire. Unless the sender and receiver clocks are precisely synchronized, the receiver may incorrectly interpret the received bit. The way the receiver keeps its clock synchronized with the sender is to detect transitions from the received signal. Whenever the signal changes on transition, the receiver recognizes a clock cycle boundary.

- The bandwidth is always equal to the data rate, since there is a transition at the middle of each bit time.
- The average DC component is $\frac{1}{2}$ of +V volts (for a string of alternating 1s and 0s), assuming the signal transitions are between 0V and +V volts. Note that an alternate form of Manchester coding has the signal transition between –V and +V volts.
- Clock synchronization can be maintained since the transition in the middle of the bit time is used to provide a clock pulse.
- The signal is polarized.

Differential Manchester

Differential Manchester, or conditional diphase (CDP) coding retains the transition at the middle of each bit time (clock cycle). A "0" bit is represented by a beginning transition; a "1" bit is represented by no beginning transition.

Differential Manchester coding has the following characteristics:

- The bandwidth is always equal to the data rate, since there is a transition at the middle of each bit time.
- The average DC component is 0V volts (for a string of alternating 1s and 0s) for signal transitions between –V and +V volts.
- Clock synchronization can be maintained since the transition in the middle of the bit time is used to provide a clock pulse.
- The signal is non-polarized.

High Density Bipolar 3 (HDB3)

The Manchester/Differential Manchester coding techniques provide clock synchronization; however, bandwidth usage is high. High Density Bipolar 3 (HDB3) is a line coding technique that relies on *bipolar violations*. In HDB3, a "0" bit is represented by a voltage level of 0V; alternate "1" bits are represented by either +V, or –V. Additionally, to improve clock synchronization, whenever there is a string of four consecutive 0s, they are changed to 000V, where V represents a bipolar violation.* A bipolar vio-

---

* A line-code technique where a "0" bit is represented by a voltage level of 0V and alternate "1" bits are represented by either +V or -V, without any bipolar violations, is called Alternate Mark Inversion (AMI). HDB3 introduces bipolar violations as an improvement over AMI to solve the consecutive 0s problem.

lation occurs because the polarity of the V is the same as the previous 1 bit. In other words, it is +V volts if the previous 1 bit was +V, and –V volts if the previous 1 bit was –V. This is considered a violation because normally the polarity of the high-level signal alternates between +V and –V. This solves the problem of consecutive 0s, but reintroduces a non-zero DC level (since we have either +V +V or –V –V). To solve this new problem, alternate V bits are of alternate polarity.

Binary eight-zero substitution (B8ZS) is a line code similar to HDB3 with an 8-bit string that is not confined to byte boundaries. After substitution, there can be no more than seven zeros in a row.

HDB3 has the following characteristics:

■ The bandwidth equals half the data rate for a string of alternating 1s and 0s.

■ The average DC component is 0V volts (for a string of alternating 1s and 0s) for signal transitions between –V and +V volts.

■ Synchronization is maintained between the sender and receiver.

■ The signal is non-polarized.

Practical Use of Baseband Line Codes

Figure 3.14 illustrates how the various baseband line codes would encode a signal for a typical bit pattern.

Figure 3.14
Baseband line code
examples

0V

NRZ/NRZI are often used on slow-speed serial links (such as RS-232) and legacy protocols (such as SDLC). A modem connected to a PC over a serial data cable would implement NRZ line coding over that link. Manchester codes are rarely used on wide-area links since they take up bandwidth; however, because of their advantages, they are often used on local area networks (such as Ethernet). Finally, codes that are based on bipolar violations are used on wide-area links since they offer the advantages of lower bandwidth while maintaining synchronization. Bipolar violation-based line codes do require additional processing steps at the receiver, so they are reserved for the slower wide area links rather than high-speed local area links. B8ZS, and the related B3ZS, are used in DS (T1) and DS3 (T3) transmission, respectively[5,6]. HDB codes are used in the European equivalents called E1 and E3. A version of pulse amplitude modulation called 2B1Q is used in ISDN and HDSL[7,8]. It will be discussed in greater detail in Chapter 5 on HDSL and HDSL2.

Passband Line Codes

An outline of some of the commonly used passband line coding techniques follows.

Quadrature Amplitude Modulation (QAM)

QAM is an example of a complex modulation technique that combines more than one of the basic modulation techniques mentioned above. QAM utilizes a signal that can be synthesized by summing amplitude-modulated sine and cosine waves. These waves are shifted 90° in phase, hence the name quadrature amplitude modulation. QAM modulates both the carrier signal's phase and its amplitude. Using appropriate modulation techniques, it is possible to increase the number of bits/baud as illustrated in Figure 3.15. The digital values can be encoded, and the corresponding phase and amplitude are represented using a constellation diagram. In such a diagram, the length of the vector from the origin to the constellation point represents the amplitude of the carrier wave (M), and the corresponding angle of the vector ($\emptyset$) represents the phase.

In Figure 3.15, each dot is known as a *symbol*. Table 3.1 shows the bit values for 4 QAM.

Table 3.1

Bit values for 4 QAM

Bit Values	Amplitude (M)	Phase (Ø)
00	1	45°
01	1	135°
11	1	225°
10	1	315°

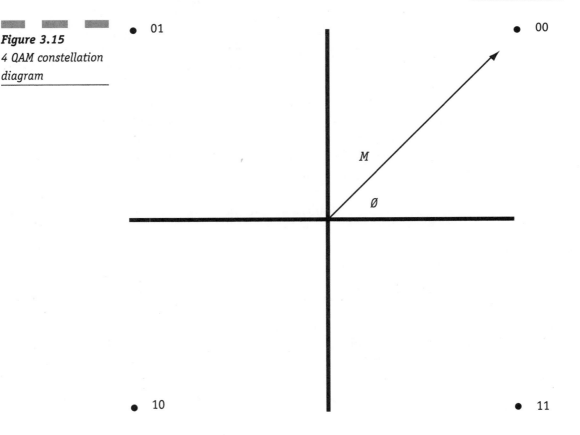

Figure 3.15
4 QAM constellation diagram

By increasing the constellation size, and hence the bit density per symbol, higher data rates can be achieved. This technique is used in V.34 modems, as discussed later.

Carrierless Amplitude Phase (CAP) Modulation

CAP modulation is closely related to QAM. As in QAM, amplitude and phase are used to represent the binary signal. The difference between the two lies in the state representation of the constellation pattern. CAP does not use a carrier signal to represent the phase and amplitude changes; rather, it uses two waveforms to encode the bits[9]. The absence of carrier gives rise to the name carrierless amplitude phase. CAP is fundamentally equivalent to QAM in performance; the difference is in the implementation. CAP is discussed in detail in Chapter 4 on ADSL and ADSL lite.

Discrete Multi-tone (DMT) Modulation

QAM and CAP are examples of single-channel passband line codes because they operate over the entire passband. By contrast, DMT is an example of a multichannel passband line code[10,11]. The basic idea behind multichannel line codes is "divide and conquer." In other words, instead of operating over the entire passband (which may have difficult transmission characteristics due to various forms of interference as noted earlier), multichannel line codes divide the passband into mini-transmission lines. The advantage of such division is that, taken individually, the characteristics of each mini-transmission line are likely to be conducive to transmission. The name DMT comes from the fact that the multiple mini-transmission lines are each treated as discrete sub-channels for purposes of transmission. DMT is discussed in detail in Chapter 4 on ADSL and ADSL lite.

Practical Use of Passband Line Codes

QAM is well known among modem engineers. It is familiar as the modulation of choice to support higher speeds as in the ITU V.32 standard for 9,600 b/s and is also used in cable modems. CAP has been used in some of the early implementations of ADSL, and DMT has been adopted as the line code of choice by the standards organizations for ADSL.

Error Control

Although error prevention is the ideal goal of transmission, it is impossible to achieve in practice due to the impairments experienced by a copper loop. Techniques such as shielding the cable or twisting the wire pair* are useful in minimizing, but not eliminating, errors. Therefore, all transmission has to presume that there will be errors on the line. Error control is the process of detecting and correcting errors in the transmission.

In general, an error control technique should:

▌ Detect more than one bit error. Impulse noise, for example, can clobber more than one bit in the transmission

---

* The twisted wire pair was invented in 1881 by Alexander Graham Bell as way of reducing the problem of crosstalk.

- Combine error detection with error correction
- Minimize (or eliminate) retransmission by the sender
- Minimize the number of overhead bits to accomplish the error detection and/or correction strategy.

There are two general techniques for error control:

- **Forward error correction (FEC).** With FEC, the strategy is error correction—introduce enough redundancy into the transmitted data so that the receiver can deduce and automatically correct the error.
- **Automatic Repeat reQuest (ARQ).** With ARQ, the strategy is error detection—introduce sufficient redundancy into the transmitted data so that the receiver can detect that an error has occurred, but not which error. The receiver can then request retransmission of the block of data in error.

Fundamental Principles of Error Control

Before discussing the details of each strategy (FEC versus ARQ), it is important to describe the basic principles of error control. Consider that the probability that a single bit can be in error is e. Therefore, the probability of successful transmission is $1-e$. If a block of data contains n bits, the probability that all bits will be transmitted successfully is $(1-e)^n$. Therefore, the probability that the block is in error is $1-(1-e)^n$. If e is very small (e.g., an error rate of 10^{-7}), the probability of an entire block being in error can be approximated as en. In other words, the probability of a block error is directly proportional to the number of bits transmitted in the block. This is the reason why small block sizes are preferred. On an error-free line, however, large block sizes are preferred since they improve the data throughput (by minimizing the packetizing overhead, thereby reducing latency). In practice, the block size is chosen to balance between these two considerations.

Consider a block of m (message) bits and r (redundant) bits for a total block length of n bits. In error control terms, the block of n bits is referred to as a *codeword*. The number of valid codewords is 2^m out of a possible 2^n combinations. If we take any two codewords of the same length, it is possible to determine the number of bits that differ between the two. This difference is known as the distance. For example, if we take two codewords 0110 and 1001, the distance between the two is four, since all four bits are different. However, the distance between codewords 0110 and 1111 is only two, since only two bits (the first and the last) are different. The codeword distance is also called the *Hamming distance* and can be computed by performing an XOR (exclusive OR) between two codewords and then adding up the resulting number of 1s. The Hamming distance represents the number of single bit errors to convert one codeword into another. Efficient line coding involves trying to maximize the Hamming distance between every pair of valid codewords—this maximizes the probability of successful error detection.

Another important principle is *parity checking*. The idea behind parity checking is simple—introduce a single redundant bit (known as the parity bit) into the block of m message bits, so that the total size of the block is $n = m+1$. The value of the parity bit is such that the total number of *1* bits in the resulting codeword is odd (known as odd parity) or even (known as even parity). For example, if the message bits are 010, then odd parity would result in a codeword of 010<u>0</u>, and even parity would result in a codeword of 010<u>1</u>. The last bit in the codeword is the parity bit. Now, if the receiver receives a block with a parity different from that expected, it immediately knows that an error has occurred. Continuing the example above, if the receiver receives the block 0001 (i.e., the second bit has been clobbered), and it is expecting even parity, then it knows an error has occurred. However, it does not know where the error has occurred, since the same situation could occur if the received block is 1101 (i.e., the first bit has been clobbered).

Forward Error Correction

Forward error correction is the preferred technique used in modems, since retransmission of data over a copper loop is not desirable.* Another advantage of FEC is that it can be implemented at the physical layer with complete transparency to the upper layers. Examples of FEC techniques are Hamming codes, Reed-Solomon codes, and Trellis codes[12,13]. The first two (Hamming and Reed-Solomon) are examples of block codes. Trellis is an example of convolutional code. The terms block codes and convolutional codes will be discussed later.

---

* If the error is caused by an impairment such as crosstalk, retransmission of data may cause the block to be in error again, in which case, the sender would be continually retransmitting the block in error until the higher layers time out.

■ **Hamming codes** are based on the principle of introducing redundancy into blocks of data (multiple bits) by using multiple parity bits. Hamming codes are a family of codes with the general property of $n=2^r-1$, $m=n-r$, and $r>2$ (implying that the minimum distance is 3). They are represented as (n,m). Consider a message consisting of 4 bits ($m=4$) such as –0101 (represented by $M_1M_2M_3M_4$). Let us take this message and compute even parity bits P_1, P_2, and P_3 as follows:

  ■ P_1 is the parity of bits 1, 2, and 3 of the source message (i.e., 010 in the example).

  ■ P_2 is the parity of bits 2, 3, and 4 (i.e., 101 in the example).

  ■ P_3 is the parity of bits 1, 3, and 4 (i.e., 001).

The resulting values are $P_1=1$, $P_2=0$, and $P_3=1$. Now, let's append these parity bits to the original message to get $M_1M_2M_3M_4P_1P_2P_3$, or 0101<u>101</u>. This example of the Hamming code is expressed as a (7,4) code. The resulting codeword length is 7, the number of message bits is 4, and the number of redundant bits is, of course, 7 − 4, or 3.

Now suppose that the received message is 1101<u>101</u> (bit 1 was clobbered). The receiver computes the parity bits P_1, P_2, and P_3 and comes up with values of $P_1=0$, $P_2=0$, and $P_3=1$. The received value of P_1 (1) is different from the computed value (0), however, P_2 and P_3 match. Therefore, bit 1 must be the one in error, since parity over positions 2, 3, and 4 is unchanged. This example illustrates how Hamming codes can be used to detect, and correct, a single bit error.

A technique called *interleaving* is used to address error bursts. A group of k codewords is arranged in a matrix. Instead of transmitting one codeword at a time, the transmitter sends one column at a time. The receiver collects the columns in order, thereby reconstructing the matrix. If an error burst clobbers k bits, only one bit in each of the k codewords will be clobbered. Since Hamming codes can effectively reconstruct the codeword with 1 bit error, the entire matrix is error corrected.

■ **Reed-Solomon (R-S) codes** are also block codes, but unlike Hamming codes they operate on symbols, rather than bits. For example, consider an R-S code of 255 symbols. Assuming that each symbol is two bits, the block size is now 2^{255} or 2,040 bits. The advantage of a large block size is that it minimizes the overhead of error control. By operating on symbols, rather than bits, R-S codes increase the block size at minimal computational cost. As with Hamming codes, interleaving* can be used with R-S codes to address error bursts. Since R-S codes operate on symbols, which result in larger block sizes, interleaved R-S codes can deal with longer error bursts than interleaved Hamming codes.

■ **Trellis codes** are examples of convolutional codes (also known as tree codes). Unlike block codes, convolutional codes operate on continuous streams of data. They implement a shift register to manipulate the symbols to form a "code tree." The size of the shift register has a direct impact on the efficiency of the code.

Trellis coding was adopted as modem data rates increased. To understand the rationale for Trellis codes, let us go back to Figure 3.15, where we can see that each symbol represents two bits of data in a 4-QAM constellation. To achieve higher data rates, we can increase the constellation size, but this introduces a problem. As the constellation size increases, the granularity of the phase and amplitude differences between different constellation points diminishes. Therefore, it becomes increasingly difficult to decipher the constellation points accurately, especially in the presence of channel noise or interference, as shown in Figure 3.16.

From Figure 3.16, we can see that it follows that increasing the distance between symbols (known as the *Euclidean distance*) increases detection probability. This is the goal of Trellis coding,† developed by Dr. G. Ungerboeck. Trellis coding introduces redundancy into a bit-stream with a convolutional code. The output of the convolutional code is then mapped to an expanded constellation with a greater number of possible points. Thus, the technique is bandwidth efficient because the symbol rate, and hence, the bandwidth, is not increased. The state evolution of the convolutional code couples with the particular constellation mapping of the Trellis

---

* There are two different interleaving techniques—block interleaving and convolutional interleaving. Convolutional interleavers require less memory (by a factor of two to four) than block interleavers, and they cut the delay by half. Therefore, convolutional interleavers are used exclusively for DSL.

† Prior to the discovery of Trellis coding, Hamming coding was popular. Today, most modems use Trellis coding because it improves the performance of QAM-based modems without requiring increased bandwidth.

code design to restrict the possibilities for the sequence of symbols at the receiver. In this manner, the receiver is able to eliminate certain symbols from consideration during detection at the receiver. However, the transmitted *sequence* of symbols must now be detected. Normally, this would be an overwhelmingly complex operation, but the Viterbi[14] algorithm provides an efficient implementation of the sequence detection algorithm.

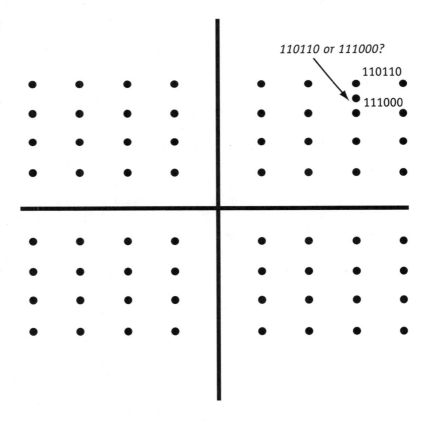

Figure 3.16
64-QAM
constellation
diagram

Automatic Repeat reQuest (ARQ)

ARQ is preferred in relatively noise-free environments where the overhead of possible retransmission is negligible in comparison to the overhead of adding redundancy for error correction. In other words, ARQ is preferred in scenarios where only error *detection*, and not error *correction*, is desirable.

There are three basic techniques of retransmission:

■ **Stop and wait ARQ.** In this technique, a block of data is numbered as a packet. The receiver checks the sequence number of the received packet. If it is in the expected sequence, it sends back an acknowledgment (ACK) and waits for the sender to send the next packet in the sequence. If the packet sequence number is not in the expected sequence, it sends back a negative acknowledgment (NAK), and the sender retransmits that packet.

■ **Go back N ARQ.** The problem with stop and wait ARQ is that it requires a positive acknowledgment for every packet. With go back N, the sender can continually send packets. The receiver can ACK or NAK the packets. Once the receiver NAKs a packet, the sender must "go back" to the packet number that was NAKed and resend all packets from that sequence number onward. For example, if the sender had previously sent packets 1 through 5, and 2 was NAKed, the sender would then resend packets 2, 3, 4, and 5.

■ **Continuous ARQ.** The problem with go back N ARQ is that it forces the sender to resend all the packets from the sequence number that was NAKed, even though only one packet may have been in error, and the remaining packets were successfully received by the receiver. With continuous ARQ, the sender only resends the single packet in error (i.e., only the sequence number that was NAKed, but not the ones following). For example, if the sender had sent previously packets 1 through 5 and 2 was NAKed, the sender would resend only packet 2. It is presumed that 3, 4, and 5 have been successfully received.

As we noted before, parity checking is one of the simplest methods of error detection. However, single-bit parity checking is insufficient to detect multiple bit errors. Let's go back to the example where a sender sends the message bits 010 with even parity resulting in a codeword of 0101. If the receiver receives the message 0001 (second bit clobbered), it knows that a parity error has occurred. However, if the receiver receives the message 1001 (first and second bits clobbered), the parity is still even and the receiver will not be able to detect the multi-bit error.

An improvement on this technique is to add a new block of data that checks the parity bits of the previous N blocks of data. This special block of data is called a block control character (BCC), since the extra character of data is used to parity-check the previous N characters that have been transmitted. However, even with BCC, it is possible for errors to go undetected.

The most reliable error-checking technique is based on cyclic redundancy code (CRC). In CRC, a polynomial code of the form $x^N + x^{N-1} + x^{N-2} \ldots x^2 + x^1 + x^0$ is used in a calculation* performed by both the sender and the receiver. If the calculations result in the same value, it is assumed that the data have been transmitted successfully. If not, the data are assumed to be in error, and a retransmission is requested (using one of the three basic techniques mentioned above).

Practical Use of Error Control Protocols

Modems generally use FEC (Hamming and Reed-Solomon codes) to avoid retransmission. In some cases, CRC checks are also performed on the data. Modems use one or more of the following standard error control protocols:

- **Microcom networking protocol (MNP) classes 2–4** are de facto standards developed by Microcom (now part of Compaq Computer Corporation). MNP framing defines several formats to be used during modem operation such as link request, link disconnect, link transfer, link acknowledgment, link attention, link attention acknowledgment, link management, and link management acknowledgment. Each of the frames performs a specific function, such as requesting link establishment, transmitting data across the link, acknowledging successful reception of data, requesting a change in the transmission data rate or requesting session termination.
- **MNP class 10** is a de facto standard developed by Microcom for adverse line conditions.
- **V.42**, the international standard for error correction, supports link access procedure for modems (LAPM) as the primary error control protocol, with fallback to MNP class 4.

Alternatively (or additionally), some modems may support a proprietary error control protocol.

---

* The calculation involves dividing the block of data by the polynomial code. This results in a remainder and quotient—the remainder portion is sent along with the data. The receiver performs the same calculation, and compares the computed remainder with that sent with the data to verify successful reception.

Data Compression

Another method for improving data throughput is to compress the input data before it is sent. The receiving modem reverses the process and decompresses the data before passing it on. Compression and decompression are done "on the fly," i.e., data compression is performed as it is received from the sending computer and decompressed just before it is transmitted to the receiving computer (with some buffering to distinguish "redundancy" in the data stream). Common techniques used in data compression are Huffman coding, run-length coding, and Lempel-Ziv coding.

Practical Use of Data Compression Protocols

Modems use one or more of the standard data-compression protocols noted below.

- **MNP class 5** is a de facto standard developed by Microcom, and can achieve a maximum compression ratio of 2:1.
- **V.42 bis**, the international standard for data compression protocols, can achieve a maximum compression ratio of 4:1.

Alternatively (or additionally), some modems may support a proprietary data compression protocol for higher compression ratios. Compression ratios, however, are highly dependent upon the type of input data. If there is no "redundancy" in the data stream (i.e., data have already been compressed), further compression will not yield any benefits.

Advanced Techniques

The evolution to modern voiceband and broadband technology closely matches corresponding advances in silicon technology. The availability of specialized processors called digital signal processors (DSPs) enabled modem engineers to implement algorithms to improve system performance economically. These advanced techniques allow V.34 modems to achieve data rates close to the theoretical limit. Some of the advanced features of V.34 modems are discussed below.

Echo Cancellation

To achieve full-duplex operation (i.e., simultaneous channels for transmitting and receiving), voiceband modems split the passband into two separate channels—one each for upstream and downstream. This is frequency division multiplexing, as discussed in Chapter 1 and in this chapter. In FDM modems, there are two separate analog carrier signals as shown in Figure 3.17.

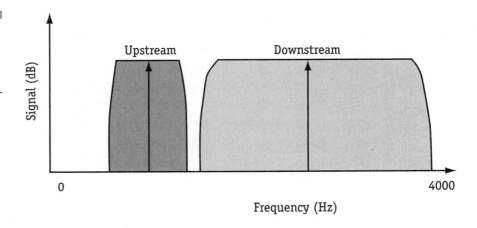

Figure 3.17
FDM technique for full-duplex operation

Modern voiceband modems employ echo cancellation techniques to achieve full duplex operation with a further increase in the data transmission rates. Echo cancellation-based modems use a single channel for both transmission and reception, as shown in Figure 3.18. Echo cancellation involves removing the effects of the modem's transmission on its reception. This allows a modem pair to use the entire available bandwidth for both directions.

Multi-dimensional Trellis Coding

In Figures 3.15 and 3.16, we can see that a way to increase the data rate is to increase the constellation size, and hence, the bit density per symbol. However, as the constellation size increases, it becomes increasingly difficult to decipher the constellation points accurately, especially in the presence of channel noise or interference. Trellis coding is used to solve this problem; it applies signal processing to extract information more reliably for a given S/N ratio.

Figure 3.18
*Echo cancellation
technique for full-
duplex operation*

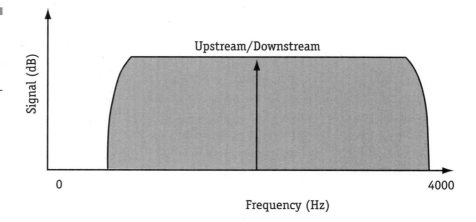

Initially, 2D Trellis coding was employed in V.32 bis modems. 2D Trellis coding introduces controlled redundancy to a bit-stream using a convolutional code. With 2D Trellis coding, introducing redundancy means adding one more bit/symbol. For example, for a 4 QAM constellation, the number of bits/symbol is two; with redundancy, the number of bits/symbol is three. So, the constellation size now increases from 2^2 to 2^3, and it becomes an 8-QAM constellation. In general, if the number of bits/symbol is N, 2D Trellis coding results in N+1 bits/symbol. Hence, the constellation size increases from 2^N to 2^N+1. In effect, 2D Trellis coding doubles the constellation size. Doubling the constellation size is not a problem for smaller constellations. However, as constellation sizes get bigger, the problem of detecting constellation points due to increased density (shorter Euclidean distances between constellation points) is reintroduced. Therefore, it is desirable to introduce redundancy without doubling the constellation size. Multi-dimensional Trellis coding provides the solution.

In 2D Trellis coding, redundancy is introduced in a 2D space. However, if the redundant bits were to be distributed in multi-dimensions (M>2), the number of bits/symbol mapped to a 2D space would be fewer. Therefore, it is possible to introduce redundancy without doubling the constellation size. For example, with 4 QAM, it is possible to achieve redundancy with 5 bits for every two symbols, i.e., $2\frac{1}{2}$ bits/symbol. So, the constellation size increases to $2^{2\frac{1}{2}}$, rather than 2^3. A practical limit on the number of dimensions for multi-dimensional Trellis code modulation (TCM) is 4 (i.e., M=4). This limit comes from striking a balance between increased complexity versus better performance. For further details on TCM, the reader is referred to [15].

Shell Mapping (Constellation Shaping)

If we refer to the constellation diagrams shown previously, we can see that as constellations get larger, placing symbols in a rectangular 2D plane requires more energy at the corners rather than at the edges. This is not very power efficient. Instead, if the outer boundary of the constellation were roughly circular (rather than rectangular), less energy would be required to place symbols at the edges. This would result in power savings of the average energy required per symbol.

If we were to take a uniform, spherical distribution of symbols and project it onto a 2D plane, we would end up with a non-uniform circular distribution of symbols, with the density of symbols highest at the center and decreasing at the edges. This reintroduces the problem that the distances between symbols are small when they are located close to the center. To overcome this problem, additional algorithms are required to maintain a balance between power reduction and adequate symbol distance. These algorithms are often implemented in DSPs because they involve complex mathematical calculations.

Fast Training

In Figure 3.3, the voice passband was depicted as linear. In actuality, the passband response is nonlinear because the higher frequencies get attenuated more than lower frequencies in the local loop. The equipment also introduces various nonlinearities. To compensate for the nonlinear behavior of the local loop, each modem's receiver must synchronize to the other's transmit signal during the initialization sequence. This is known as *training*. Training involves sending a known signal down the wire that the receiver can use to compare the received signal against the original non-distorted known signal. The receiver can then calculate the distortion effects of the channel and compensate accordingly.

Due to large error bursts on the line, it is possible for modems to perform training after the connection has been up for a while. This is known as retraining. Training/retraining is a complex process that takes time. During the time period when modems are training/retraining, no data can be sent. Hence, it is desirable to reduce this period. Unfortunately, estimated training times become longer for higher data rates such as the 33.6 Kbps supported by V.34 modems. The necessary reduction of training times is achieved by (a) changing the training sequence to one based on receiver timing rather than transmitter timing and (b) better error recovery procedures to avoid complete reinitialization.

Evolution of Modem Technology and Standards to DSL

The original modems developed in the 1950s used proprietary modulation protocols. Examples were the Bell 103 (300 b/s U.S. standard) and the Bell 212A (1200 b/s U.S. standard), based on radio frequency (RF) technology. In the 1960s, the International Telecommunications Union (ITU)* began to ratify standards-based modem protocols beginning with V.21 (300 b/s standard outside the United States). This marked an important milestone in the development and adoption of international standards for modem technology. In the 1980s, the next major improvements came with the addition of echo cancellation and Trellis coding. Echo cancellation allowed modem pairs to use the entire available bandwidth for both upstream and downstream movement. Trellis coding made it possible to implement error correction in modems that resulted in the ability to extract information more reliably for a given S/N ratio. This enabled data rates to increase to 9.6 Kbps with V.32 and 14.4 Kbps with V.32bis.

In early 1990, work began on raising the data rate beyond 14.4 Kbps to an initial goal of 19.2 Kbps. With the adoption of several key technologies—line probing, precoding, multi-dimensional Trellis coding, shell mapping, and warping—a top speed of 28.8 Kbps was achieved in 1994. By 1996, V.34 standard-based modems could achieve 33.6 Kbps, which is close to the theoretical limit. This was followed by the introduction of 56-Kbps modems based on the V.90 standard. V.90 modems do not violate Shannon's law because communication is not between

---

* The ITU was formerly known as the CCITT.

two modems, but rather between a 56K modem and a digital system. This results in reduced quantization noise because only one analog-to-digital conversion must take place. Table 3.2 summarizes some of the significant milestones in modem history.

Table 3.2

Some milestones in modem history

Year	ITU Recommendation	Data Rate (b/s)	Bandwidth	Modulation
1968	V.26	2,400	1200	4-PSK
1972	V.27	4,800	1600	8-PSK
1976	V.29	9,600	2400	16-QAM
1984	V.32	9,600	2400	2D TCM
1994	V.34	28,800	3400	4D TCM

In the 1980s and early 1990s, simultaneously with the advances in modem technology, researchers began studying the problems of copper-loop access with a view to improving data rates and removing the restrictions of the 4-kHz narrow bandwidth. Copper can natively support much higher bandwidth (up to about 2 MHz) without significant attenuation; however, filters in the loop plant constrain the bandwidth to 4 kHz for voice transmission. Bellcore engineers and others began to consider improvements in transmission that could be achieved by removing the filters and other restrictions on the loop to allow the complete bandwidth to be utilized. The first DSL technology was ISDN, which provided an end-to-end symmetric digital connection all the way to the subscriber. Symmetric DSL technology also evolved as a replacement for T1 lines. Asymmetric DSL also evolved out of Bellcore research in the late 1980s. Many of these technologies utilize, and improve upon, the techniques developed for voiceband modems. For example, ADSL modems use Reed-Solomon FEC, echo cancellation, multi-dimensional Trellis coding, and shell mapping. Even more advanced techniques based on both CAP (e.g. symmetric and asymmetric transmission profiles) and DMT (e.g., Zipper) have been proposed for VDSL. A discussion of these techniques is reserved for Chapter 6 on VDSL.

Simultaneous POTS and DSL Service

Certain DSL technologies, such as ADSL, permit simultaneous POTS and DSL service, since the associated DSL frequencies are located above the voiceband and do not overlap with voiceband frequencies. This is by design, since researchers realized that it would be necessary to provide mechanisms to support legacy POTS devices, including voiceband modems. To support *independent* simultaneous POTS and DSL service, devices known as *splitters* can be deployed at both the CO side and the premises (residence) side, as shown in Figure 3.19. A splitter consists of a high-pass filter (HPF) and a low-pass filter (LPF) to ensure that POTS and DSL do not interfere with each other.

Figure 3.19
*Function of
a splitter*

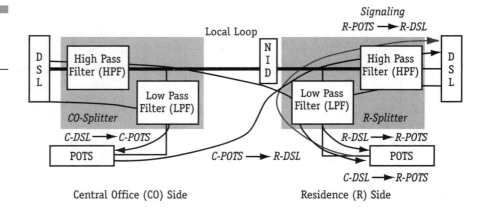

The following observations can be made about Figure 3.19:

- **C-DSL into C-POTS:** CO-side DSL service interference with CO-side POTS is prevented by LPF incorporated into the CO splitter.
- **R-DSL into R-POTS:** Premises-side (residential) DSL service interference with premises-side POTS is prevented by LPF incorporated into the premises splitter.
- **C-DSL into R-POTS:** CO-side DSL service interference with premises-side POTS is prevented by LPF incorporated into the premises splitter.
- **C-POTS into R-DSL:** CO-side POTS interference with premises-side DSL service is prevented by HPF incorporated into the premises splitter.

■ **R-POTS into R-DSL:** Premises-side POTS interference with premises-side DSL service is prevented by HPF incorporated into the premises splitter.

Splitters are important devices that provide necessary isolation, and were originally considered mandatory in architectures that support simultaneous POTS and DSL service. However, as we discuss later in this book, the industry organized itself in 1998 to make at least the premises splitter optional for a variety of important reasons.

Summary

The PSTN access architecture has strongly influenced copper-loop transmission and, consequently, voiceband and DSL modem technologies. Although copper is capable of supporting greater bandwidth, the bandwidth of the loop is constrained to 4 kHz to support voice. Additionally, since the copper loop is a noisy environment, modem engineers had to develop ingenious ways to improve the S/N ratio within the restricted bandwidth. Some of the major advances came with improved error correction and modulation techniques such as Reed-Solomon coding, multi-dimensional Trellis coding, and shell mapping. The coding gains realized with these techniques have enabled V.34 and later modems to achieve almost the theoretical limit of data rates imposed by Shannon's theorem. These techniques have since found application and further improvement in digital subscriber line technologies.

References

1. Nyquist, H. "Certain Factors Affecting Telegraph Speed," *Transactions A.I.E.,E.* (1924).
2. Shannon, C. E. "A Mathematical Theory of Communication," *Bell Systems Technical Journal*, Vol. 27 (1948).
3. Stallings, W. *Data and Computer Communications*, 5th edition (New York: Prentice-Hall, 1996), Chapter 4.

4. Starr, T., Cioffi, J., and Silverman, P. *Understanding Digital Subscriber Line Technology* (New York: Prentice-Hall, 1999), Chapter 6.

5. "DS1 Metallic Interface," *ANSI Standard T1.403* (1995).

6. "DS3 Metallic Interface," *ANSI Standard T1.404* (1994).

7. "Integrated Services Digital Network (ISDN)—Basic Access Interface for Use on Metallic Loops for Application on the Network Side of the NT (Layer 1 Specification)," *ANSI Standard T1.601* (1992).

8. ANSI Technical Report TR28 (on HDSL).

9. Im, G. H. and Werner, J. J. "Bandwidth-Efficient Digital Transmission over Unshielded Twisted-Pair Wiring," *IEEE Journal on Selected Areas in Communications*, Vol. 12, No. 9 (December 1995).

10. Cioffi, J. M. "A Multicarrier Primer," *ANSI Contribution T1E1.4/91-157* (November 1991).

11. Bingham, J. "Multicarrier Modulation for Data Transmission: An Idea Whose Time Has Come," *IEEE Communications Magazine*, Vol. 28, No. 5 (May 1990).

12. Lin, S. and Costello, D. J. *Error Control Coding: Fundamentals and Applications* (New York: Prentice-Hall 1983).

13. Wilson, S. G. *Digital Modulation and Coding* (New York: Prentice-Hall, 1996).

14. Viterbi, A. J. "Error Bounds for Convolutional Codes and an Asymptotically Optimum Decoding Algorithm," *IEEE Transactions on Information Theory*, Vol. 13 (April 1967).

15. Ungerboeck, G. "Trellis Coded Modulation with Redundant Signal Sets: Parts I and II," *IEEE Communications Magazine*, Vol. 25, No. 2 (February 1987).

ADSL and ADSL Lite

Background

History

As described in Chapter 3, the problem of running data over legacy voice infrastructure severely limits the maximum available data rate using traditional modem technology. By the 1980s, the issues were well understood. Researchers began to study ways of utilizing the bandwidth that an ordinary twisted pair of copper wire could natively support. A study of typical loops between the customer premises and a central office showed that, for loop lengths up to approximately 18,000 feet, it is possible to achieve enough signal strength at the receiver to code data bits meaningfully into symbols at frequencies well above the voice passband. These efforts gave rise to the family of technologies known as XDSL.

Asymmetric digital subscriber line (ADSL), an important member of the XDSL family, was developed in 1989 out of research work done at Bellcore by Joseph Lechleider. Two key features distinguish ADSL:

■ It allows POTS to coexist on the same wire. To achieve this, researchers proposed using a guard band to separate the voiceband and broadband frequencies. They further proposed utilizing the broadband frequency spectrum to transmit digital data (carrying any combination of voice, video, and data in digital format), thus increasing the intrinsic bandwidth capability of a copper loop.
■ ADSL gets its name from its inherent asymmetry. The upstream bandwidth is less than the downstream bandwidth.*

Bellcore originally envisioned ADSL for video on demand (VoD) applications. The motivation for Bellcore's research in the early 1990s was the need for telcos to respond to competition by cable companies in delivering VoD services to the home. Bell Atlantic conducted the initial VoD trials using ADSL in northern New Jersey; about the same time Time Warner conducted VoD trials using cable in Orlando, Florida. In early VoD trials, the downstream rates topped at approximately 1.5 Mbps—sufficient for the delivery of MPEG-1 video streams. The

---

* *Downstream* means the data rate from the network to the subscriber, *upstream* is the reverse direction.

upstream rates, at around 64 Kbps, were sufficient to allow users to send simple commands up to a video server (i.e., commands to select the movie and VCR-like commands to pause, play back, fast forward, and rewind).

By the mid 1990s, the market for VoD all but dried up due to lack of demand. The primary reason was not the ADSL technology itself; rather, it was the fact that the capital expense of deploying expensive video services translated into high monthly subscriber charges. Consumers were unwilling to pay these charges, when the trip to the local video store was so much cheaper. When the VoD market contracted, both cable companies and telcos began eyeing new opportunities for their technologies.

The need for high-speed Internet access gave new life to ADSL. The inherent asymmetry of the technology made it well suited for Web browsing applications, where the downstream content information generally required greater bandwidth than upstream requests. The upstream and downstream rates improved to 640 Kbps and approximately 6–8 Mbps,* respectively (depending on loop length and conditions). This meant that downstream rates were high enough to support MPEG-2 video. The ratio of 1:10 for upstream-to-downstream bandwidth was deliberately chosen as the optimum value to suit TCP/IP traffic. Another improvement, rate adaptation, allowed two DSL modems to adjust their upstream and downstream rates based on loop conditions. Yet another benefit of ADSL for Internet access was the "always on" feature.

Using the local loop for both analog voice and digital data required more efficient line codes. The discrete multi-tone (DMT) line-coding technique, originally developed at Bellcore, was refined and implemented around 1987 as the result of work performed by Professor John M. Cioffi at Stanford University, who founded Amati Corporation† in 1992. Amati developed an ADSL modem called Prelude using off-the-shelf components. Prelude was tested by telcos throughout the world to evaluate the basic technology. Amati subsequently incorporated lessons learned from these tests into its next-generation ADSL transceivers and modems. Globespan Corporation promoted a competing line-code tech-

---

* This version of ADSL (offering 6–8 Mbps downstream rates) is also referred to as full-rate ADSL.
† Amati Corporation was subsequently sold to Texas Instruments in 1997.

nique called *carrierless amplitude phase modulation*. A third proposed option for line coding, quadrature amplitude modulation, was closely related to CAP in its basic technology. Some groups endorsed both CAP and QAM because they took advantage of the established knowledge base among voiceband modem engineers and manufacturers.

In the early 1990s, Bellcore evaluated various line code options for ADSL and agreed upon DMT, based on the successful demonstration of the technology by Amati. Though CAP proponents disagreed, DMT was believed to perform better than the competing CAP line code in the presence of line noise. The DMT camp was able to convince the appropriate standards committees in the United States and Europe about DMT's merits. On the other hand, the early trials and deployments of ADSL were based on CAP. Each line coding technique demonstrated specific advantages/disadvantages over the other, depending on the line conditions. Therefore, technical debates raged in the industry about the appropriate line code to use for ADSL[1,2]. These debates became quite charged at times, leading to line-code "holy wars" in the marketplace. DMT proponents pointed to the adoption by the standards bodies; CAP proponents pointed to actual deployments. To this argument, the DMT proponents shot back that the deployment numbers at that time were insignificant. Finally, the vendor community, with prodding from the service providers responsible for deployment, agreed to support DMT as the line code of choice for ADSL. The growing momentum of international standardization efforts by the International Telecommunications Union (ITU) behind DMT was clearly a strong influence in this choice.

While standardization efforts and line-code debates continued at the physical layer, other groups were forming to develop the link, network, and transport layers into deployable architectures. In early 1994, the first steps toward the formation of an industry group focused on enabling deployment of copper access technologies took place in New York City. At an initial meeting between Gorham & Partners, consultants to the International Copper Association, the need for an industry forum to bring together individuals, companies, and organizations with interests in developing and promoting ADSL was recognized. This idea gained momentum and, in October 1994, the first meeting of the ADSL Forum was held in London. Approximately 50 people representing 43 organizations from 14 countries attended that first meeting. The meeting ended with the nomination of an Interim Steering Group, and a commitment to hold the first full meeting in San Antonio, Texas three months later. Since then, regular quarterly meetings have been held, alternating between locations in North America and Europe, until 1998 when the first meeting in Asia took place in Singapore. By 1999, the attendance at these meetings had grown to more than 500 people, representing numerous companies and countries.

By charter, the ADSL Forum has stayed line-code neutral. Instead, it has chosen to focus its efforts on solving problems above the physical layer. Over the years, the ADSL Forum has developed liaisons to other working groups and standards bodies to resolve many issues relating to end-to-end architectures that incorporate copper-loop access. It has also released several public documents detailing its work based on technical contributions.

In early 1998, another significant event in the history of ADSL occurred—the formation of the Universal ADSL Working Group (UAWG). The shift in focus from video on demand to high-speed Internet access required the creation of a new ADSL standard for mass-market deployment. This was driven by factors such as reduced cost to consumers and the removal of barriers to widespread deployment such as network provider installation and new wiring. Industry proponents recommended decreasing the total cost to consumers by focusing on a reduced set of ADSL features that would be optimized for data access at a maximum downstream rate of 1.5 Mbps. This 1.5 Mbps rate to enable Internet access to consumers was chosen for some key reasons:

- This rate was close to what service providers were comfortable with for the initial ADSL service roll out.
- The rate closely matched the Internet backbone capability. In other words, increasing the rate to 6–8 Mbps would not have material impact on the overall throughput since the Internet backbone would have throttled the overall rate back down to around 1.5 Mbps.
- The rate versus cost to consumers could be made competitive with technologies such as cable.

Therefore, the term ADSL lite came to be associated with the new standard, although, with typical industry confusion, various terms have since been associated with this standard for a variety of reasons.* The older standard, providing 6 Mbps to 8 Mbps downstream bandwidth, came to be known as full-rate ADSL or ADSL "heavy."

In a unified response to the critical need for a standard that would enable large-scale deployment of high-speed Internet access over copper loops to consumers, the computer, telecommunications, and networking industries took the unprecedented step of combining their technological and marketing clout. This culminated in the formation of the UAWG to expedite the international standardization of ADSL lite.

The roles of the ADSL Forum and the UAWG have been complementary. The ADSL Forum has focused on full-rate ADSL and architectural issues above the physical layer. The UAWG has focused primarily on the necessary optimization of the physical layer to enable mass-market high-speed data access.

Basic Network Architecture

The basic network architecture, shown in Figure 4.1, is important as a backdrop for the discussion of the technology. A detailed discussion of the architecture is reserved for a later chapter in this book.

Figure 4.1

Basic network architecture for ADSL and ADSL lite (Source: Texas Instruments)

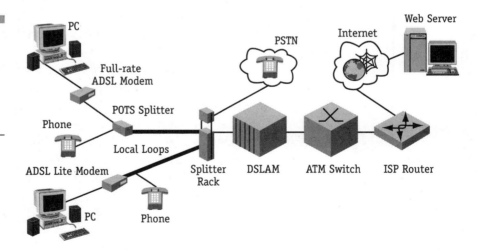

---

* The official ITU designation for the standard is G.992.2. However, since this designation does not convey the idea behind the standard, most people in the industry refer to the standard by its working title, "G.lite." Prior to G.lite standardization, there were a multiplicity of vendor implementations that conveyed the motivations behind the standard, such as "Splitterless ADSL" and "Consumer DSL."

Between every subscriber premises* and the central office is a twisted-pair copper wire commonly referred to as the local loop. For full-rate ADSL, splitters are at both ends of the local loop[†] to isolate POTS and ADSL from each other.

At the premises, a splitter may[‡] be installed at the demarcation point[§] ("demarc") between the telco and the premises wiring. At the demarc, another device called the network interface device (NID) provides the necessary isolation between the telco and the premises. The splitter is installed on the premises side "behind" the NID[||] and two separate wire pairs exit from there. The first wire pair, which is usually the existing wire, provides voice services. Fanning out into tree-and-branch wiring, the branches terminate at wall jacks where POTS devices, such as telephones and FAX machines, are typically connected. The second wire pair, which may be new wiring, leaves the splitter, to provide ADSL service.[¶] This ADSL wire then connects to an ADSL modem at the premises called an ADSL termination unit, remote (ATU-R).

Splitters are also installed on the central office side where the local loop is first terminated on a main distribution frame (MDF). This is the central point at which copper loops that fan out to the premises are terminated. For ADSL, a wire pair then connects each copper loop into a CO side splitter (actually a bank of splitters, one for every loop that has ADSL service). As in the case of the premises splitter, two wire pairs leave the CO splitter. The first connects into a voice switch such as a Lucent 5ESS or a Nortel DMS 100 to provide traditional POTS. The second wire pair connects into the CO counterpart of the ATU-R, known as the ADSL termination unit, central (ATU-C). In short, the ATU-R and ATU-C are the modems at either end of the ADSL wire. For

---

* Typically, a premises is a residence, and the two terms are sometimes used interchangeably. In actuality, a premises could be any number of buildings such as a residence, a multi-family dwelling, an office complex, etc.

† One of the simplifications of ADSL lite is to make the premises splitter optional. This eliminates the need for the service provider to send out an installer to install the premises splitter, i.e., it eliminates the need for a "truck roll."

‡ As noted earlier, the premises splitter is required for full-rate ADSL and is optional for ADSL lite.

§ The demarcation point varies from country to country. The scenario described here is applicable for U.S. installations.

|| A splitter is sometimes incorporated into the NID.

¶ When the premises splitter is eliminated, the copper-loop wire pair fans out into the typical tree-and-branch wiring used to provide POTS; one of the branches then connects to an ADSL lite modem (rather than a POTS device). Note that this potentially eliminates the need for new wiring; the ADSL lite modem can be plugged into any existing wall jack.

efficiency reasons, a bank of ATU-Cs is combined with a multiplexing function to form a DSL access multiplexer (DSLAM) in the central office, and this connects to a service provider's network.

The architectural impact of an optional premises splitter for ADSL will be discussed later in this book. For now, it is sufficient to understand that the possible elimination of the customer premises splitter implies that POTS and ADSL service may exist concurrently without isolation—in other words, they could interfere with each other. Therefore, ADSL lite technology incorporates mechanisms to mitigate the impact of such interference.

Applications

The inherent asymmetry of ADSL makes it well suited to almost any application that requires high downstream bandwidth, but lower upstream bandwidth. As noted earlier, VoD was the initial driver for ADSL; however, Internet access quickly became the primary driver for both full-rate ADSL and ADSL lite. The following represent the types of applications being developed and deployed for these technologies:

▌ **Telecommuting.** Telecommuting allows people to work from their homes and connect to resources at the workplace. To the degree that telecommuting implies data-only access, both full-rate ADSL and ADSL lite can support this application. However, telecommuters increasingly need remote access to both data and voice services. For example, there is a need for a telephone at home to perform as a logical extension off a corporate PBX. Companies are developing products that utilize the bandwidth of full-rate ADSL to support multiple virtual phone lines that can make ordinary POTS devices (e.g., telephones and fax machines) act as extensions of a corporate PBX. Supporting combined voice and data services may require migration to full-rate ADSL with guaranteed QoS.

▌ **Video streaming or real-time information.** ADSL enables delivery of real-time, bandwidth-intensive applications such as news, stock tickers, and weather. This is closely tied to "push technology" where a viewer can subscribe to channels of interest that can be regularly updated by taking advantage of the "always on" connection that both ADSL and ADSL lite provide.

- **Distance learning.** Full-rate ADSL with guaranteed QoS can support an MPEG-2 video stream, thus allowing a training center to simulcast training videos to multiple sites and communicate with trainees at those sites.
- **Telemedicine.** Doctors, including radiologists, can diagnose and recommend procedures using X-rays and other video images sent from another location. Typically, the doctor is a specialist in a hospital, and the remote location is a rural area. This has the advantage of bringing specialized consulting services to rural medicine. In one application developed by the University of Alabama, a radiologist can remotely control a microscope to view a patient's slide and perform functions such as panning and zooming utilizing a full-rate ADSL connection.
- **Videoconferencing.** In the initial analysis, videoconferencing might appear to be mismatched for ADSL because it requires bandwidth symmetry. However, with full-rate ADSL and guaranteed QoS support, it is possible to dedicate an H0 channel (i.e., 384 x 384 Kbps) out of the available ADSL bandwidth to a videoconferencing application, while still leaving adequate bandwidth for other applications. This is an example of how certain applications with symmetric bandwidth requirements can be adapted to ADSL.

Standards and Working Group Activities

As with other mass-market technologies, standardization of ADSL is important to enable vendors to build modems that interoperate. The International Telecommunications Union is responsible for global modem standards at the physical layer. The work of the ITU is assisted by other standards bodies—the most significant among them being the American National Standards Institute (ANSI) and the European Telecommunications Standards Institute (ETSI). In addition, the ADSL Forum and the UAWG are industry work groups that also provide input into the standards process at the ITU.

The ADSL Forum has been responsible for developing recommendations for both packet and cell transport over ADSL. The UAWG has been responsible for accelerating the standardization of ADSL lite through the ITU process.

American National Standards Institute (ANSI) Committee T1

In the United States, committee T1[3] is accredited with ANSI and has the responsibility for developing national telecommunications standards and influencing international standards through the ITU. Within committee T1, the T1E1 subcommittee is responsible for developing standards relating to the interfaces, power, and protection of networks. T1E1.4 is the working group in T1E1 that is focused on DSL. This committee developed the original "mother" ADSL standard based on DMT line coding—known as T1.413. This standard has gone through a couple of versions termed *issues*—the latest version is known as T1.413 Issue 2, or simply T1.413i2.[4]

Issue 1 defines several optional capabilities and features, including echo cancellation, Trellis code modulation, and dual latency (to support both delay-sensitive traffic such as voice and video- and delay-insensitive traffic such as data). Issue 2 defines additional options such as transport of a network timing reference, transport of STM and/or ATM, and a reduced overhead framing mode.

European Telecommunications Standards Institute (ETSI)

ETSI[5], the European counterpart to the ANSI committee T16, is responsible for developing European telecommunications standards and influencing international telecommunications standards through the ITU. Within ETSI, the Technical Committee on Transmission and Multiplexing (TM) is responsible for standardization of the functionality and performance of transport networks and their elements. This includes working group TM6 that focuses on DSL. ETSI TM6 has increased the value of the work of the ANSI T1.413i2 standard by adapting it to satisfy European requirements.

International Telecommunications Union (ITU)

The ITU[6] is the standards organization that oversees the development of global modem standards, known as recommendations. Just as the "V series" of ITU recommendations are relevant for voiceband modems, the "G series" of recommendations are relevant for XDSL modems. ITU recommendation G.902 is a framework for the architecture and functions of access networks. It describes access types, management, and service node aspects. The lead Study Group for access networks is

Study Group XV (SG 15), which has the responsibility for developing recommendations on XDSL. Figure 4.2 illustrates how various standards organizations coordinate their work to input into the ITU for ADSL.

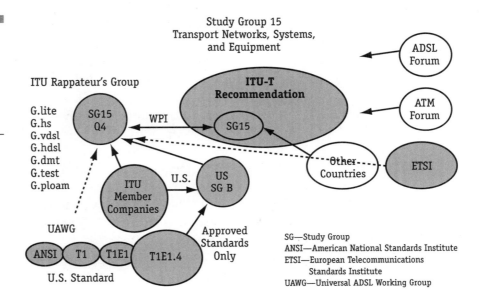

Figure 4.2

ITU process for the "G series" of standards for ADSL (Source: Texas Instruments)

Within SG 15, the following projects have been approved for standardization with respect to ADSL:

■ **G.992.1 (G.dmt)**[7]—Internationalized version of T1.413i2 with annexes for full-rate ADSL

■ **G.992.2 (G.lite)**[8]—ADSL lite standard to support mass market Internet access and other consumer applications

■ **G.994.1 (G.hs)**[9]—Handshake procedures to facilitate interoperation between G.992.1and G.992.2 modems (similar to the V.8 bis procedure for voiceband modems)

■ **G.996.1 (G.test)**—Testing procedures for XDSL modems

■ **G.997.1 (G.ploam)**—Physical layer operations, administration, and maintenance for XDSL modems.

ADSL Forum (ADSLF)

The ADSL Forum[10] was formed to help telcos and their suppliers realize their respective business objectives with ADSL. Its technical efforts have concentrated on the new copper-loop access systems required to deploy ADSL. These systems necessitate protocols and connections for home networks and terminals, access networks that concentrate traffic and route signals to appropriate destinations, and network management to install, configure, maintain, and migrate ADSL systems. Therefore, the ADSL Forum's technical projects have focused on system features above the physical layer.

The ADSL Forum's technical work is divided into seven areas, each with its own working group:

- ATM over ADSL (including transport and end-to-end architecture aspects)
- Packet over ADSL (this work is now complete)
- Customer premises equipment (CPE) and central office (CO) configurations and interfaces
- Operations
- Network management
- Testing and interoperability
- Support to the VDSL study group.

Each work group develops technical reports through a technique called "Working Texts," documents that capture and organize work in progress. When the work is completed, the Working Text becomes a Technical Report upon membership approval. This is then made public and distributed by the Forum to interested parties. In many cases, when the Forum identifies a particular requirement for ADSL that is outside its scope, it advises another standards body of this requirement through liaisons to coordinate the work effort. The ADSL Forum has established formal liaisons to key standards bodies and working groups, including the ITU, ATM Forum, ANSI T1E1.4, ETSI TM6, and the UAWG.

Universal ADSL Working Group (UAWG)

In January 1998, the UAWG[11] was officially formed with defined goals to:

- Promote a universal, single standard by developing a common specification backed by strong industry support and adoption of the specification as the global standard known as G.992.2
- Simplify installation by reducing additional device and/or wiring requirements at the premises
- Deliver downstream bandwidths at speeds up to 25 times faster than voiceband modems
- Provide an always-connected service that avoids time-consuming connection procedures and enables new classes of applications.

A key decision among the participating members of the UAWG was to use the ANSI T1.413i2 standard and to modify it as necessary to develop a new standard optimized for mass-market Internet access and other consumer applications. This "lighter" version of the ADSL standard traded off higher speed for extended reach and reduced cost, while providing adequate speeds in support of Internet access (up to 1.5 Mbps downstream and 512 Kbps upstream). Just 10 months after the formation of the UAWG the members made several technical contributions to the ITU, which resulted in accelerating the ITU's determination of G.992.2 (G.lite) by October 1998.

It is important to note that the UAWG is not a permanent body. It has a "sunset clause" that will eventually result in the Work Group's ceasing to exist. The work efforts of the UAWG, especially on interoperability testing, will make a logical transition to the ADSL Forum.

Technology Discussion

Modulation Schemes

Both DMT and CAP are efficient line codes designed to take advantage of the large amount of high-frequency bandwidth that lies above the analog voice passband. However, DMT and CAP are fundamentally different in their implementation; hence, it is not possible for a DMT transceiver to interoperate with a CAP-based one. CAP is based on quadrature amplitude modulation, a popular line coding technique for analog modems described in Chapter 3. DMT and CAP line coding are discussed in detail in the following sections.

DMT

DMT-based[12,13,14] ADSL modems can be thought of as many (usually 256) "mini-modems," 4 kHz each, that run simultaneously. DMT uses many carriers that create subchannels, with each subchannel carrying a fraction of the total information. The subchannels are independently modulated with a carrier frequency corresponding to the center frequency of the subchannel and processed in parallel. Each subchannel is modulated using QAM and can carry between 0 and a maximum of 15 bits/symbol/Hz. The number of actual bits carried per subchannel depends upon the line characteristics. Certain subchannels can be left unused because of external interference. For example, an AM radio station causing radio frequency interference in a particular subchannel can cause that subchannel to be unused. DMT is illustrated in Figure 4.3.

Figure 4.3
*Discrete Multi-Tone (DMT)
(Source: Texas Instruments)*

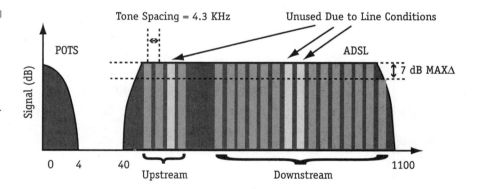

The theoretical maximum upstream bandwidth is 25 channels × 15 bits/symbol/Hz/channel × 4KHz = 1.5 Mbps.

The theoretical maximum downstream bandwidth is 249 channels × 15 bits/symbol/Hz/channel × 4KHz = 14.9 Mbps.

DMT offers several advantages as the line-coding technique for ADSL. Among them are:

■ **Evolution from V.34 modem technology.** As we discussed in Chapter 3, V.34 modems use several advanced techniques to maximize the data rates on noisy lines. DMT-based ADSL modems represent a natural evolution from V.34 modem technology. DMT modems employ QAM (each of the subchannels in DMT implements QAM), echo cancellation,* multi-dimensional Trellis coding, and constellation mapping.

■ **Performance.** DMT increases modem performance because independent subchannels can be manipulated individually with consideration of line conditions. DMT measures the S/N ratio separately for each subchannel and assigns the number of bits carried by the subchannel accordingly. Typically, the lower frequencies can carry more bits because they are attenuated to a lesser extent than higher frequencies. As a result, this procedure increases the overall throughput, even under adverse conditions.

■ **Robustness to line impairments.** During initialization, DMT monitors the line conditions and computes the bit carrying capacity of each subchannel based on its S/N ratio. If a subchannel is experiencing external interference such as radio frequency interference (RFI) and crosstalk, then it may not be used at all in favor of other subchannels.

■ **Rate adaptation.** DMT can dynamically adapt the data rate to line conditions. Each subchannel carries a certain number of bits depending on its S/N ratio. By adjusting the number of bits per channel, the DMT can automatically adjust the data rate.

CAP

Like QAM, CAP uses both multilevel amplitude modulation (i.e., multiple voltage levels per pulse) and phase modulation, resulting in constellations, as shown in Figure 4.4.

The difference between CAP and QAM is really in the implementation. With QAM, two signals are combined in the analog domain. However, since the carrier signal does not carry information, CAP implementations do not send the carrier at all. The signal modulation is done digitally using two digital filters with equal amplitude characteristics that differ in phase response.[†15] The advantage over QAM is the digital,

---

* Echo cancellation is even more important in a DMT modem than in a voiceband modem. Without echo cancellation the upstream and downstream bandwidths are treated separately, with the lower frequencies used for upstream bandwidth. However, lower frequencies get attenuated less than higher frequencies, hence the bit-carrying capacity of subchannels that span these frequencies is greater. Therefore, echo canceling modems are able to provide greater downstream bandwidth by utilizing the lower frequencies—something that is not possible in FDM modems.

† These filters are known as a *Hibert pair*.

rather than analog, signal modulation—this results in cost savings. The absence of carrier makes the implementation "carrierless" or "carrier suppressed." However, there is a small price to pay for suppressing the carrier. With QAM, the constellation is fixed; with CAP, the constellation is free to rotate since there is no carrier to fix it to an absolute value. To compensate for this, a CAP receiver must include a rotation function to detect the relative position of the constellation. Fortunately, the cost of implementing the rotation function in silicon is minor.

Figure 4.4

64-CAP constellation

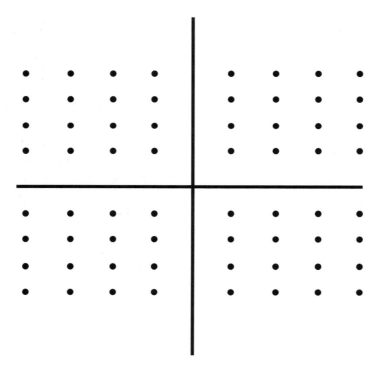

In contrast to DMT, CAP uses the entire available bandwidth (except for the analog voice passband), so there are no subchannels in CAP. In other words, although DMT and CAP are both based on QAM, the important difference is in the frequency range over which the QAM-like technique is employed. DMT uses QAM in each subchannel; CAP (like traditional QAM) evenly distributes energy across the entire range of frequencies. CAP systems employ frequency division multiplexing (FDM) to separate the frequencies into an upstream channel and a downstream channel.

CAP offers the following advantages:

■ **Mature technology evolving from V.34 modems.** Since CAP is directly based on QAM, it is a well-understood, mature technology and, due to the absence of subchannels, is simpler to implement than DMT.

■ **Rate adaptation.** In CAP, rate adaptation can be achieved either by changing the constellation size (4-CAP, 64-CAP, 512-CAP, and so on), or by increasing/decreasing the range of the frequency spectrum utilized.

The Line Code Debates

The merits of DMT versus CAP have been hotly contested in the standards working group discussions, in technical journals, and in the media. Technically speaking, both line codes have their advantages and disadvantages—after all, if one technology were clearly superior, there would not be any debate. In the final analysis, however, the standards organizations gave the nod to DMT largely due to the general agreement (at least among the members who developed the standards) that DMT is more robust in the presence of line impairments. In practice, CAP has also been shown to be very robust. However, the results depend on a variety of factors influencing line conditions. For now, the debate has been settled in favor of DMT, at least for full-rate ADSL and ADSL lite. Unfortunately, the debate continues around VDSL.

ANSI T1.413 Standard Overview

ANSI T1.413[16] is the "mother" standard upon which the ITU standards for full-rate ADSL (G.992.1 a.k.a. G.dmt) and ADSL lite (G.992.2 a.k.a. G.Lite) are based. The following are some of the important specifications in T1.413:

■ DMT line code and spectral composition of the signals transmitted by the ADSL modems at both ends of the wire

■ The transmission technique used to support the simultaneous transport of voiceband services and both simplex (unidirectional) and duplex (bidirectional) digital channels on a single twisted-pair

■ Electrical and mechanical specifications of the network interface

■ Organization of transmitted and received data into frames
■ Functions of the operations channel.

For the remainder of this technology discussion reference is primarily made to the T1.413i2 standard, with the exception of the fast retraining and power management topics unique to G.lite.

ADSL System Reference Model

As mentioned earlier, the ADSL Forum stepped up to the task of defining the reference architecture around the basic physical-layer technology. The system reference model shown in Figure 4.5 illustrates the functional blocks required to provide ADSL service.

Figure 4.5
System reference model for full-rate ADSL (Source: ANSI Contribution T1.E14/97-007R6, July 1998)

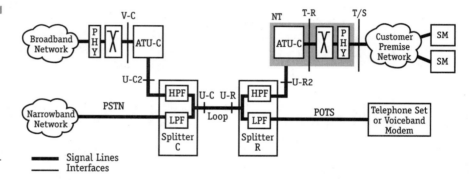

To simplify discussion, only the full-rate ADSL system reference model is shown. The system reference model for ADSL lite is similar to the one shown in Figure 4.5, except that the premises splitter is optional.

Contrasting Figure 4.5 with Figure 4.1, we can see that the system reference model incorporates the basic architecture. However, since it is used as the basis for standardization efforts, the standard interfaces have specific labels, as shown in Figure 4.5.

With reference to Figure 4.5, the following interfaces are defined:

Splitter C	Interface between PSTN and splitter—CO side
Splitter R	Interface between PSTN and splitter—Remote side
U-C	U interface—CO side
U-C2	U interface—CO side from splitter to ATU-C

U-R	U interface—Remote side
U-R2	U interface—Remote side from splitter to ATU-R
V-C	V interface—CO side from access node to network interface

ADSL Bearer Channels

An ADSL system may transport up to seven bearer channels simultaneously.* The data rates of all bearer channels can be programmed in any combination of multiples of 32 Kbps; i.e., 1.536 Mbps (North America) or 2.048 Mbps (Europe and elsewhere). The 32 Kbps comes from the granularity of DMT.

There can be up to four independent downstream simplex (unidirectional) bearer channels that are labeled AS0 to AS4.

1. Bearer channel AS0 supports data rates from 32 Kbps up to 6.144 Mbps[†] (at all multiples of 32 Kbps).
2. AS1 supports the range from 32 Kbps to 4.608 Mbps (4.096 Mbps in Europe and elsewhere).
3. AS2 supports the range from 32 Kbps to 3.072 Mbps (2.048 Mbps in Europe and elsewhere).
4. AS3 supports the range from 32 Kbps to 1.536 Mbps.[‡]

As with AS0, the ranges for AS1 to AS3 are in steps of integral multiples of 32 Kbps. AS0 support is mandatory; support for the other channels is optional.

There can be up to three duplex (bidirectional) bearer channels[§] that are labeled LS0 to LS2. Bearer channel LS0 supports data rates of 16

---

* These channels are logical channels, i.e., bits from all channels are multiplexed over the same physical link.
† The actual upper bound of the channel data rate is dependent on loop conditions. Note that T1.413i2 supports even higher downstream data rates, so the AS0 support for an upper-bound data rate of 6.144 Mbps is only a convenient marker (6.144 Mbps is both a multiple of 1.536 Mbps for North American systems and 2.048 Mbps for European systems). In other words, a modem vendor is required at least to support an upper bound of 6.144 Mbps for the AS0 bearer channel. However, depending upon loop conditions and vendor implementations, AS0 can support even higher data rates.
‡ AS3 is for North America only and is not applicable for Europe and elsewhere.
§ The three duplex bearer channels may alternatively be configured as independent unidirectional simplex bearer channels, and the rates of the bearer channels in the two directions do not need to match.

Kbps* plus the range from 32 Kbps to 640 Kbps (at all multiples of 32 Kbps). LS1 and LS2 support the range from 32 Kbps to 640 Kbps† (at all multiples of 32 Kbps). LS0 support is mandatory; support for the other channels is optional. Note that although the duplex channels are bidirectional, they are generally used for upstream in actual implementation.

It is important to note that the ADSL data multiplexing format is flexible enough to allow other transport data rates that are non-integer multiples of 32 Kbps. This is useful for ADSL deployments that need to interact directly with data rates that are non-integer multiples of 32 Kbps, such as T1 rates of 1.544 Mbps. This interaction is accomplished by carrying the additional bits in an ADSL overhead channel shared among the bearer channels.‡

Support for data rates that are non-integer multiples of 32 Kbps is optional, because it is deployment dependent. The net data rate is the total data rate minus ADSL system overhead, some of which is dependent on configurable options and some of which is fixed. Therefore, support of data rates that are non-integral multiples of 32 Kbps requires the ADSL overhead channel to have sufficient capacity left over after all configuration requirements have been met.

ADSL Transport Classes

Although nothing precludes a bearer channel from transporting any multiple of 32 Kbps (with the upper bound dependent upon the total carrying capacity of the link for specific loop conditions), as a practical matter for interoperability considerations, it is necessary to establish common transport classes.

The following transport classes have been established for North American implementations, using 1.536 Mbps as the baseline multiple for downstream data rates.

---

* Support for the 16-Kbps data rate with LS0 is an exception to the 32-Kbps multiple rule. It comes from the need to support a special mandatory control channel called the "C" channel. The C channel is used to transport signaling messaging for selection of services and call setup, much like the D channel in ISDN.

† The upper bounds of data rates for LS1 and LS2 have been revised upward from Issue 1 of the T1.413 standard.

‡ The shared overhead channel is used to transport bits that maintain synchronization. The overhead channel may have additional capacity (to carry the extra bits of a bearer channel) that exceeds the integral multiple of 32 Kbps. Exactly how much additional capacity is available depends upon configuration options that contribute to the overhead.

Transport Class 1. This transport class is mandatory and is intended for the shortest loops. The following configurations are allowed, each totaling up to 6.144 Mbps downstream:

- One simplex bearer channel at a downstream rate of 6.144 Mbps (using AS0 (mandatory)
- One simplex bearer channel at 4.608 Mbps and the other simplex bearer channel at 1.536 Mbps
- Two simplex bearer channels, each at 3.072 Mbps
- One simplex bearer channel at 3.072 Mbps, and two simplex bearer channels at 1.536 Mbps
- Four simplex bearer channels, each at 1.536 Mbps.

Transport Class 2. This transport class is optional and is intended for medium loops. The following configurations are allowed, each totaling up to 4.608 Mbps downstream:

- One simplex bearer channel at a downstream rate of 4.608 Mbps
- One simplex bearer channel at 3.072 Mbps and the other simplex bearer channel at 1.536 Mbps
- Three simplex bearer channels, each at 1.536 Mbps.

Transport Class 3. This transport class is optional and is intended for medium loops. The following configurations are allowed, each totaling up to 3.072 Mbps downstream:

- One simplex bearer channel at a downstream rate of 3.072 Mbps
- Two simplex bearer channels, each at 1.536 Mbps.

Transport Class 4. This transport class is mandatory and is intended for long loops. The following is the only configuration allowed, and it is mandatory.

- One simplex bearer channel at a downstream rate of 1.536 Mbps (using AS0).

The following transport classes have been established for European (and elsewhere) implementations, using 2.048 Mbps as the baseline multiple for downstream data rates.

Transport Class 2M-1. The following configurations are allowed, each totaling up to 6.144 Mbps downstream:

- One simplex bearer channel at a downstream rate of 6.144 Mbps (using AS0)
- One simplex bearer channel at 4.608 Mbps and the other simplex bearer channel at 2.048 Mbps
- Three simplex bearer channels, each at 2.048 Mbps.

Transport Class 2M-2. The following configurations are allowed, each totaling up to 4.096 Mbps downstream:

- One simplex bearer channel at a downstream rate of 4.096 Mbps
- Two simplex bearer channels, each at 2.048 Mbps.

Transport Class 2M-3. The following configuration is the only one allowed:

- One simplex bearer channel at a downstream rate of 2.048 Mbps (using AS0).

ADSL Overhead

ADSL Overhead Channel (AOC)

An AOC shared among all bearer channels is available to carry the extra bits that exceed the non-integer multiple of 32-Kbps capacities of bearer channels. The purpose of the ADSL overhead channel, however, is primarily to exchange necessary operational information among the ATUs at either end of the copper loop. The following is an example of information that might be exchanged over the AOC using an aoc protocol.*

- **Bit swapping.** Bit swapping enables an ADSL system to change the number of bits assigned to a DMT subcarrier or change the trans-

---

* Every AOC message has a header that identifies the type and length of the message. It is repeated five consecutive times—the receiver is required to respond to an AOC message only if it has received three identical messages in a time period spanning five of that particular message. A sender must insert at least 20 blanks ("stuffing" messages) between two consecutive groups of five concatenated and identical messages.

mit energy of a subcarrier without interrupting the data flow. It tries to equalize (as far as practical) the error rate of each subcarrier and maintain this over time by continually moving bits away from carriers with high error rates to those with lower error rates.[17] Either ATU may initiate a bit swap. The swapping procedures in the upstream and downstream channels are independent,* and may take place simultaneously. The ATU that initiates the bit swap transmits a request message and expects to receive an acknowledgment message from the other end.

Embedded Operations Channel (EOC)

The ATU-C and ATU-R communicate over an EOC for in-service and out-of-service maintenance and for the retrieval of a limited amount of ATU-R status information and ADSL performance monitoring parameters. This is known as the embedded operation channel since the EOC bits are carried along with the user data bits as part of the ADSL superframe, as described later.

The following are examples of information that might be exchanged over the EOC using an EOC protocol:†

■ **Perform ATU-R self test.** This message exchange is initiated by the ATU-C to request the ATUR to perform a self-test. The result of the self-test is stored in a register at the ATUR, which the ATU-C can read.

■ **"Dying gasp."** Whenever an ATU-R detects loss of power (when electrical power has been shut off), it inserts emergency priority eoc messages into the ADSL upstream data to implement a dying gasp indicator to the ATU-C. The ATU-R attempts to send at least six

---

* There can be a maximum of one downstream bit swap request and one upstream bit swap request outstanding at any time.

† Unlike AOC, where either end may initiate an AOC message exchange, EOC message exchange is always initiated by the ATU-C. The only exception to this rule is the "dying gasp" message where an ATU-R may autonomously notify the ATU-C via EOC. The ATU-C sends EOC (command) messages to the ATU-R to perform certain functions. Some of these functions require the ATU-R to activate changes in the circuitry. Other functions are to read from and write into data registers at the ATU-R. Some of these commands are "latching," meaning that a subsequent command will be required to release the ATU-R from that state. A separate command, "return to normal," is used to unlatch all latched states. This command is also used to bring the ADSL system to a known state, the idle state, when no commands are active in the ATU-R location. To maintain the latched state, the command "hold state" will be sent to bring the ADSL system to a known state, the idle state.

contiguous dying gasp EOC messages. The ATU-C detects the loss of power at the ATU-R when it receives at least four messages. This information can then be used by the service provider to determine that a customer's ADSL modem has been shut off.

■ **Power management.** Currently, power management is an important feature that is specified in ADSL lite but not in full-rate ADSL. Power management capability information, which will be discussed in detail in a later section, is exchanged over the EOC.

ADSL Framing

At the lowest level, the line codes (whether DMT or CAP) carry a number of bits per symbol. The bits are organized into frames, which are further organized into superframes, much like T1 frames and superframes.

In ADSL, 68 consecutive frames (labeled frame 0 through 67) form one superframe, as shown in Figure 4.6. Each frame is encoded and modulated into a DMT symbol.

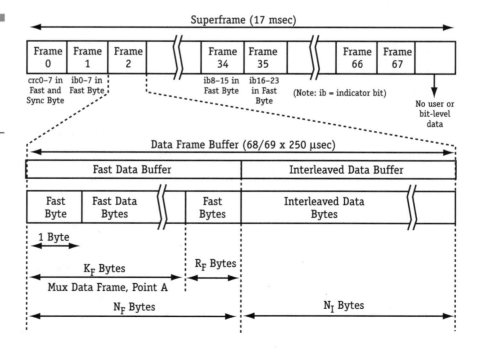

Figure 4.6
ADSL superframe (Source: ANSI Contribution T1.E14/97-007R6, July 1998)

Certain frames have special significance:

■ Frame 0 carries error control information.
■ Frame 1 carries indicator bits (discussed later in this section).
■ Frames 34 and 35 carry other indicator bits.
■ Sync frame is not part of the ADSL superframe, but immediately follows every ADSL superframe, i.e., it is the 69th frame transmitted after the 68 frames of the ADSL superframe. The purpose of the sync frame is to maintain synchronization and to equal one DMT symbol.

An ADSL frame is sent every 250 μsecs; therefore, it takes 17 milliseconds to transmit a superframe (250 μsecs × 68 frames). With full-rate ADSL, the 250 μsecs of the ADSL frame are further broken up into two parts of 125 μsecs each:

■ **Fast data from a fast data buffer.** This is for delay-sensitive, but error-tolerant traffic such as audio and video. In other words, this data has to be transmitted with minimum latency, but does not have to be error corrected. If there is still an error, it may be possible to compensate for the loss of a particular frame algorithmically or by skipping the frame. Fast data do incorporate forward error correction as an attempt to provide some measure of error protection without the need to retransmit frames. The first byte of each frame is designated as the fast byte; however, actual usage of the fast byte depends on the frame number, as we will soon see.
■ **Interleaved data from an interleaved data buffer.** This is for delay-insensitive, but error-intolerant traffic such as pure data applications. In other words, a certain amount of latency is acceptable, but the traffic has to be sent error free. In this case, it is acceptable to retransmit the frame. Interleaved data use a cyclic redundancy check as the error protection mechanism.

As mentioned before, the fast byte of frame 0 carries CRC bits (CRC 0–7) of the superframe. The fast bytes of frames 1, 34, and 35 carry indicator bits. The purpose of the indicator bits and the frames they are carried in are shown in Table 4.1.

Table 4.1	Indicator bit	Definition	Frames carried in
Indicator bits	Ib 0–7	Reserved	Frame 1
functions	Ib 8	Febe-I (Far-end block error on interleaved data)	Frame 34
(Source: ANSI	Ib 9	Fecc-I (Forward-error correction code on interleaved data)	Frame 34
Contribution			
T1.E14/97-007R6,	Ib10	Febe-NI (Far-end block error on non-interleaved data)	Frame 34
July 1998)	Ib 11	Fecc-NI (Forward-error correction code on non-interleaved data)	Frame 34
	Ib 12	Los (Loss of signal). This indicates when a pilot signal in the opposite direction drops below a certain threshold	Frame 34
	Ib 13	Rdi (Remote defect indication) to indicate reception of a severely errored frame (sef).	Frame 34
	Ib 14–15	Reserved	Frame 34
	Ib 16–23	Reserved	Frame 35

The fast byte in other frames (i.e., 2–33 and 36–37) is assigned in even-frame/odd-frame pairs either to the EOC or to synchronization control of the bearer channels assigned to the fast buffer, as shown in Figure 4.7.

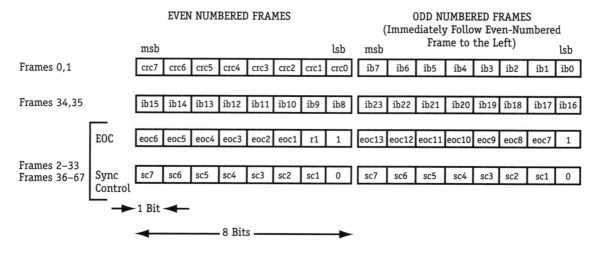

Figure 4.7 *Usage of fast byte (Source: ANSI Contribution T1.E14/97-007R6, July 1998)*

Full-rate ADSL supports both the fast and interleaved data paths,* and it is referred to as dual-latency ADSL (i.e., it supports both delay-sensitive and delay-insensitive traffic). ADSL lite, by contrast, only supports single latency with interleaved data. However, fast-byte usage, depending on which frame number carries the CRC, indicator, and EOC bits, is the same as full-rate ADSL.

Initialization

Initialization is required for an ATU-R and ATU-C pair to establish a communications link and may be initiated by either side. To maximize throughput and reliability, the transceivers need to determine certain relevant attributes of the connecting channel and establish transmission and processing characteristics suitable to that channel. The time line of Figure 4.8 provides an overview of this process.

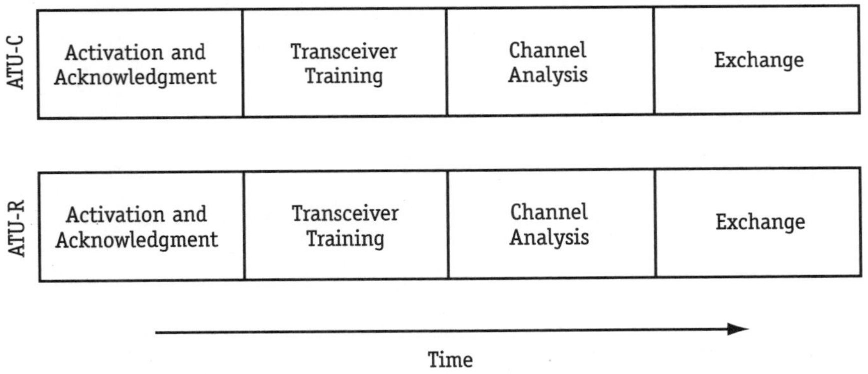

Figure 4.8
Overview of initialization (Source: ANSI Contribution T1.E14/97-007R6, July 1998)

ATU-C	Activation and Acknowledgment	Transceiver Training	Channel Analysis	Exchange

ATU-R	Activation and Acknowledgment	Transceiver Training	Channel Analysis	Exchange

Time →

In Figure 4.8, each receiver determines the relevant attributes of the channel through the transceiver training and channel analysis procedures. Certain processing and transmission characteristics can also be established at each receiver at this time. During the exchange process, each receiver shares with its corresponding far-end transmitter certain

---

* Although full-rate ADSL supports dual latency, it merely provides the transport mechanism. The standard does not specify the criteria by which the fast and interleaved buffers are filled—this is left up to the vendor. Furthermore, the specification only provides the time length (125 us) for the fast and interleaved buffers; the buffer sizes are dependent upon the data rate.

transmission settings it expects to see. Specifically, each receiver communicates to its far-end transmitter the number of bits and relative power levels to be used on each DMT subcarrier, as well as any messages and final data rate information. For highest performance, these settings should be based on the results obtained through the transceiver training and channel analysis procedures.

Activation and Acknowledgment

There are two options for initialization during the activation and acknowledgment phase:

- Using the "legacy" T1.413i2 method
- Using G.994.1 (G.hs) method.

The Legacy T1.413i2 Method

Figure 4.9 shows the timing diagram for the activation and acknowledgment phase 4.

Figure 4.9

Timing diagram for activation and acknowledgment (Source: ANSI Contribution T1.E14/97-007R6, July 1998)

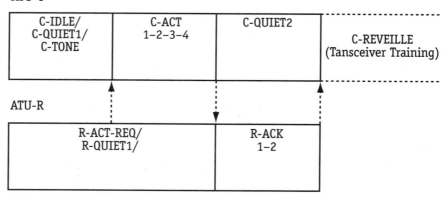

From the perspective of the ATU-C, after power-up and an optional self-test it enters a quiet state referred to as C-QUIET1 first. Once in this state, if it either receives a command from the host controller or detects a tone from the ATU-R (RACTREQ*), the ATU-C enters an active state. The ATUC will transmit an activated signal (generally referred to as C-ACT) to establish a communication link with the remote ATU. To facilitate interoperability among different implementations of frequency division and echo cancellation systems, four activate signals—C-ACT1 to C-ACT4—are defined. These four mutually exclusive signals are used to distinguish different system requirements for loop timing and the use of a pilot tone. Consequently, an ATU-C will transmit only one of these acknowledgment signals (generally referred to as C-ACK).

Loop timing is defined as the combination of "slaving"[†] an ADC[‡] clock to the received signal (i.e., to the other transceiver's DAC clock) and tying the local DAC and ADC clocks together. This can be performed by only one of the two transceivers and is always active during any ADSL connection. The ATU-C decides which transceiver will perform the loop timing and informs the ATU-R via the choice of the activation tone used. After acknowledgment occurs, the ATU-C enters a second quiet state (C-QUIET2) while it waits for an acknowledgment from the ATU-R in the form of an R-ACK1 or R-ACK2. The purpose of the second quiet state is to allow the detection of R-ACK1 or R-ACK2 without the need to train the ATU-C echo canceller. Once the ATU-C receives an acknowledgment from the remote ATU, it enters one of three states:

■ **C-REVEILLE.** If the ATU-C detects R-ACK, it enters the state C-REVEILLE in preparation for the next phase of transceiver training.

■ **C-ACT.** If the ATU-C fails to detect R-ACK, and the C-ACT state has not been entered more than twice, the ATU-C enters the C-ACT state again.

■ **C-QUIET1.** If the ATU-C does not detect R-ACK after returning to C-ACT twice, it returns to C-QUIET1.

From the perspective of the ATU-R, after power-up, it begins by transmitting an activate request signal or R-ACT-REQ. The ATU-R will

---

* An ATU-C can prevent an ATU-R from sending R-ACT-REQ by sending C-TONE.
† A term used to describe a signal that depends on another signal called the master, i.e. its characteristics are derived from the master signal source.
‡ ADC refers to the Analog-to-Digital Converter section of the transceiver, and DAC refers to the Digital-to-Analog Converter section.

remain in this state until it receives an activation signal from the ATU-C. It then responds with either an R-ACK1 or R-ACK2 (a third response called R-ACK3 is reserved, but left unused). Like the various versions of C-ACK, the versions of R-ACK (R-ACK1 and R-ACK2) are used to indicate options that have to do with the length of the quiet period required prior to entering transceiver training.

The G.994.1 Method

For the ATU-C to initiate a startup, the following sequence occurs[9]:

- The ATU-C begins by transmitting tones.
- When these tones are detected by the ATU-R, it responds by sending its tones from one signal family.
- When the ATU-C detects this, it responds by sending x'81' GALF characters (GALF is FLAG* spelled backwards) on modulated carriers.
- The ATU-R responds to the GALF characters by transmitting x'7E' FLAG characters on modulated carriers.
- When the ATU-C receives the FLAG characters, it responds by sending FLAG characters, as well.
- When the ATU-R receives the FLAGs back from the ATU-C, it can begin the first transaction.

For the ATU-R to initiate a startup, the following sequence occurs:

- The ATU-R begins by entering state R-SILENT1 on power-up.
- Upon command from the host controller, it initiates handshaking by transmitting tones (generally referred to as R-TONES-REQ or a tones request).
- When these tones are detected by the ATU-C, it responds by sending its tones to the remote ATU.
- When the ATU-C's tones have been detected by the ATU-R, it transmits silence for some period, followed by its tones from only one signal family.
- When the ATU-C detects this, it responds by sending x'81' GALF characters on modulated carriers.

---

* FLAG and GALF characters are used to maintain synchronization between the transmitter and the receiver when there are no other data to send.

■ The ATU-R responds to the received GALF characters by transmitting x'7E' FLAG characters on modulated carriers.

■ When the ATU-C receives the FLAG characters, it responds by sending back FLAGs.

■ When the ATU-R receives the FLAGs from the ATU-C, it can begin the first transaction.

Interoperability Between the Two Methods

The G.994.1 method is preferred; therefore, for interoperability between the two methods, the following functions are specified within the G.994.1 text[9]:

■ On the receiver side, the ATU-C monitors for a signal called R-ACT-REQ (indicating that the ATU-R attempting to connect is using the T1.413i2 method).

■ On the receiver side, the ATU-R monitors for the following signals: C-ACT1-4 and C-TONE.

■ On the transmitter side, the ATU-R alternates between G.994.1 and T1.413i2 initialization[4] as follows:

   ▪ Transmit R-TONES-REQ for two seconds (to attempt to start the G.994.1 initialization sequence).

   ▪ Transmit silence for 100 ms.

   ▪ Transmit R-ACT-REQ for two seconds (to attempt to begin the T1.413i2 initialization sequence).

   ▪ Transmit silence for 100 ms.

   ▪ Repeat sequence starting with R-TONES-REQ.

Transceiver Training

Figure 4.10 shows the timing diagram for the transceiver-training phase[4]. Synchronization of mutual training begins with the transmission of a signal called R-REVERB1 by the ATU-R. This allows the ATU-C to:

■ Measure the upstream power to adjust the ATU-C transmit power downstream

■ Adjust its receiver gain control

■ Synchronize its receiver and train its equalizer.

Figure 4.10

Timing diagram for transceiver training (Source: ANSI Contribution T1.E14/97-007R6, July 1998)

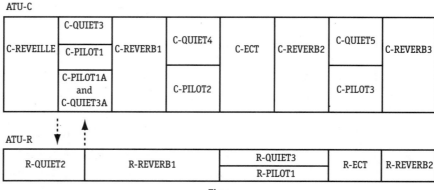

Synchronization is maintained throughout training by both transceivers' counting the number of symbols starting from R-REVERB1. During training, there is provision for implementation with echo cancellation to train the echo cancelers at both ends (if they are supported by both the ATU-C and the ATU-R).

Channel Analysis

Figure 4.11

Timing diagram for channel analysis (Source: ANSI Contribution T1.E14/97-007R6, July 1998)

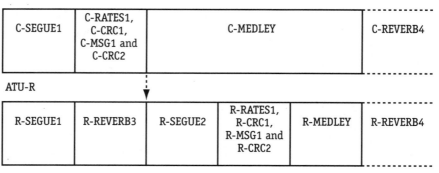

Figure 4.11 shows the timing diagram for the channel analysis phase[4]. After transceiver training, the ATU-C and ATU-R begin channel analysis by transmitting signals C-SEGUE1 and R-SEGUE-1, respectively. During this analysis, several functions are performed both downstream and upstream. In the downstream direction:

- The ATU-C sends a signal called C-RATES1. The purpose of C-RATES1 is to communicate options for data rates and formats to the ATU-R (i.e., how the AS and LS channels will be used), as well as the Reed-Solomon, FEC, and interleaver parameters.
- C-RATES1 is followed by C-CRC1 as a check for error detection in the reception of C-RATES1.
- This is followed by C-MSG1 whose purpose is to communicate to the ATU-R the vendor ID information, the ATU-C transmit power level used, and the echo canceler option. C-CRC2 follows as a check for error detection in the reception of C-MSG1.
- Next, the ATU-C transmits C-MEDLEY, a signal used by the ATU-R to estimate downstream S/N (signal-to-noise) ratio.
- Following C-MEDLEY, C-REVERB4 is transmitted in preparation for entering the final phase of exchange.

In the upstream direction:

- The ATU-R sends a signal called R-RATES1 that is similar to C-RATES1. As with C-RATES1, R-RATES1 is followed by R-CRC1 as a check for error detection in the reception of R-RATES1.
- R-MSG1, similar to C-MSG1, follows. As with C-MSG1, R-MSG1 is followed by R-CRC2 as a check for error detection in the reception of R-MSG1.
- After this, the ATU-R transmits R-MEDLEY, a signal used by the ATU-C to estimate upstream signal-to-noise ratio.
- Following R-MEDLEY, R-REVERB4 is transmitted in preparation for entering the final phase of exchange.

Exchange

Figure 4.12 shows the timing diagram for the exchange phase[4]. During exchange, there are two events that can cause either end to revert to the activation and acknowledgment phase—timeouts or error detection by a CRC checksum. However, assuming things are normal, exchange proceeds to get the two modems into a steady state called SHOWTIME. During this time, several functions are performed in both the downstream and the upstream directions. The primary purpose, however, is to begin a second exchange of rates because the new rates will, in general, be closer to the optimum bit rate for the channel than the previously specified rates. The channel information received in the previous C-MSG1 (or R-MSG1) is used to calculate the new optimized data rates.

Figure 4.12
*Timing diagram
for exchange
(Source: ANSI
Contribution
T1.E14/97-007R6,
July 1998)*

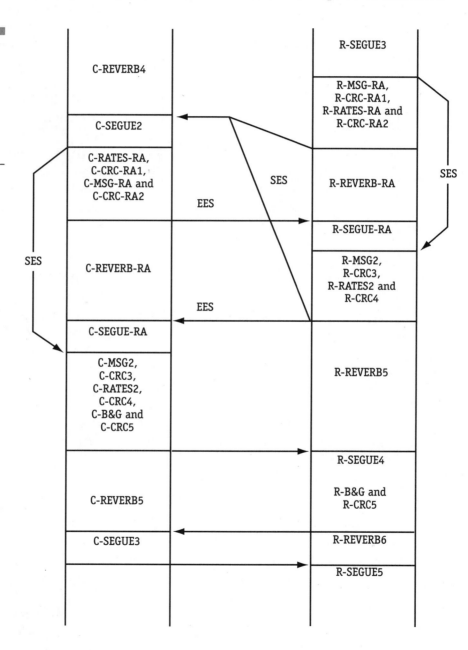

From the perspective of the ATU-C, in the downstream direction:

- The ATU-C sends a signal called C-RATES-RA to communicate options for data rates and formats to the ATU-R (i.e., how the AS and LS channels will be used), as well as the Reed-Solomon, FEC, and interleaver parameters. This is followed by C-CRC-RA1 as a check for error detection in the reception of C-RATES-RA.

- This is followed by C-MSG-RA, which is used to communicate to the ATU-R the new minimum required SNR margin, and the maximum and minimum noise margin in steady state. C-MSG-RA is followed by C-CRC-RA2 as a check for error detection in the reception of C-MSG-RA.

- After this, the ATU-C transmits C-REVERB-RA and C-SEGUE-RA. These signal intervals are used to let the ATU-R process the C-RATES-RA and C-MSG-RA information.

- Then the ATU-C transmits C-MSG2 to communicate such information as the total number of bits per symbol supported, the estimated upstream loop attenuation, and the performance margin with the selected rate option. C-MSG2 is followed by C-CRC3 as a check for error detection in the reception of C-MSG2.

- This is followed by C-RATES2, which is the ATU-C's reply to R-RATES-RA that would have been sent by the ATU-R. This combines the selected upstream and downstream options, i.e., the final decision on rates that will be used in both directions. As usual, C-RATES2 is followed by C-CRC4.

- The last important piece of information to be exchanged is called the C-B&G, and represents the bits and gains table. This contains the bits and gains information to be used on the upstream DMT subchannels. For any subchannel 'i', b_i represents the number of bits to be coded by the ATU-R transmitter, and g_i represents the scale factor relative to the gain that was used for that carrier. For subchannels where no data are to be transmitted (for example, due to the presence of AM interference), both b_i and g_i will be zero. For subchannels where no data are currently transmitted, but where bits could be allocated later (for example, due to an improvement in the signal-to-noise ratio), b_i will be zero, but g_i will have a value. C-B&G is followed by another CRC check—C-CRC5.

■   Following C-CRC5, the ATU-C transmits C-REVERB5 until it is
      prepared to transmit according to the conditions specified in the R-
      B&G signal sent by the ATU-R. When the ATU-C is ready, it trans-
      mits C-SEGUE3 to notify the ATU-R that it is about to enter the
      signaling steady-state called C-SHOWTIME.

In the upstream direction, the sequence occurring from the perspective of the ATU-R is similar (although the signal names begin with the
letter 'R' rather than 'C').

Fast Retraining for ADSL Lite

Motivation for Fast Retraining

A fast retrain procedure is defined in ADSL lite[8] to adapt transmission
characteristics to changing line conditions. The primary motivation for
fast retraining is splitterless operation with ADSL lite, which implies
that POTS and ADSL service can be concurrent on the wire without
any isolation between them. Under these circumstances, when a POTS
device such as a telephone goes off hook, the change in impedance can
cause dramatic change in line conditions. In turn, this causes the ATU-
C and ATU-R to retrain. Since any retraining results in temporary disruption of ADSL service, it is desirable to retrain as quickly as possible—hence, the need for fast retrain.

Procedure

Fast retrain is based on the concept of stored profiles. A minimum of
two profiles (profiles 0 and 1) must be supported and the number can
go as high as 16 profiles, with the highest profile number being 15. Profiles contain the following information at a minimum:

■   B&G tables
■   FEC parameters for Reed-Solomon
■   Interleaver depth D.

The flow chart for fast retrain is shown in Figure 4.13.

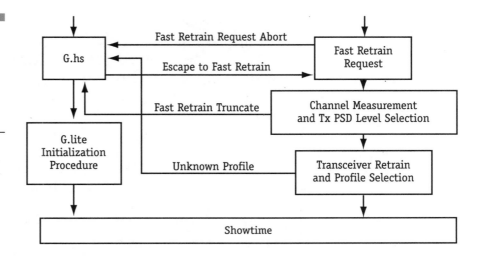

Figure 4.13
Flow chart for fast retrain (Source: Draft ITU Recommendation, COM 15-134)

Fast Retrain (ATU-C Initiated)

Fast-retrain Request

- The ATU-C initiates fast retrain when it enters the C-RECOV state and sends this signal to the ATU-R.
- The ATU-R leaves R-SHOWTIME on request from the ATU-C within R-T1 symbols after reception of C-RECOV and goes to R-RECOV.
- After the reception of R-RECOV, the ATU-C sends C-RECOV for a minimum period of 32 symbols after which it starts to transmit C-REVERB-FR1.

Channel Measurement

- The ATU-R detects a minimum of 256 symbols of C-REVERB-FR1 before switching from R-RECOV to R-REVERB-FR1.
- The ATU-C switches to C-PILOT-FR1 within 32 symbols after reception of the transition of R-RECOV to R-REVERB-FR1. The R-REVERB-FR1 is sent for 512 symbols followed by 10 symbols of R-SEGUE-FR1.
- The ATU-R follows the R-REVERB-FR1 signal with 512 symbols of R-LINE PROBE. This signal allows the ATU-R to measure the echo/reflections to set the upstream TX power appropriately.

Selection of Transmit Power Levels

- After R-LINE-PROBE, the ATU-R sends a message indicating the new downstream and upstream PSD levels via the sequence of 16 R-REVERB-FR2 symbols, 10 R-SEGUE-FR2 symbols, and 32 R-MSG-FR1/R-CR-FR1 symbols. It is followed by R-QUIET-N, which is terminated 750 ± 6 symbols after the start of R-SEGUE-FR1, after which R-REVERB-FR3 is transmitted at the new power level.

- At the ATU-C, the C-PILOT-FR1 signal is terminated 512 symbols after reception of R-SEGUE-FR1, after which the ATU-C sends a message indicating a fast retrain truncate bit, via a sequence of 64 C-REVERB-N symbols, 10 C-SEGUE-N symbols, and 32 C-MSG-N/R-CR-N symbols. Next, the C-REVERB-FR2 is transmitted at the new power level.

Selection of Transmit Power Levels

- At the ATU-R, the R-REVERB-FR3 signal is sent for 1,248 symbols, followed by 1,536 symbols of R-QUIET-FR1 and 32 symbols of R-REVERB-FR4. The final symbol of R-REVERB-FR4 may be shortened by any number of samples (using a sampling rate of 552 kHz) that is an integer multiple of four in order to accommodate the transmitter-to-receiver frame alignment. 512 symbols of R-ECT-FR, a minimum of 128 and a maximum of 160 symbols of R-REVERB-FR5, and 10 symbols of R-SEGUE-FR3 immediately follow R-REVERB-FR4.

- At the ATU-C, the C-REVERB-FR2 signal is sent for 192 symbols, followed by 1,024 of C-PILOT-FR2, 32 symbols of C-REVERB-FR3, 512 symbols of C-ECT-FR, 1,060 symbols of C-REVERB-FR4, 512 symbols of C-PILOT-FR3 or C-QUIET-FR, depending on whether R-ACK2 or R-ACK1 was received during a preceding normal initialization procedure, 128 symbols of C-REVERB-FR5, and 10 symbols of C-SEGUE-FR1.

- The ATU-C after C-SEGUE-FR1 and the ATU-R after R-SEGUE-FR3 introduce the cyclic prefix and switch to transmitting the MEDLEY signal. The signals C-MEDLEY-FR and R-MEDLEY-FR are transmitted for 1,024 symbols. This part of the fast retrain is intended for signal-to-noise ratio measurement.

Profile Exchange

■ Following R-MEDLEY-FR, the ATU-R line profile selection is exchanged using 16 symbols of R-REVERB-FR5, followed by 10 R-SEGUE-FR4 symbols, and 32 R-MSG-FR2/R-CRC-FR2 symbols.

■ Following C-MEDLEY-FR, the ATU-C line profile selection is exchanged using 134 symbols of C-REVERB-FR5, followed by 10 symbols of C-SEGUE-FR2, and the C-MSG-FR2/C-CRC-FR2 (32 symbols). Both the ATU-C and the ATU-R have independent profile selections for upstream and downstream, respectively.

■ The ATU-C sends the final sequence consisting of 400 symbols of C-REVERB-FR6, followed by 10 symbols of C-SEGUE-FR3, after which the C-SHOWTIME is resumed with restart of the superframe counter.

■ The ATU-R sends the final sequence consisting of 518 symbols of R-REVERB-FR6, which is followed by 10 symbols of R-SEGUE-FR5. After this, the R-SHOWTIME is resumed with restart of the superframe counter.

Fast Retrain (ATU-R initiated)

Fast-retrain Request

■ The fast-retrain request is started by the ATU-R leaving R-SHOW-TIME autonomously and going to R-RECOV.

■ After the reception of R-RECOV, the ATU-C leaves C-SHOWTIME and starts C-RECOV within C-T1 symbols after reception of R-RECOV. C-RECOV is sent for a period of 64 symbols, after which it starts to transmit C-REVERB-FR1.

■ The ATU-R detects a minimum of 256 symbols of C-REVERB-FR1 before switching from R-RECOV to R-REVERB-FR1.

■ The ATU-C switches to C-PILOT-FR1 within 32 symbols after reception of the transition of R-RECOV to R-REVERB-FR1.

■ The rest of the procedure is identical to the ATU-C initiated case.

Power Management for ADSL Lite

Motivation for Power Management

The primary motivation for enabling power management functionality was the recognition that ADSL lite modems would be used extensively in consumer PCs that have stringent requirements for power management. Certain regulatory programs* require that PCs be able to enter power-down modes in which the total power consumption is 1 W or less. To achieve this, ADSL lite modems must be able to enter a low-power state. This state will generally correspond to a similar state for the rest of the PC (in which most of the system will be suspended).

ADSL Link States

ADSL link states[8] allow an ATU to enter a low-power state without totally disconnecting the link. These are stable states and are generally not expected to be transitory. An ATU must support the ADSL link states shown as mandatory in Table 4.2.

Table 4.2

ADSL link states (Source: Draft ITU Recommendation, COM 15-134)

State	Name	Support	Description
L0	Full on	Mandatory	The ADSL link is fully functional
L1	Low power	Optional	The L1 state maintains full L0 state functionality at a lower net data rate (except for power management transitions). Power reduction in L1 can be achieved by methods provided in the exchange entry procedure (e.g., reduced data rate, reduced number of tones, and reduced power per tone). The reductions are implementation specific.
L2			Reserved
L3	Idle	Mandatory	There is no signal transmitted at the U-C and U-R reference points. The ATU may be powered or unpowered in L3.
L4-L127			Reserved by the ITU-T.
L128-L255			Reserved for vendor specific implementations.

---

* These programs include the U.S. Energy Star and the EC Blue Angel programs.

Link-state Transitions

Link-state transitions are initiated by various events. The following events are identified as potentially leading to link transitions:

- *Grant* is a negotiated event that results from a successful eoc handshake (described later).
- *Command* results from an unconditional request to change states.
- *Change in line condition* results when the receiver detects that conditions have changed sufficiently to merit an initialization or fast-retrain procedure.
- *Failure* is one of the defined failure conditions, e.g., loss of power.

EOC handshake is used for power management coordination between the ATU-C and ATU-R (and vice versa). A successful handshake is defined as a grant event used to enable a power-management state transition. An unsuccessful result does not trigger a state transition, and the power-management state remains unchanged.

EOC Handshake (ATU-C Initiated)

The ATU-C initiates the handshake using the following procedure:

- The ATU-C writes the value of the new ADSL link state into the link-state data register using the EOC write protocol.
- The ATU-C ends the handshake sequence by issuing the GNTPDN eoc command protocol. If the echo of the eoc command is returned, the handshake was successful. If an "unable to comply" (UTC) message is received, then the handshake was unsuccessful.

If the ATU-R cannot support the granted link state for some reason (e.g., the ATU-R does not support the ADSL link state), then it will respond to the GNTPDN command with a UTC message using the eoc command protocol.

EOC Handshake (ATU-R Initiated)

The ATU-R initiates the handshake as follows:

■   The ATU-R writes the value of the requested ADSL link state into the link-state data register.

■   The ATU-R then sends a REQPDN EOC autonomous message.

■   After receiving a "request power down" (REQPDN) message from the ATU-R, the ATU-C responds by reading the requested power-down state from the link-state data register using the EOC read protocol.

■   After receiving the REQPDN message, the ATU-C may optionally propose an alternate ADSL link state by writing a different value into the link-state data register using the EOC write protocol.

■   After determining that it can grant the state transition request, the ATU-C will issue the grant power down command using the eoc command protocol. If the EOC command echo returns, the handshake ends successfully. If the UTC message is returned in response to the EOC command, the handshake was unsuccessful.

If no response to the REQPDN autonomous message is received from the ATU-C within five seconds, then the ATU-R resends the REQPDN message. If the ATU-R is in the middle of a multibyte read or write EOC protocol sequence, then the timeout count does not begin until the multibyte sequence ends. Upon timeout, the ATU-R may send the REQPDN message up to four more times, after which the handshake procedure terminates with an unsuccessful result.

If the ATU-C is in the middle of a multibyte read or write sequence when a REQPDN message is received, then the ATU-C may choose to terminate the multibyte sequence or delay response until the end of the sequence.

If the ATU-C cannot grant the power-down request for some reason (e.g., it does not support the requested state), it sends the REJPDN EOC command. After the ATU-R receives the REJPDN eoc command, the handshake ends unsuccessfully.

If the ATU-R cannot support the granted link state for some reason, it responds to the GNTPDN command with the UTC eoc message using the EOC command protocol. This could happen if, for example, the ATU-C responded via the optional write with a different ADSL link state, or the ATU-R no longer needed to go into the granted power-down state.

Differences between T1.413i2, G.dmt, and G.lite

As noted earlier, both G.dmt and G.lite are based on T1.413i2. Consequently, there are a lot of similarities among the three. However, there are certain important differences. The key distinctions between G.lite and T1.413i2/G.dmt can be summarized as follows:

- G.lite supports a maximum of 1.5 Mbps downstream and 512 Kbps upstream. T1.413i2/G.dmt supports a theoretical maximum of 14.9 Mbps downstream (though practical limits are in the neighborhood of 6 Mbps to 8 Mbps downstream) and 1.5 Mbps upstream.
- G.lite does not require a premises splitter; T1.413i2/G.dmt does.
- G.lite only supports ATM transport. T1.413i2/G.dmt supports both ATM and STM transport.
- G.lite supports single latency only. T1.413i2/G.dmt supports dual latency.
- G.lite incorporates newer features such as fast retraining and power management. However, it is likely that these features may be incorporated into newer versions of G.dmt.

Given the key distinctions listed above, G.dmt and T1.413i2 are a lot closer in their texts, but with some exceptions noted below:

- G.dmt provides electrical characteristics for both North American and European standards. T1.413 Issue 2 only provides North American standards.
- G.dmt provides separate sections for the requirements of ADSL systems operating in (a) the frequency band above POTS, (b) the frequency band above ISDN, and (c) in the same cable as ISDN. ANSI T1.413 Issue 2 only addresses systems operating at the frequency band above POTS.
- The initialization section of G.dmt is identical to that of T1.413 Issue 2, except that the activation and acknowledgment subsection is replaced with a handshake procedure, as defined in G.994.1.

Current State of the Technology

Deployment Issues and Possible Solutions

Since ADSL and ADSL lite are technologies that are just beginning to be deployed, it is impossible to predict all the issues that might come up. Nevertheless, industry "wisdom"[18] suggests the following issues and possible solutions to mitigate their impact.

Loop Reach versus Distance

Issue

Loop reach is of paramount importance for new service coverage from a telco. It depends on the routing factor (a function of the radius of loop length, wire gauge, and the type/location of a telco's CO and the surrounding area). The longer the loop reach, the bigger the circle of coverage, and hence the higher probability of serving new customers from a single central office. On the other hand, ADSL data rates are distance sensitive. This means the achievable data rate drops off with longer distances, as shown in the example in Figure 4.14.

Figure 4.14

Typical downstream data rate versus distance from the CO (24-gauge wire) (Source: Texas Instruments)

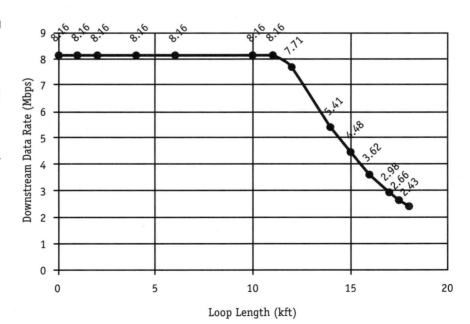

As Figure 4.14 illustrates, the downstream data rate at 18 Kft of 24-gauge wire is around 2.4 Mbps. However, factors such as mixed gauges of wire can reduce actual data rates, and achievable data rates beyond a certain distance are too low to be of practical use.

Possible Solutions

Data rates drop off with distance due to attenuation of the signal (one of the important factors influencing attenuation is loop length). It follows that receiver implementations that can recover the strongest signal despite attenuation loss can achieve higher data rates. Digital signal processing techniques are often used to implement such algorithms. However, since these algorithms can be processor intensive, it is important to have sufficient CPU processing headroom to improve performance.

Furthermore, sufficient MIPS headroom allows for implementation of digital filters in the DSP itself. The motivations to implement digital filters include:

■ Analog filters are subject to variations depending on factors such as temperature and humidity. Digital filters are free from such variations.
■ Digital filters can be designed within very tight tolerances, unlike analog filters.
■ Reprogramming the DSP can alter the characteristics of digital filters. By contrast, once an analog filter is manufactured, its characteristics (e.g., passband frequency range) cannot be easily altered.

Line Conditions

Issue

ADSL transmission over telephone lines is subject to various line impairments such as bridged taps, crosstalk[19], and radio frequency interference[20]. Additionally, a loop may experience gradual changes in channel capacity over time as a result of temperature and moisture changes during the day.

Possible Solutions

DMT adapts well to line impairments. Figure 4.15 shows that DMT divides the channel into carriers and utilizes only those carriers that have acceptable error rates for bit loading.*

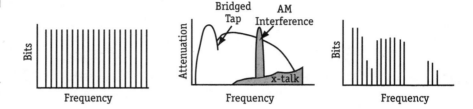

Figure 4.15
DMT's dynamic adaptation to line conditions

As described earlier, bit swapping tries to equalize (as far as practical) the error rate of each subcarrier and maintain this over time by continually moving bits away from carriers with high error rates to those with lower error rates. This technique allows for gradual changes in loop conditions and achieving optimal performance at a given data rate over time.

Premises Conditions (ADSL Lite)

Issue

ADSL lite poses a challenge for concurrent operation (i.e., simultaneous POTS and ADSL). POTS equipment such as telephones, fax machines, and voiceband modems going from on-hook state to off-hook state can cause transients on the line that could have an impact on ADSL service.

In addition, the impedance seen at the ADSL modem can change quite significantly during on/off-hook transitions. Telephones offer a wide variety of impedances ranging from linear to nonlinear. For example, a phone may appear as a high-impedance inductive load on hook and a nonlinear diode mixer when it is off hook. Some have integrated circuits; others are purely passive electromechanical devices.

---

* DMT employs a bit-loading algorithm to determine the number of bits a carrier can carry based on the S/N ratio.

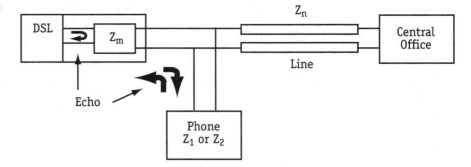

Figure 4.16

Impedance change due to on-hook to off-hook transition

If we refer to Figure 4.16, when the phone is on hook, the impedance value is Z_1 and the echo is approximately linear. However, when the phone goes off hook, the impedance changes to Z_2, and the echo can be nonlinear as a result of components in the phone. The problem with these impedance changes is that they cause the modems to retrain. During the modem retraining period, no data can be passed end to end; therefore, it is desirable to avoid (or at least minimize) modem retraining during the concurrent operation of POTS and ADSL.

British Telecommunications plc conducted a study of several phones in 1998[21]. Their results were presented as contributions to the ITU, ADSL Forum, and the UAWG. Figure 4.17 and 4.18 show a sample measurement taken from this study that illustrates the dramatic change in impedance of a phone going from on hook to off hook.

Figure 4.17

British Telecom study—modulus of impedance seen by ADSL modem, "Tribune" phone on hook (Source: ADSL Forum Contribution 98-032, March 1998)

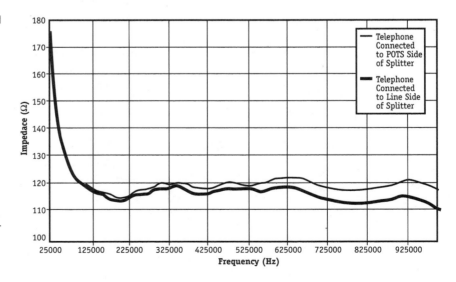

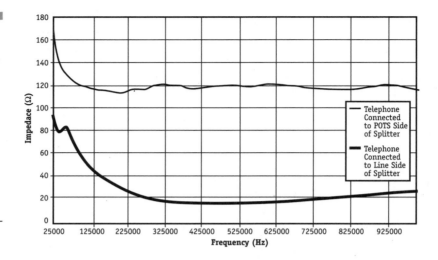

Figure 4.18
British Telecom Study—modulus of impedance seen by ADSL modem, "Tribune" phone off hook (Source: ADSL Forum Contribution 98-032, March 1998)

Possible Solutions

Concurrent operation requires a fast error-recovery procedure to minimize service disruption during on/off-hook transitions due to the impedance changes seen at the modem. Therefore, G.922.2 (G.lite) modems implement a fast-retraining algorithm, in addition to the other error recovery procedures incorporated into the T1.413i2 standard (i.e., resync and full initialization).

Interference between POTS and ADSL Service (ADSL Lite)

Issue

Installing and operating an ADSL modem at the customer premises without a POTS splitter raises at least two issues (discussed below), since POTS and ADSL services can interfere with each other as shown in Figure 4.19.

■ **ADSL interference with POTS.** POTS quality is affected by the presence of both the upstream and downstream ADSL signal (the R-DSL into R-POTS and C-DSL into R-POTS interference shown in Figure 4.19). For example, when the R-DSL signal interacts with the nonlinearities in the POTS device that is in off-hook state, the R-DSL signal will be mixed down into the voiceband frequencies. This interference is audible as noise during a telephone conversation. Since POTS quality includes all voiceband services such as fax and voiceband modems, it is necessary to quantify both electrical and physiological limitations on the interference that will bleed from the ADSL signal band into the voiceband.

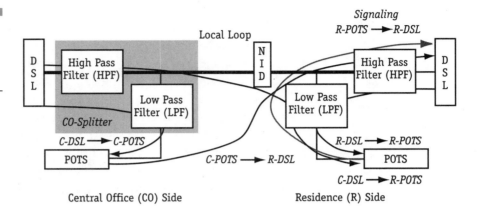

Figure 4.19
Interference caused by removing premises splitter

- ■ **POTS interference with ADSL.** POTS interference with ADSL comes from several areas:
  - ■ C-POTS into R-DSL and R-POTS into R-DSL interference, as shown in Figure 4.19.
  - ■ Ring-trip, the impulse created by the interruption of the low frequency ringing voltage by the customer's phone going off hook. This particular waveform, when convolved with the impulse response of the subscriber loop, can cause serious short-term impairments to the ADSL signal. The amplitude and duration of the impulse depends on the exact moment in the signaling when the transition takes place. This interference puts an additional burden on error correction over normal noise problems.
  - ■ Pulse dialing, which has a serious impact on ADSL service because of lengthy error bursts on the line
  - ■ Premises wiring and on/off-hook impedance changes.

Possible Solutions

- ■ **Power cutback.** Cutting back power is an effective technique to address ADSL interference with POTS. The basic idea is to reduce power so that the voltage level of the ADSL signals across the nonlinear components in the POTS device circuitry is within the linear region. It is important to avoid the nonlinear region because the peaks become clipped and/or distorted, resulting in noise at the POTS device. Another impact of nonlinearity is that clipping shows up as distortion across the entire frequency range. This in turn, has an impact on upstream and downstream performance.* Power cutback is easy to implement, though it has an impact on reach.

■ **Peak to average ratio (PAR) reduction.** Peak to average ratio (PAR) reduction/clip mitigation techniques[22,23] are an effective way to counter the negative impact of power cutback on reach, while retaining the benefit of reduced POTS interference. The basic principle of PAR reduction/clip mitigation is shown in Figure 4.20. The idea is to eliminate the peaks, thereby avoiding the nonlinearity. Since the average power using PAR is more likely to stay within the linear region, the amount of power required is lower.

Figure 4.20
PAR reduction/clip mitigation

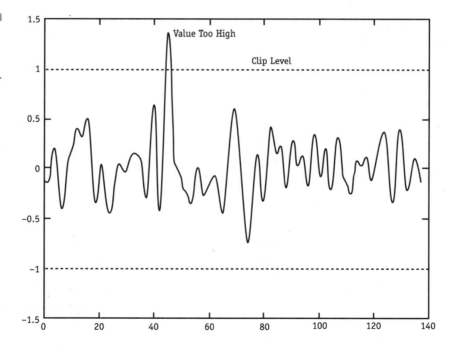

---

* In actuality, upstream performance is more severely degraded because of the greater likelihood of impact of the nonlinearity as the distance between the peak signals and diodes is very short. Therefore, the noise does not have a chance to be attenuated by the loop length. This is unlike the case of downstream bandwidth.

■ **Additional solutions.** The following solutions have also proven effective:

   ■ *R-DSL into R-POTS.* Normally, the LPF in the premises splitter prevents this interference. The impact of this interference caused by removing the premises splitter can be reduced using the power cutback and PAR reduction techniques discussed above. For certain conditions, installing in-line (low-pass) filters at every POTS device can eliminate this interference, as shown in Figure 4.21. Symmetrical in-line filters are desirable because they function in the same manner regardless of the way in which they are installed with respect to a POTS device—an important consideration for consumer installations.

Figure 4.21
Benefit of in-line filters

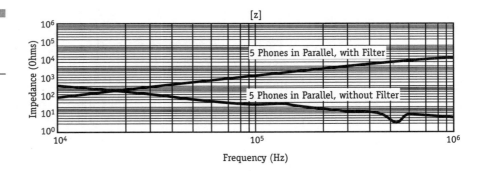

   ■ *C-DSL into R-POTS.* Normally, the LPF in the premises splitter and the HPF in the CO splitter prevent this interference. Without a premises splitter, increased HPF capability at the CO splitter and CO power cutback should minimize the effect of this interference for longer loops.

   ■ *R-POTS into R-DSL, C-POTS into R-DSL, and signaling.* Normally, the HPF in the premises splitter prevents these forms of interference; however, when the premises splitter is eliminated, moving the HPF functionality into the ADSL modem minimizes the problem. Note that certain forms of interference, such as ring-trip are inherently wideband, so an HPF at the modem will not completely eliminate the problem. The impact will be reduced data rate.

Interoperability Efforts

The Need for Interoperability

Achieving interoperability between one vendor's DSLAM and another vendor's CPE modem is a critical milestone toward enabling broad market deployment of ADSL. This will allow service providers to deploy DSLAMs of their choice in their central offices, while enabling consumers to purchase modems, based on price points and feature sets, from local retail outlets. It will also enable computer manufacturers to include DSL-based broadband capabilities as part of their offerings on a large scale without thought to a particular technology or vendor implementation that might be prevalent in certain geographical areas. In short, interoperability eliminates risk for the service provider, the consumer, and the computer manufacturer.

For this reason, the industry working groups (UAWG and ADSL Forum) have recently focused their efforts on enabling interoperability around the G.992.1 (G.dmt) and G.992.2 (G.lite) standards for full-rate ADSL and ADSL lite, respectively. However, a question may arise—if there are ADSL standards, why is there a need to "enable" interoperability? After all, if everyone follows the standard, shouldn't interoperability be automatic? The answer lies in some of the practical issues:

■ A way to generate confidence in a particular implementation of the standard is to prove, through adequate testing, that it interoperates with other implementations and conforms to the standard.

■ Vendor interpretations of the standard can differ, especially if there are ambiguous areas in the text. While every effort is made to avoid and/or remove ambiguities in the standard, sometimes these ambiguities surface only when two vendors try to connect to each other.

■ ADSL standards, like any others, are based on a collection of technical contributions from domain experts that are debated and discussed among peers. These contributions are sometimes based on theoretical reasoning. Other times, they are based on results gained from trials and lab simulations. However, especially in the early stages of the standard, they are not put to the test of real world experience. Real world tests on a large scale are necessary to validate the development of the standard. Interoperability tests are a way to validate and clarify the new standard text.

■ Vendors may inadvertently fail to implement (or may misimplement) portions of the standard. Certification that a product meets the standard can be determined only through testing.

Approaches to Interoperability

There are two approaches to ADSL interoperability:

■ **The "plugfest" approach.** This approach is typically favored by the data communications industry. The idea is for companies involved in developing products for a standard to gather in a single location (often, before a standard is fully ratified) to attempt communication between devices as a means of validation, and to some degree, vendor debugging. The test goals and procedures are decided a priori to the event, so that all participants are in agreement. Plugfests are repeated during the standards development process until the standard is final, and a small number of devices are at a point where they work together under certain circumstances. The drawback is that, as new devices are added to the testing matrix, it must successfully connect to all other existing devices to be considered interoperable. As the number of devices increases, the matrix grows exponentially, so that it eventually becomes impossible to accomplish satisfactory testing at a single event.

■ **The "certification" approach.** This approach is favored by the telecommunications community, in large part due to their history of ensuring high levels of reliability for the PSTN. The telcos prefer certification because they are often regulated entities and therefore, answerable to the public about the reliability of the services they offer. This mentality extends to ADSL as well. Certification involves a stable standard against which a set of interoperability tests can be established and agreed upon by the industry. Vendors then submit their equipment to a test house that performs these tests and sends notification of success or failure. The drawback of certification is that it can turn into a time-consuming (and often expensive) process.

The UAWG has chosen to adopt the plugfest approach to begin interoperability efforts for G.992.2 (G.lite) in the early cycle of standards development. The ADSL Forum has chosen to adopt the certification approach of developing test cases against the standards for both full-rate ADSL and ADSL lite. The UAWG is cooperating with the ADSL Forum, however, on the joint definition of test cases for ADSL lite.

UAWG Efforts

The UAWG has recently completed several "plugfests" among multiple vendors. The objectives of these plugfests are to:

- Validate the text of the G.992.2 (G.lite) standard
- Enable manufacturers to gain confidence in their implementations.

The UAWG has put its interoperability test plan in the public domain. The test plan covers the following areas:

- **Test 1—Initialization procedures.** The purpose of this test is to verify the interoperability of the initialization procedures. The test is divided into four separate subtests, corresponding to the four major phases of the initialization sequence: 1) activation and acknowledgment, 2) transceiver training, 3) channel analysis, and 4) exchange. Successful completion of this test occurs when the two ATUs under test simultaneously reach the SHOWTIME state.
- **Test 2—Basic data transmission operation.** The purpose of this test is to verify the interoperability of basic data transfer between the ATU-C and ATU-R operating within the G.992.2 (G.lite) terms of reference. Tests of basic data transmission are divided into three subtests: 1) fixed data rate operation with no forward error correction, 2) fixed data rate operation with specific FEC parameters, and 3) vendor-optional connection over a 9,000-foot 26 AWG wire loop (CSA Loop 6).
- **Test 3—ADSL overhead channel (AOC) transmission and processing, and fast-retrain procedure under momentary off-hook state.** The purpose of this test is to verify the fast retraining procedure, and, implicitly, AOC operation (since fast retraining results in profile exchanges over the AOC).
- **Test 4—G.994.1 (G.hs) initialization procedures.** The purpose of this test is to verify handshaking procedures during initialization. Handshaking is used among modems to determine the capabilities at either end.

■ **Test 5—L3-L0 transition and embedded overhead channel (EOC) transmission and processing.** The purpose of this test is to verify sections of the power management implementation (in particular the L3-L0 transition), and, implicitly, EOC operation since this involves message exchanges.

■ **Test 6—Interoperability between a G.992.2 (G.lite) ATU-R and a full-rate (T1.413 i2) ATU-C.**

ADSL Forum Efforts

Within the ADSL Forum, the Testing and Interoperability Working Group is charged with developing detailed test plans with the goal of generating standards that would ensure interoperability among various implementations. As with other ADSL Forum efforts, the interoperability test standards are contributed by industry sources. When they are approved, the ADSL Forum proposes to make these test plans publicly available so that independent test houses can come forward with proposals to certify products. A common set of test plans would ensure uniformity in the certification process.

The Forum proposes three areas of testing to ensure that products meet the specifications and that they can interoperate without observable problems under different load conditions—conformance testing, static interoperability, and dynamic interoperability. Each of these test areas is independent of the others, and one is not necessarily a prerequisite for the other two. However, the combination of all three types of testing will provide the highest degree of confidence that successfully tested products will provide reasonable performance when connected to similar successfully tested equipment and when deployed "in the field" on reasonable copper loops.

■ **Conformance.** Conformance testing attempts to evaluate an implementation against a specification. There are three general levels of conformance testing:
  ■ Electrical conformance
  ■ Physical conformance; electrical and physical conformance are closely related. Examples of areas evaluated during electrical and physical conformance testing include DC characteristics, voiceband characteristics, POTS splitter characteristics, and ADSL band characteristics.

■ Higher layers of conformance; protocols supported between the ATU-R and ATU-C above the physical layer may need to be tested to validate conformance with architectural specifications such as ADSL Forum TR-002 24 and TR-003 25. Areas for which tests may need to be specified include (a) PPP over ATM on ADSL or (b) frame-based architectures (e.g., FUNI on ADSL or HDLC on ADSL).

■ **Static interoperability.** Static interoperability testing verifies the operation of a pair of modems in a benign, stable laboratory environment (i.e., NULL loop and no noise intrusions). It involves testing the capabilities and behavior of two implementations and validating satisfactory communication. The objective is to confirm the degree to which they communicate with each other. For example, an ATU-C and an ATU-R may implement the same mandatory features and functions, yet differ in their optional implementations. Sometimes, the ability to interoperate depends on these optional features.

■ **Dynamic interoperability.** Dynamic interoperability testing attempts to evaluate an implementation in a real network environment under varying conditions to see how it performs. The following are examples of performance test conditions used during dynamic interoperability testing:

■ Loop characteristics: length, wire gauge, bridged taps, attenuation, etc.

■ Customer premises wiring

■ PSTN conditions: ringing, ring-trip, battery feed, signaling, etc.

■ Co-channel noise interference: BR-ISDN, HDSL, ADSL, etc.

■ Other interference: impulse noise, RFI, etc.

The following are examples of performance metrics used during dynamic interoperability testing:

■ Upstream and downstream transmission rates

■ Noise margin

■ ADSL line status

■ Transmitted blocks

■ Corrected blocks

■ Uncorrectable blocks

■ Counters for current and previous loss of signal, loss of frame, loss of power, and errored seconds

- Bits/carrier
- Interleave delay
- Rate adaptation (e.g., dynamic rate repartitioning, dynamic rate adaptation, and fast retrain).

The following are examples of end-to-end performance metrics used during dynamic interoperability testing (these metrics are of importance to network providers):

- Service quality
- Latency
- Bandwidth and data rate availability.

Trials and Deployments

ADSL began to move out of the trial phase into actual deployments in 1998. Although availability is still spotty and prices vary, there is encouraging news ahead, as ILECs move to stave off competition from cable, CLECs, and IXCs. In third quarter of 1998, TeleChoice conducted a survey of several major service providers in North America to assess the rate of ADSL deployments. TeleChoice requested information about the number of subscribers, number of DSL-equipped COs, and number of potential customers from DSL-equipped COs. Although the number of subscribers is closely guarded by the service providers, the survey response on the number of DSL-equipped COs and potential customers was positive enough to derive meaningful conclusions. According to TeleChoice, the survey shows approximately 925 DSL-equipped COs in North America (United States and Canada) and approximately 20 million potential customers. Trials and deployments are also under way in Europe, Latin America, and the Asia-Pacific region.

The following are some representative deployments:*

- Singapore Telecom was one of the first to deploy ADSL service in 1997 to deliver a full suite of multimedia services, including Internet access, VoD, news, shopping, games, and music on demand. The service is branded as Magix and is delivered over the Singapore One network using ATM transport.

---

* The ADSL Forum maintains an ongoing list of current ADSL trials and deployments worldwide.

▪ In May 1998, US WEST* announced the availability of MegaBit ADSL services in several metropolitan areas of the western United States. MegaBit Services include:

    ▪ MegaHome—Gives standard Internet users 256 Kbps access

    ▪ MegaOffice—Telecommuters and small businesses can get 512 Kbps

    ▪ MegaBusiness—Heavier-use business customers can get 768 Kbps

    ▪ MegaPak—Combines MegaHome and US WEST.net Internet service

    ▪ MegaBit—Intensive business users and cyber-surfers can get 1 Mbps to 7 Mbps.

▪ Deutsche Telekom announced in October 1998 that it would deploy ADSL initially in eight major German cities. Both ATM and Ethernet interfaces would be available to the customer premises equipment. The Deutsche Telekom deployment is one of the first that allows coexistence of ADSL and ISDN on a single phone line.

Future Directions

As ADSL and ADSL lite move into the deployment phase, the work effort has shifted to areas necessary to make these technologies work in a real-world environment. The following is representative of some of the work in progress:

▪ **Interoperability.** As we discussed earlier, industry work groups such as the UAWG and the ADSL Forum are developing test suites that independent test houses can use to certify products. In the standards arena, the ITU SG 15 is reviewing G.996.1 (G.test) entitled *Test Procedures for Digital Subscriber Line Transceivers*. The following is a brief description of items under review:

    ▪ Test setup both with and without splitters at the ATU-R end

    ▪ Tests to study the impact of crosstalk, impulse, and POTS interference

---

* Since this announcement, other U.S. telcos, such as BellSouth and Southwestern Bell Communications (which includes Pacific Bell), have announced service availability.

- POTS quality testing during concurrent operation of ADSL and POTS at the ATU-R end
- Test loops (e.g., typical North American and European loops to be used in testing)
- Premises wiring models to study the impact of premises wiring during concurrent operation of ADSL and POTS without a splitter at the ATU-R end.

■ **Operations, administration, maintenance, and provisioning (OAM&P).** At the technology level, the ITU SG 15 is reviewing G.997.1 (G.ploam) for "physical layer OAM." At the architecture level, several efforts are underway to develop a framework for automatic service provisioning and configuration, as well as for billing systems. These efforts are described in Chapter 8 on ADSL architectures.

■ **Network management.** In most cases, OAM&P tools rely on network management agents that collect information about a particular managed entity in the management information base (MIB). The ADSL Forum is reviewing line-code specific MIBs for CAP and DMT. The ADSL Forum is also reviewing the network management architecture for ADSL systems.

Summary

The fundamental idea behind ADSL is utilization of the native bandwidth of copper loops and eliminating the constraint of a narrow 4-kHz spectrum. This enhances the investments that the telcos already have in terms of connections to both consumers and businesses. The inherent asymmetry (by trading off upstream bandwidth in return for higher downstream bandwidth) allows ADSL to support a variety of applications.

ADSL has evolved significantly since the days of the early VoD trials. The explosive growth of the Internet, the pent-up demand for high-speed access, and the need for integrated services that can be delivered over an "always-on," high-speed pipe have all served to galvanize the industry toward enabling the technology to realize its full potential. Telcos have been energized to deliver value-added services to the same markets that are being coveted by cable companies. The computer industry has become excited about the possibility of developing new applications and hardware that would take advantage of bandwidth to the premises. All these factors have contributed to an increasing number of strategic alliances and the formation of industry work groups focused on ADSL and ADSL lite. The vision is that ADSL will satisfy the demand of reliable, integrated value-added services, whereas ADSL lite will offer an intermediate step for those consumers hungry for relief from Internet access over a modem. The ADSL Forum and the UAWG have made significant contributions in the industry to advance the deployment of these technologies.

ADSL operates in the frequency spectrum above the POTS band, allowing ADSL and POTS to coexist on the same copper wire. With full-rate ADSL, the two services are totally independent because a POTS splitter is installed at both the premises and the central office. With ADSL lite, the splitter at the premises is optional. This allows easier rollout of the service because a truck roll to install the premises splitter can be eliminated. However, ADSL lite trades off higher bandwidth of full-rate ADSL in return for this benefit.

Some of the issues that have throttled aggressive deployment of ADSL technologies until recently are:

▮ Maximizing the data rate for specific loop distances (to increase service availability to potential customers)
▮ Line conditions
▮ Spectral compatibility
▮ Premises conditions (for ADSL lite)
▮ Interference between POTS and ADSL (for ADSL lite).

In 1998, the industry stepped forward with innovative solutions and workarounds to these key issues. There has also been growing momentum in the industry to enable interoperability among multiple vendors to reduce the risk of deployment for both service providers and customers. In turn, this promotes increased availability and reduces total cost of service delivery.

The first steps toward serious deployment occurred in 1998, with service providers moving beyond the trial phase and announcing service availability in selected metropolitan areas worldwide. As deployment gathers steam, the focus is shifting away from the base technology into areas that are necessary to make ADSL and ADSL lite work in a real-world environment. The ADSL Forum, with the help of the UAWG, is continuing to lead the way in developing solutions for copper loop-based access architectures. In the standards arena, the International Telecommunications Union has been moving forward to enable ratification of XDSL-related standards not only for the base technology, but also for testing and operational issues.

References

1. Smithwick, L. "Carrierless Amplitude/Phase Modulation (CAP) vs. Discrete Multi-Tone Modulation (DMT)," (January 1997).
2. Baines, R. "Discrete Multi-Tone (DMT) vs. Carrierless Amplitude/Phase (CAP) Line Codes," *Analog Devices White Paper,* (May 1997).
3. Information is available at the official Website at **www.tl.org/tlel/el4home.htm**.
4. "Issue 2 ADSL Standard T1.413," *ANSI Contribution T1.E14/97-007R6*, as provided for default ballot (July 1998).
5. Information is available at the official Website at **www.etsi.org**.
6. Information is available at the official Website at **www.itu.int**.
7. "G.992.1, Asymmetrical Digital Subscriber Line (ADSL) Transceivers," *Draft ITU Recommendation*, COM 15-131.
8. "G.992.2, Splitterless Asymmetrical Digital Subscriber Line (ADSL) Transceivers," *Draft ITU Recommendation*, COM 15-136.
9. "G.994.1, Handshake Procedures for Digital Subscriber Line (DSL) Transceivers," *Draft ITU Recommendation*, COM 15-134.
10. Information is available at the official Website at **www.adsl.com**.
11. Information is available at the official Website at **www.uawg.org**.
12. Cioffi, J. M. "A Multicarrier Primer," *ANSI Contribution T1E1.4/91-157* (November 1991).
13. Bingham, J. "Multicarrier Modulation for Data Transmission: An Idea Whose Time has Come," *IEEE Communications Magazine* Vol. 28, No. 5, (May 1990).
14. Chow, P,. Bingham, J., and Coiffi, J. "DMT-Based ADSL: Concept, Architecture and Performance," *Colloquium Digest IEE 1994*, Issue 192.

15. Im, G. H. and Werner, J. J. "Bandwidth-Efficient Digital Transmission over Unshielded Twisted-Pair Wiring," *IEEE Journal on Selected Areas in Communications*, Vol. 12, No. 9 (December 1995).

16. "Asymmetric Digital Subscriber Line (ADSL) Metallic Interface," *ANSI Standard T.413-95* (1995).

17. Cioffi, J. "The Essential Merit of Bit-Swapping," *ANSI Contribution T1E1.4/88-318* (November 1988).

18. Warrier, P. "Universal DSL deployment of G.Lite—Issues and Solutions," *Texas Instruments Application Report SPAA007A* (September 1998).

19. Carbonelli, M., Seta, D. D., Perucchini, D., and Petrini, L. "Evaluation of Near-End Crosstalk Noise Affecting ADSL Systems," *International Conference on Communication Systems,* (November 1994).

20. Foster, K. T., and Cook, J. W. "The Radio Frequency Interference (RFI) Environment for Very High-Rate Transmission Over Metallic Access Wire-Pairs," *ANSI Contribution T1E1.4/95-020* (February 1995).

21. Costello, E., and Laidler, G. "Telephony CPE Impedance Measurements for Splitterless ADSL," *ADSL Forum Contribution 98-032* (March 1998).

22. Tellado, J., and Cioffi, J. "PAR Reduction with Minimum or Zero Bandwidth Loss and Low Complexity," *ANSI Contribution T1E1.4/98-173* (June 1998).

23. Gatherer, A., and Polley, M. "Controlling Clipping Probability in DMT Transmission," *1997 Asilormar Conference*, (November 1997).

24. "ATM over ADSL Recommendations," *ADSL Forum, TR-002.*

25. "Framing and Encapsulization Standards for ADSL: Packet Mode," *ADSL Forum TR-002.*

HDSL
and HDSL2

Background

History

The origin of high-speed digital subscriber line (HDSL) is rooted in T1 and ISDN technology, and the motivation for HDSL came from the success of T1 deployments. The original T1 commercial deployments (around 1984) used a line-coding method called bipolar alternate mark inversion (AMI). Devices called channel service units (CSUs) and digital service units (DSUs), at either end of the T1 link from the customer premises to the serving central office, were used to multiplex 64-Kbps voice channels and encode the bits for transmission over the link. The link itself consisted of two pairs of copper wire—a sender and a receiver—each carrying 1.544 Mbps. Since the length of the T1 link from the customer premises to the CO could span several thousand feet, repeaters were necessary to regenerate the digital signal along the link.* Typically, repeaters were placed every 6,000 feet (i.e., 6 Kft), the same distance as loading coils—a legacy of conditioning copper loops to carry analog voice channels. Therefore, by replacing the loading coils with repeaters,† and by utilizing two copper loops, the telco could now provide the customer with capacity equivalent to 24 digital voice channels, i.e., 1.536 Mbps.‡ This increase in capacity proved to be very attractive to businesses since T1 service was often priced so that it became cost effective to replace approximately six leased lines (dedicated copper loops) with a single T1 link. Furthermore, since T1 offered raw transport, businesses could now use the increased capacity to cost-effectively carry a combination of voice and data over the same link. These factors contributed to the growing popularity of T1 and became the basis for delivery of ISDN primary rate interface (PRI) service.

---

* T1 transmission based on AMI required use of a broad frequency spectrum, including high frequencies. As discussed in Chapter 3, since high frequencies attenuate rapidly with distance, it became necessary to introduce repeaters at intervals to regenerate the signal and counter the effects of attenuation.

† As discussed in Chapter 3, loading coils act as filters and are therefore fundamentally incompatible with transmissions that utilize the higher frequencies such as T1 and DSL technologies. Therefore, to provision T1 service over an existing analog voice-grade loop, a telco had first to eliminate any loading coils, and then use repeaters to regenerate the digital signal.

‡ Although the T1 rate is 1.544 Mbps, 8 Kbps are used for signaling, so the capacity to the user is 1.536 Mbps.

However, as T1 deployments grew, it soon became obvious that the need to have repeaters approximately every 6 Kft would curtail the rate of deployment. Competitive pressures also prompted service providers to reduce the lead times to deploy the new service. The repeaters also proved to be troublesome during line troubleshooting since they did not run any network management software* and were often located in hard-to-reach places such as conduits. Hence, the telcos prompted Bellcore engineers to come up with T1 alternatives where the same usable line capacity of 1.536 Mbps could be provided without the need for repeaters. Therefore, "repeaterless T1" became the basis for research that led to the development of HDSL.

To enable repeaterless T1, engineers had to adopt a line code more efficient than the existing bipolar AMI used for T1. Fortunately, engineers had experience with a line code called 2 Binary, 1 Quaternary (2B1Q). 2B1Q, which provided the ability to encode 2 bits/baud rather than 1 bit/baud, was the line code used to deliver ISDN basic rate interface (BRI). Bellcore engineers proposed a new technology, HDSL, that utilized 2B1Q line coding for repeaterless T1. As before, HDSL utilized two copper wire pairs, except now the need for repeaters was eliminated. Each pair carried 784 Kbps or half the total "raw" T1 rate of 1.544 Mbps. Instead of CSUs/DSUs at either end of the link, the new technology required the use of an HTU-R and an HTU-C at the premises and CO ends, respectively. Over time, HDSL became cost effective enough for service providers to replace existing T1 links. It is important to note that these changes occurred only from a technology and network perspective. As far as customers were concerned, they still received T1 service equal to 1.536 Mbps of usable capacity.

To cater to European requirements, early versions of repeaterless E1 using HDSL required three wire pairs because the equipment manufacturers simply reused their 784 Kbps transceivers to push the additional up capacity to 2.048 Mbps. Later, with advances in transceiver technology, it became possible to support both T1 and E1 speeds using only two twisted pairs and 2B1Q line coding remained. Today, HDSL delivers either 1.544 Mbps or 2.048 Mbps over two twisted pairs at distances of up to 12 Kft.

---

* Repeaters are simple, low-cost devices without any intelligence for network management.

In the mid 1990s, the demand for eliminating the second wire pair increased. Service providers, especially in the United States, came under pressure to maximize the use of their existing infrastructure to achieve economies of scale and to reduce costs in an increasingly competitive environment. Copper pairs became increasingly more valuable as U.S. regulations required ILECs to unbundle pairs and lease them to competitors. In addition, customers began to demand more services and lines, which often required additional wiring. Then, as pairs were utilized for other services, it became increasingly difficult to find multiple free pairs between two locations—an important requirement for two-pair HDSL. For these reasons, telcos were motivated to turn to researchers and the industry to find ways to conserve copper by maximizing the number of services over a single pair. Vendors could also derive a direct benefit from going to a single pair. It meant fewer components,* and a framer could be simplified by eliminating the need for mapping and inverse multiplexing. These benefits translated to lower power consumption and board space—especially attractive for digital loop carrier (DLC) implementations.

In response to telco demands, the industry initially developed a variety of proprietary technologies and related terminology that, unfortunately, had the impact of confusing and fragmenting the market. Symmetric (or single-pair) digital subscriber line or SDSL was one of the early single-pair technologies. However, its typical reach topped out around 10 Kft. Other implementations, sometimes known as medium-speed digital subscriber line or MDSL, increased the loop length to around 22 Kft, but at the expense of line capacity (restricted to a maximum of 784 Kbps). The telcos were not satisfied with these offerings. Further, they demanded interoperable solutions from the vendors so they would not be locked into a particular vendor. In response to telco requirements, the industry quickly moved toward standardization of the next generation of HDSL, known as HDSL2. Today, HDSL2 is on its way to becoming formally standardized in the United States as a result of the ANSI T1E1.4 committee efforts. When HDSL2 is deployed, the goal of the HDSL2 standard is to provide full T1 line rate capability at the loop lengths served by HDSL over a single pair.

---

* Only one transceiver and half the number of line protection and remote power feeding units compared with a two-pair implementation.

Basic Network Architecture

The basic HDSL network architecture for T1 is shown in Figure 5.1. This architectural diagram is important as a backdrop for the discussion of the technology. A detailed discussion of the architecture is reserved for a later chapter in this book.

Figure 5.1

Basic network architecture for HDSL for T1

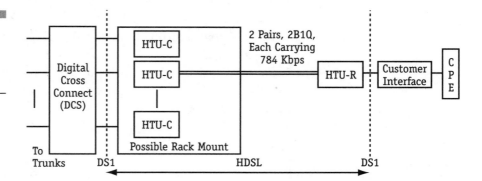

As with ADSL, HDSL architectures are based on copper loops between the subscriber premises and the central office. However, aside from the obvious differences between symmetry and data rates, HDSL is architecturally different from ADSL in two respects:

■ ADSL assumes simultaneous POTS and ADSL service. HDSL does not include support for POTS; indeed, HDSL utilizes frequencies that overlap the voice passband. For this reason, HDSL architectures are inherently "splitterless."

■ ADSL requires one, and only one, copper pair. HDSL requires two (and in some legacy implementations, up to three) pairs. However, HDSL2 will operate over a single pair.

On the premises side, a device called an HDSL termination unit-remote (HTU-R) connects the customer premises equipment to the copper loop pairs that form the HDSL link. Typically, the CPE is a router with a serial interface that connects directly into a port on the HTU-R. As far as the router is concerned, its serial port is connected to a T1. An HTU-R consists of a transceiver, a mapper (to map the T1 bits into the

HDSL frame structure and vice-versa), and an interface module to accept a standard T1 interface. Another term used for an HTU-R is line termination unit (LTU). On the central office side, a device called the HDSL termination unit-central (HTU-C) terminates the other end of the copper loop pairs. Another term used for an HTU-C is network termination unit (NTU). HTUs replace the traditional CSUs and DSUs used for T1 and perform pretty much the same functions. As in the case of ADSL, for efficiency reasons, a bank of HTU-Cs are combined with a multiplexing function to form a DSL access multiplexer (DSLAM)* that connects to a service provider's network.

The basic HDSL network architecture for E1 is similar to that for T1, except that three copper loop pairs were originally used, instead of two. This is shown in Figure 5.2.

Figure 5.2
Basic network architecture for HDSL for E1

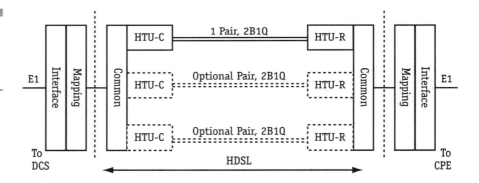

As mentioned earlier, with improvements in transceiver technology that enabled each pair to carry up to 1.168 Mbps, it became possible to support both T1 and E1 line rates with just two pairs. Today, three-pair implementations of HDSL for E1 have been almost phased out.

Applications

It is important to note that both HDSL and HDSL2 (and related technologies such as SDSL and MDSL) all provide *symmetric* bandwidth.

---

* Often, the same DSLAM chassis can simultaneously support both ATU-C and HTU-C line cards for ADSL and HDSL, respectively.

In other words, since the upstream and downstream rates are the same, these technologies are suitable to applications where symmetry is the predominant requirement (i.e., where ADSL would not be an option). Of course, nothing precludes a customer from also running asymmetric bandwidth applications on top of HDSL/HDSL2 (and related technologies) as long as there is sufficient capacity.* This flexibility is what makes HDSL/HDSL2 (and related technologies) generally more suitable for business use. In practice, factors such as deployment, pricing, and customer applications dictate the use of a particular technology in a given situation. With that in mind, the following are some typical symmetric applications that can be supported using HDSL (and related technologies):

- **Private T1 line.** This can satisfy a variety of scenarios:
  - Frame relay or ATM line from business site to data switch
  - Frame relay or X.25 link between business sites
  - Videoconference line to business site
  - Voice trunk to wireless cell site.
- **Integrated services to businesses.** Businesses need to integrate voice and data services in order to achieve economies of scale. Most businesses today, particularly small businesses, maintain separate infrastructures for voice and data, which is inefficient and expensive. HDSL2 is an economical way to provide integrated voice and data access over a single copper loop to a business using multiline derived digital voice lines.
- **Web hosting.** Most businesses today have embraced the Web by establishing a Web "storefront," i.e., a home page. For businesses that choose to maintain and host their own Web pages on a server at their premises, HDSL/HDSL2 provides the necessary symmetric bandwidth to allow multiple users simultaneous access.
- **Videoconferencing.** HDSL/HDSL2 can easily support multiple H0 channels of 384×384 Kbps to deliver high-quality videoconferencing.
- **LAN-LAN interconnect.** The symmetric bandwidth of HDSL/HDSL2 allows companies to interconnect their LANs for sharing files and other resources.

---

* It is important to note that HDSL/HDSL2 line rates are fixed, and not rate adaptive.

- **Telecommuting.** Unlike telecommuting with ADSL, HDSL/HDSL2 can support high-speed bidirectional communications (such as bulk file transfers) that might be required for certain classes of telecommuters.
- **Shared Internet access.** Multiple PCs on a LAN at a business can share the bidirectional high-speed bandwidth.
- **"Reverse ADSL" option.** Certain applications (such as uploading "end-of-the-day" transactions to corporate sites) require high upstream bandwidth and low downstream bandwidth. It would seem that merely reversing the direction of an ADSL system could satisfy this requirement. However, "reverse ADSL" is not practical given current technology and deployment constraints. "Reverse ADSL" in the same binder group as traditional ADSL would result in significant degradation of service due to the crosstalk induced by opposing systems. Therefore, symmetric DSL services like HDSL and HDSL2 offer the only real choice in such situations.

Standards Activities

Many of the organizations involved in the standardization of ADSL have also been active in the HDSL/HDSL2 standardization process. Indeed, there is considerable overlap between the people who attend meetings associated with ADSL and HDSL technologies. As with ADSL, the initial standardization effort for HDSL came out of the ANSI T1E1.4[1] technical committee. For Europe, ETSI TM6[2] has been involved in the HDSL/HDSL2 standards activities; internationally, the ITU[3] determined the G.991.1 standard (G.hdsl) in 1998.

In 1994, ANSI T1E1.4 published a technical report presenting the results of an investigation into the feasibility and advisability of bidirectional HDSL lines that operate over twisted-wire pairs in the outside loop plant with a payload capacity substantially greater than the 160 Kbps of ISDN basic access DSLs (IDSL). The report described an HDSL system that could provide an alternative to T1-repeatered lines to transport DS1-based services at a rate of 1.544 Mbps over loop lengths that met the carrier service area (CSA) guidelines. The report recommended a frame structure and some operations features for 2B1Q-based systems. Vendors subsequently used this report to develop their own HDSL systems. However, since the ANSI report was not intended to be a full specification of a working HDSL system, vendor implementations were not guaranteed to be interoperable. The process toward international standardization would take several more months and significant effort by the ANSI TE1.4 and ETSI TM6 committees. It was only in October 1998 that the ITU determined the G.991.1 standard.

At the time of this writing, the ANSI T1E1.4 committee had not yet finalized HDSL2. However, it is expected that the HDSL2 standard could be approved after letter ballot review by T1E1.4 sometime in 1999. Meanwhile, at the October 1998 ITU meeting in Geneva, Study Group 15 agreed to begin work on single-line HDSL (G.shdsl), which will be based on work done in ANSI T1E1.4 and ETSI TM6 committees.

The ADSL Forum has not been as active in the area of HDSL/HDSL2 standards as it has been on other DSL technologies such as ADSL and VDSL. However, aside from the symmetric bandwidth offered over the loop and lower maximum line rates, there are no significant changes to the higher layers between ADSL and HDSL/HDSL2. As previously noted, there is generally nothing precluding applications that run on top of ADSL working well over HDSL. In other words, most of the end-to-end architectures for ADSL are also applicable for HDSL/HDSL2.* For this reason, the ADSL Forum has not focused its efforts on HDSL/HDSL2.

Technology Discussion

Modulation Schemes

AMI line coding has an efficiency of 1 bit/baud. Therefore, to transmit 1.544 Mbps for T1, AMI requires use of the frequency spectrum from zero to 1.544 kHz, which is well into the high-frequency range where attenuation versus distance becomes an important factor. Since high frequencies attenuate rapidly with distance, repeaters are required to ensure enough signal quality at the receiving end. It follows that a mechanism to eliminate repeaters involves reducing the usage of high frequencies—i.e., limiting frequency use to those with enough signal

---

* At least one commercial deployment for telecommuting offers customers the choice of ADSL or SDSL (the precursor to HDSL2) as the loop technology, with no difference in the backbone architecture of the service provider.

strength to carry it across a loop without repeaters. Limiting high frequencies, in turn, implies contraction of the usable frequency spectrum (smaller bandwidth). The way to carry the same line rate with smaller bandwidth is to increase the efficiency of the line coding or, put another way, to improve the bits/baud.

As we discussed in Chapter 3, advanced modulation techniques can be used to increase the bits/baud. The quaternary line communications technique takes a pair of bits and codes them into a four-level symbol using two levels of positive voltage and two levels of negative voltage. 2B1Q is a simple modulation technique that increases the line code efficiency to 2 bits/baud by requiring the receiver to detect both the amplitude and the polarity of the signals. In effect, 2B1Q cuts the frequency spectrum usage by half to carry the same line rate as AMI. For T1, the highest frequency utilized is only 772 kHz—well within the range of frequencies that can be modulated without significant attenuation across a loop without repeaters.

Single-pair implementation of symmetric DSL requires even more efficiency in line coding. Multiple line codes were proposed for HDSL2, however, the two predominant ones discussed in the standards organizations were CAP[4] and PAM[5]. CAP has already been described in Chapter 4 on ADSL, and that discussion will not be repeated here. PAM stands for pulse amplitude modulation and is a simple line-code technique that uses amplitude and polarity—just like 2B1Q. In fact, some documents refer to 2B1Q as "4-level PAM," With a slight modification to the basic 2B1Q method, the line-code efficiency can be improved to 3 bits/baud (up from 2 bits/baud).

As with ADSL, there was a lot of debate in the industry about the choice of line codes for HDSL2. Eventually, the proponents of PAM won, largely on the basis of its simplicity and close ties to the established 2B1Q modulation for HDSL.

The remainder of this technology discussion focuses on the HDSL2 draft standard[6].

Overview of the Draft HDSL2 Standard

The standard provides the HDSL2 transceiver requirements and guidelines to assist manufacturers, providers, and users of products connected to the telephone network. Therefore, an HDSL2 transceiver is considered standards compliant when the requirements are met. Guidelines, as informative appendices, provide technical details for assisting manufacturers and users, but are not requirements for compliance.

To ensure interoperability between transceivers, the standard specifies the frame structure, a transmitter physical layer, a startup procedure, electrical characteristics, and other considerations. In the standard text, test loop and text circuit examples are included in the appendices.

HDSL2 System Reference Model

The system reference model shown in Figure 5.3 illustrates the functional blocks required to provide HDSL2 service.

Figure 5.3
System
reference model

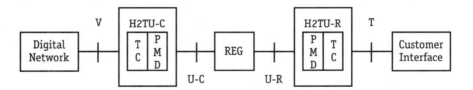

What follows are some notes on the reference model:

- The V and T interfaces are defined in terms of logical not physical, functions.
- The V interface may consist of interface(s) to one or more network elements.
- More than one T interface may be defined.
- Implementation of the V and T interfaces is optional when interfacing elements are integrated into a common element.
- Due to the asymmetry of the signals on the line, the transmitted signals must be distinctly specified at the U-C and U-R reference points.
- Use of one or more signal regenerators (REG), i.e., repeaters, is optional.
- Due to the asymmetric nature of the HDSL2 power spectra, upstream and downstream transmit directions are defined, where upstream is the direction from the H2TU-R to the H2TU-C, and downstream is the direction from the H2TU-C to the H2TU-R.

H2TU Transceiver Reference Model

The H2TU architecture is divided into a transmission convergence (TC) and a physical medium dependent (PMD) layer. The f(n) interface is a logical division between TC and PMD and does not preclude implementation efficiencies when the two are implemented on one device. The respective functions for these two blocks are identified in Figure 5.4. The transmission system is based on a PMD-independent framing structure, which provides for embedding DS1 payloads into HDSL2 frames. The reference model allows for the implementation of alternate service-specific transmission convergence blocks, and the usage of unique overhead bytes that may carry link-management statistics, encryption data, and resource requests.

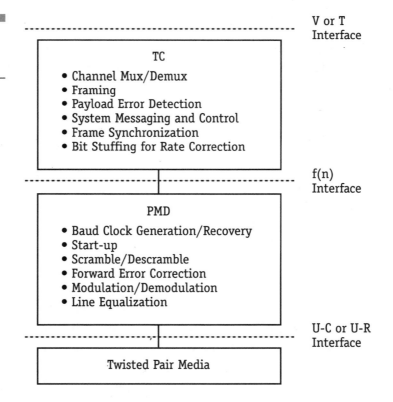

Figure 5.4

HDSL2 transceiver reference model

H2TU DS1 TC Layer Functional Characteristics

The HDSL2 frame structure for T1 service is shown in Table 5.1. The transmitted frame is 9,310 bits long when the stuff bits are not present, or 9,314 bits long when stuff bits are present.*

Table 5.1

HDSL2 frame structure

Time	Frame Bit #	HOH Bit #	Abr. Name	Full Name
0 ms	1-10	1–10	FSW 1-10-	Sync word
	11-2326	—	B1	Payload block 1
	2327	11	crc1	Cyclic redundancy check
	2328	12	crc2	Cyclic redundancy check
	2329	13	sbid1	Stuff bit id copy 1
	2330	14	losd	DS1 Loss of signal defect
	2331-2338	15–22	EOC01-EOC08	EOC bit 1 through bit 8
	2339-4654	—	B2	Payload block-2
	4655	23	crc3	Cyclic redundancy check
	4656	24	crc4	Cyclic redundancy check
	4657	25	uib	Unspecified indicator bit
	4658	26	sega	Segment anomaly
	4659-4666	27–34	EOC09-EOC16	EOC bit 9 through bit 16
	4667-6982	—	B3	Payload block 3
	6983	35	crc5	Cyclic redundancy check
	6984	36	crc6	Cyclic redundancy check
	6985	37	sbid2	Stuff bit id copy 2
	6986	38	segd	Segment defect
	6987-6994	39–46	EOC17-EOC24	EOC bit 17 through bit 24
6 ms – 2/1552 ms	6995-9310	—	B4	Payload block 4
6 ms – 1/1552 ms	9311	47	sb1	Stuff bit 1
6 ms nominal	9312	48	sb2	Stuff bit 2
6 ms + 1/1552 ms	9313	49	sb3	Stuff bit 3
6 + 2/1552 ms	9314	50	sb4	Stuff bit 4

---

* The presence or absence of stuff bits will be in accordance with the procedures specified in the standard.

The frame structure includes the following:

- **Frame sync word.** The frame sync word consists of a sequence of 10 bits, transmitted leftmost bit first.
- **Payload blocks.** The frame contains four payload blocks, each consisting of 2,316 bits or the same number of bits as in 12 DS-1 frames.
- **CRC generation.** A cyclic redundancy check (CRC) is generated for each HDSL2 frame and transmitted on the following HDSL2 frame. The six CRC bits ($c1$ to $c6$) are the coefficients of the remainder polynomial after the message polynomial, multiplied by X^6, is divided by the generating polynomial. The message polynomial consists of all the bits in the HDSL2 frame except for the synchronization word, CRC bits, and the stuff bits.* The message bits are ordered just as they are in the HDSL2 frame itself, i.e., m_0 is the first bit, m_1 is the second bit, etc.
- **Stuff bit IDs.** Bits *sbid1* and *sbid2* are used for stuff bit identification. Both bits are set to one if the four stuff bits are present at the end of that frame. Both bits are set to zero if there are no stuff bits at the end of the current frame.
- **Indicator bits.** The frame contains four fixed indicator bits to allow rapid detection of DS1 and HDSL2 faults. They are:
  - Far-end loss of DS1 signal (LOSD) to represent the status of the far-end DS1 input
  - Segment anomaly (SEGA) to indicate a crc error on the incoming frame
  - Segment defect (SEGD) to indicate a loss of synchronization (LOSYN) on the incoming frame
  - Unspecified indicator bit (UIB), reserved.
- **Embedded operations channel.** The EOC channel consists of 24 bits per frame, resulting in a nominal data rate of 4,000 bits per second.
- **Stuff bits.** The stuff bits are inserted to compensate for jitter in the frame clock.

---

* Therefore, there are 9,294 message bits in a frame covered by the CRC check.

Initialization

The block diagram of the startup mode PMD layer of an H2TU-C or H2TU-R transmitter is shown in Figure 5.5.

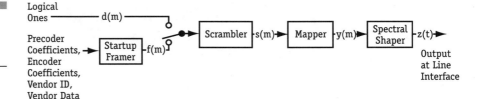

Figure 5.5

Startup PMD reference model

The time index m represents the symbol time, and t represents analog time. Since startup uses 2-PAM modulation, the bit time is equivalent to the symbol time. The output of the framer is framed information bits and is mathematically represented as $f(m)$. The output of the scrambler is $s(m)$. The output of the mapper is represented as $y(m)$, and the output of the spectral shaper at the line interface is $z(t)$. The function $d(m)$ is an initialization signal that will be logical ones for all m.

The timing diagram for the startup sequence is given in Figure 5.6, and the state transition diagram for the startup sequence is given in Figure 5.7.

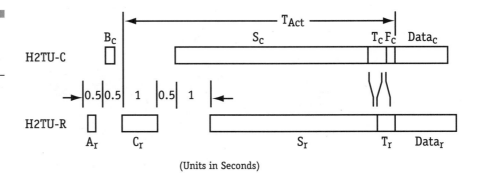

Figure 5.6

Timing diagram for startup sequence

The startup frame format is shown in Table 5.2.

Figure 5.7
H2TU-C and H2TU-R transmitter state transition diagram

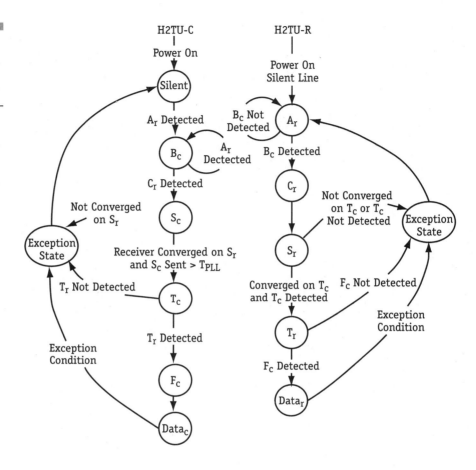

Table 5.2
Startup frame format

Startup Frame Bit LSB:MSB	Definition
1:13	Frame sync for T_c and T_r: 1111100110101, where the leftmost bit is sent first in time
	Frame sync for F_c: 1010110011111, where the leftmost bit is sent first in time
14:35	Precoder coefficient 1: 22-bit signed two's complement format with 17 bits after the binary point, where the LSB is sent first in time
36:57	Precoder coefficient 2

continued on next page

Table 5.2

continued

Startup Frame Bit LSB:MSB	Definition
58:3951	Precoder coefficients 3–179
3952:3973	Precoder coefficient 180
3974:3994	Encoder coefficient A: 21 bits where the LSB is sent first in time
3995:4015	Encoder coefficient B: 21 bits where the LSB is sent first in time
4016:4017	Size of country code in octets: 2 bits where the LSB is sent first in time
4018:4019	Size of provider code in octets: 2 bits where the LSB is sent first in time
4020:4051	Country code word: 32 bits max where the LSB is sent first in time, any unused bits follow MSB and are set to zero
4052:4083	Provider code word: 32 bits max where the LSB is sent first in time, any unused bits follow MSB and are set to zero
4084:4147	Vendor data: 64 bits of proprietary information
4148:4155	HDSL2 version: 00000001, where leftmost bit is sent first in time
4156:4211	Reserved: 56 bits set to logical zeros
4212:4227	CRC: Bit 15 sent first in time, bit 0 sent last in time

With respect to Figure 5.7, the following signals are defined:

■ **Signal A_r.** Upon power-up on a silent line, the H2TU-R sends A_r. Waveform A_r is generated by connecting the signal $d(m)$ to the input of the H2TU-R scrambler.
■ **Signal B_c.** The H2TU-C transmits B_c if, and only if, A_r has been detected. Waveform Bc is generated by connecting the signal $d(m)$ to the input of the H2TU-C scrambler.
■ **Signal C_r.** After detecting B_c, the H2TU-R sends C_r. Waveform C_r is generated by connecting the signal $d(m)$ to the input of the H2TU-R scrambler.
■ **Signal S_c.** After detecting C_r, the H2TU-C sends S_c. Waveform S_c is generated by connecting the signal $d(m)$ to the input of the H2TU-C scrambler. If the H2TU-C does not converge while S_c is transmitted, it enters the exception state.
■ **Signal S_r.** The H2TU-R sends S_r after the end of C_r. Waveform S_r is generated by connecting the signal $d(m)$ to the input of the H2TU-

R scrambler. If the H2TU-R does not converge or detect T_c while S_r is transmitted, it enters the exception state. The method used to detect T_c is vendor dependent.

■ **Signal T_c.** Once the H2TU-C has converged, it sends T_c. Waveform Tc contains the precoder coefficients and other system information. T_c is generated by connecting the signal f(m) to the input of the H2TU-C scrambler as shown in Figure 5.7, where f(m) is the start-up frame information. If the H2TU-C does not detect T_r while sending T_c, it enters the exception state. The method used to detect T_r is vendor dependent.

■ **Signal T_r.** Once the H2TU-R has converged, it sends T_r. Waveform T_r contains the precoder coefficients and other system information. T_r is generated by connecting the signal f(m) to the input of the H2TU-R scrambler, where f(m) is the startup frame information. If the H2TU-R does not detect T_c while sending T_r, it enters the exception state. The method used to detect T_c is vendor dependent.

■ **Signal F_c.** Once the H2TU-C has detected T_r, it sends F_c. Signal F_c is generated by connecting the signal f(m) to the input of the H2TU-C scrambler as shown in Figure 5.5. The signal f(m) is the startup frame information shown in Table 5.2 unless the frame sync word is reversed in time and the payload information bits are set to arbitrary values. The payload information bits correspond to the precoder coefficients, encoder coefficients, Vendor ID, Vendor data, and Reserved. The CRC is calculated on this arbitrary-valued payload. The signal F_c is transmitted for exactly two startup frames, and as soon as the first bit of F_c is transmitted, the payload data in the T_r signal is ignored.

■ **Data$_c$ and Data$_r$.** Within 200 symbols after the end of the second frame of F_c, the H2TU-C sends Data$_c$, and the H2TU-R sends Data$_r$.

If activation is not achieved within T_{act} seconds or if any exception condition occurs, the exception state is invoked. During the exception state, the unit remains silent for $T_{silence}$ seconds, then waits for transmission from the far end to cease before returning to the corresponding initial startup state. The H2TU-R begins sending A_r, and the H2TU-C remains silent and waits for the detection of A_r.

Data Mode

The block diagram of the data mode PMD layer of an H2TU-C or H2TU-R transmitter is shown in Figure 5.8.

Figure 5.8

Data mode PMD reference model

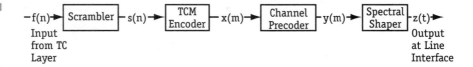

$-f(n) \rightarrow$ | Scrambler | $\vdash s(n) \rightarrow$ | TCM Encoder | $\vdash x(m) \rightarrow$ | Channel Precoder | $\vdash y(m) \rightarrow$ | Spectral Shaper | $\vdash z(t) \rightarrow$

Input from TC Layer

Output at Line Interface

The following is an explanation of the terms used in the reference model:

Time Index n	Represents bit time
Time Index m	Represents symbol time
Time Index t	Represents analog time
f(n)	Input from the TC layer
s(n)	Output from the scrambler
x(m)	Output of the Trellis coded modulation encoder
y(m)	Output of the channel precoder
z(t)	Analog output of the spectral shaper at the line interface

When transferring K information bits per one-dimensional PAM symbol, the symbol duration is K times the bit duration, so the K values of n for a given value of m are {mK+0, mK+1,..., mK+K-1}.

The PMD data rate is 1.552 Mbps (the extra 8 Kbps over the T1 line rate represents the HDSL overhead for management and control). The modulation scheme is Trellis code PAM, which results in 3 bits/symbol.

Operations, Administration, Maintenance, and Provisioning (OAM &P)

As mentioned earlier, the frame carries bits that convey problems such as far-end loss of DS1 signal (LOSD), segment anomaly (SEGA), segment defect (SEGD), and loss of sync (LOSYN). Either side can query the status of these bits to initiate recovery procedures, if needed.

Like ADSL, HDSL2 has an embedded operations channel (EOC) that the H2TU-R and H2TU-C use to communicate maintenance and control information. The eoc carries messages in an HDLC-like format beginning with the usual 0x7E sync byte and a two-byte frame check sequence (FCS). The EOC is used to carry performance parameter messages such as the signal-to-noise ratio, or informational messages such as version information and vendor ID. It can also be used to carry messages for provisioning.

Current State of the Technology

Deployment Issues and Possible Solutions

Spectral Compatibility and Binder Group Management

Issue

Multiple pairs of copper wire can be grouped in "pair units," for example, groups of 12, 13, or 23 pairs. A multiunit consists of pair units that have been assembled into a collection of 50 or 100 pairs. A binder group is a pair unit or a multiunit that has been assembled and bound with colored binder tape for identification. For example, a 50-pair multiunit can consist of two 12-pair units and two 13-pair units, and a 100-pair multiunit can consist of four 25 pair units. The most common binder group sizes are 12, 13, 25, 50, or 100 pairs. Pair units, multiunits, and binder groups are shown in Figure 5.9.

Figure 5.9

Pair units,
multiunits and
binder groups

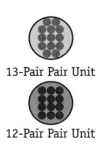

13-Pair Pair Unit

12-Pair Pair Unit

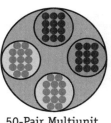

50-Pair Multiunit
(Two 13 Pair Units
Two 12 Pair Units)

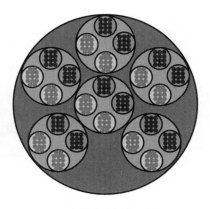

300 Pair Binder Group
(Six 50 Pair Multiunits)

Within the loop plant, pairs that carry energy are close together; therefore, they can induce crosstalk in the neighboring pairs. As discussed in Chapter 3, the crosstalk can be either near-end crosstalk (NEXT)* or far-end crosstalk (FEXT). When there are multiple loop technologies in a binder, the chances of one service interfering with another are very high. For example, if T1 service exists on a loop in the same binder pair that carries HDSL, the T1 service can interfere with HDSL and vice versa. In fact, interference is possible even among two loops carrying the same type of service (e.g., HDSL)—a phenomenon known as "self interference." Spectral studies have shown that T1 services (based on AMI line code) are among the services most likely to interfere with other services such as ADSL and HDSL. This impact is shown in Figure 5.10.

Figure 5.10
Impact of interference

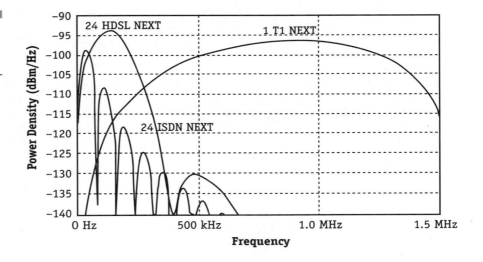

The obvious solution is to isolate the services that are the likely "offenders" and put them in a different binder group where they are least likely to cause interference. This is known as *binder group separation*.

The problem is that binder group separation works well in theory, but not in practice. The loop plant typically consists of large feeder cables near the central office, with successive cables getting smaller and smaller as they get further away—the smallest cable feeds directly into the premises. As a result, a typical loop consists of many splices along

---

* In a symmetric environment, NEXT is usually the worst source of interference that limits reach.

the way. Unfortunately, since the loop plant was originally designed to support voice services, there were no requirements to maintain the relationship of binder groups when loop cables were spliced together. In other words, binder groups that were nonadjacent in one cable section may become adjacent in the next cable section after passing a splice point. As the cable gets smaller, the likelihood of loop pairs remaining in nonadjacent binder groups decreases (see Figure 5.11).

Figure 5.11
The problem with binder group separation

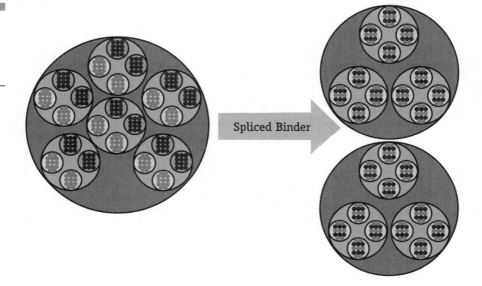

Spliced Binder

Therefore, binder group separation is almost impossible in practice. In the loop environment, it is rare to find two binder groups serving the same premises that would qualify as being truly "nonadjacent." Even if they are nonadjacent binder groups, the provider has no means of knowing because no historical records were kept.

Although spectral compatibility and binder group management is an issue for all DSL technologies (including ADSL and VDSL), it is of particular importance for HDSL/HDSL2. Since a primary use of HDSL/HDSL2 is to replace T1, it is highly likely that T1 services will exist in the same binder group as HDSL/HDSL2. And, as we've already noted, T1 is among the services most likely to interfere with other services!

Possible Solutions

The best solution is to assume binder group separation does not exist, and therefore, some interference is likely to occur. The next best option is to minimize the extent of interference. If the effects of crosstalk are not significant, it is possible for transceivers to reconstruct the signal. However, when the crosstalk is high, it is impossible for the receiver to interpret the signals correctly. Therefore, a key goal is to reduce the crosstalk to acceptable levels.

One technique to accomplish this is to define power spectral density (PSD) masks. Simply put, a PSD mask ensures that the power level of a signal transmitted at a particular frequency is low enough to minimize the possibility of energy bleeding over (and interfering with) another signal that may be transmitting at the same frequency. Therefore, the PSD mask defines acceptable power levels for the entire frequency range of the signal. The proposed power spectral density mask for an H2TU-C is shown in Figure 5.12. Notice the sharp dropoff of power beyond 400 kHz to avoid transmitting too much energy at the higher frequencies.

Another technique is to use separate frequency ranges for transmission and reception with frequency division multiplexing. FDM systems eliminate NEXT and reduce FEXT (because the origin of the FEXT signal is at the distant end of the loop). It would seem that FDM systems are the ideal choice, but the benefit comes at a significant price. The loop reach is reduced, and complex filters are needed to separate the transmit and receive frequencies (thus increasing the power and real estate of the transceiver). By contrast, systems that employ echo cancellation (EC) offer greater loop reach because they efficiently utilize the entire available spectrum. The choice of FDM versus EC systems is often an implementation decision that is left to the service provider and depends on the particular circumstances, e.g., what other services are likely to coexist in the same binder?

The best solution to manage the crosstalk issue is to research "typical" services that could be deployed in the same binder and avoid those services that introduce significant crosstalk. This can be done by conducting real-world simulations to ensure that new loop technologies such as HDSL2 are spectrally friendly to existing technologies that are likely to be in the same binder (e.g., T1, ADSL, HDSL, etc.). Figure 5.13 shows a typical test setup to simulate and analyze HDSL2 performance in the presence of HDSL disturbers.

Figure 5.12
*PSD Mask for z(t)
at H2TU-C line
interface*

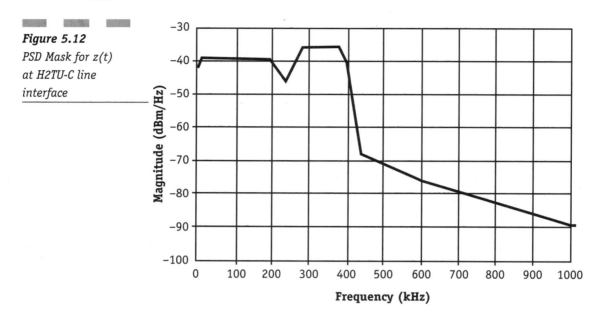

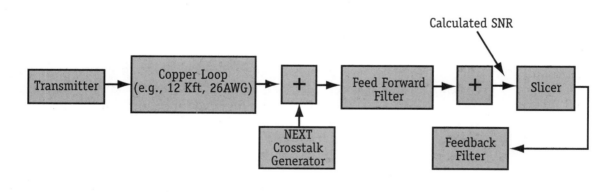

Figure 5.13 *Test setup to analyze the impact of HDSL disturbers on HDSL2*

Vendors and standards bodies have spent considerable effort simulating and analyzing the power spectral density masks of various technologies to ensure compatibility among them, as far as possible. Early research shows that:

■ The performance of T1 (based on AMI line coding) is not degraded by HDSL2 transmitting at 1.168 Kbps, 1.552 Kbps, and 2.320 Kbps any more than by other colocated T1 lines.

■ The performance of HDSL (based on 2B1Q line coding) is not degraded by HDSL2 transmitting at 1.168 Kbps, 1.552 Kbps, and 2.320 Kbps any more than it is by other colocated HDSL lines.

■ The performance of HDSL2 transmission suffers more from crosstalk "self interference" than by other colocated T1 or HDSL lines.

In summary, T1 and HDSL are spectrally compatible with HDSL2. However, as the research confirms, the problem can be mitigated, but never eliminated.

Loop Reach

Issue

As mentioned in Chapter 4 in the discussion of ADSL, loop reach is of paramount importance for new service coverage from a telco. On the other hand, the data rates for HDSL and related technologies such as ADSL, are distance-sensitive.

Possible Solutions

As in the case with ADSL, HDSL, receiver implementations that can recover the strongest signal despite attenuation loss can achieve higher data rates. Digital signal processing (DSP) techniques are often used to implement such algorithms. Other ways of increasing loop reach are to increase the number of pairs or reintroduce repeaters.* Both of these are generally considered as "avenues of last resort;" however, they are relevant in cases were coverage is otherwise impossible.

---

* Although reintroduction of mid-span repeaters may seem like taking a technological step backward, single-pair repeaters for HDSL2 are far simpler than the dual-pair repeaters for HDSL. Recent improvements in electronics design make it possible to reduce the number of repeaters required by increasing the maximum possible span between repeaters. In many cases, only one mid-span repeater may be required. Therefore, single-pair repeaters can be designed cost effectively, and they take up a lot less space—making them convenient to install. Nevertheless, they do require a line technician to go and install them.

Trials and Deployments

Beginning in 1997, competitive LECs (CLECs) in the United States began taking advantage of the Telecommunications Act of 1996 to lease unbundled copper loops from the ILECs and sell their services to businesses using HDSL (and related technologies). Of the family of related symmetric DSL technologies, SDSL was the most commonly adopted because of its advantages of single-wire implementation with T1/E1 speeds, despite the fact that encoding were proprietary. However, since the CLEC target markets were the major metropolitan areas with a high concentration of businesses, interoperability among multiple vendors was not a major concern. In the future, as the HDSL2 standard matures and CLECs expand their offerings, it is expected that most SDSL deployments will be converted to HDSL2 implementations.

The following are representative deployments (based on SDSL but expected to migrate to HDSL2):

■ In March 1998, NorthPoint Communications announced the availability of its NorthPoint DSL^{SM} service in the San Francisco Bay Area and Silicon Valley and later expanded to additional areas. To deliver its SDSL technology-based services, NorthPoint partnered with several national and regional ISPs. Their service delivers always-on bandwidth at a choice of 160 Kbps, 416 Kbps, 784 Kbps, or 1.04 Mbps, depending on user requirements and location. As the customer's needs change, service levels can be moved to faster premium services with a single phone call to a local service provider, without requiring a hardware swap or site visit. Typical applications are Internet access and multimedia use.

■ In early 1998, Covad Communications Company began offering its TeleSpeed^{SM} service using IDSL, ADSL, or SDSL to telecommuters in northern California, which was later expanded to several metro areas on both coasts. For the telecommuting application, Covad provisions an ATM PVC (permanent virtual circuit) between the teleworker's premises and an ATM switch located at its regional network facilities. The connection between the premises and the serving central office is over a copper pair leased from the local telco. The copper pair is terminated at a Covad-owned DSLAM, colocated at the telco's central office. ATM uplinks connect the DSLAM into an ATM switch at the regional facilities. On the back end of the ATM switch, connections fan out to the various corporations that have signed agreements with Covad to provide telecommuting access to their employees.

Future Directions

Although HDSL itself is a mature technology, HDSL2 is still in the standardization process. Therefore, much of the industry effort is focused on the physical layer. Typical requirements being discussed for G.sdhsl, the ITU standard for HDSL2, include:

- End-user payload symmetric bit rates of 256 Kbps, 384 Kbps, 768 Kbps, 1.536 Mbps, 2.048 Mbps, and 2.32 Mbps.
- One-way latency less than 500 microseconds for bit rates of 1.5 Mbps and above (in support of TCP/IP acknowledgments).
- Spectral compatibility with G.991.1 (G.hdsl), G.992.1 (G.dmt), G.992.2 (G.lite), ANSI T1.601 (ISDN BRI), and ANSI T1.403 (T1 carrier)
- A bit error rate of better than 10^{-7} for all test loops with the noise models and the defined noise margin.

There is also evidence that vendors are already looking ahead to improvements beyond HDSL2. The standards activities of organizations such as ANSI T1E1.4 have focused on the single-pair application of HDSL2. However, by going back to a two-pair deployment scenario, it is possible to deliver T1/E1 speeds at reaches well beyond 12 Kft. A two-pair implementation of HDSL2 could be used to provide T1/E1 line capacities to customers who cannot be served today with traditional HDSL or single-pair HDSL2. Current estimates are that the loop lengths served would be around 16 Kft.

Summary

HDSL was originally developed out of the need for "repeaterless T1" to accelerate T1 deployments. HDSL, based on the 2B1Q line coding technique, allowed improvements in the usage of the frequency spectrum over the legacy T1 line coding based on AMI. This improvement allowed HDSL to provide the same line rates as T1, but without any repeaters. However, HDSL utilized two copper wire pairs just like the traditional T1 service.

Around the mid 1990s, there was a strong push in the industry to eliminate the second wire pair for a variety of reasons. This prompted the need for an alternative to the 2B1Q line coding to increase the efficiency. Initial proprietary implementations gave way to efforts to standardize the next generation of HDSL, known as HDSL2. There was a lot of debate in the standards groups regarding the line code choice—CAP or PAM? However, the PAM proponents won the day, and this has now been settled upon as the line code scheme for HDSL2. The HDSL2 standard, however, is still in the process of standardization—initially within the United States (and concurrently in Europe), with follow-up efforts for an international standard through the ITU. However, the standardization process has not slowed down the deployment of symmetric DSL technologies, particularly by CLECs catering to the increased demand by business customers for cheap, reliable, and symmetric high-speed services. Despite the challenges (such as binder group separation), service providers have successfully demonstrated the viability of single-pair, symmetric DSL technology by deploying commercial services based on SDSL. It is expected that these deployments will evolve to HDSL2 when HDSL2 is ratified and available.

References

1. Information is available at the official Website at **www.tl.org/tlel/el4home.htm**.
2. Information is available at the official Website at **www.etsi.org**.
3. Information is available at the official Website at **www.itu.int**.

4. Im, G. H. and Werner, J. J. "Bandwidth-Efficient Digital Transmission over Unshielded Twisted-Pair Wiring," *IEEE Journal on Selected Areas in Communications*, Vol. 12, No. 9 (December 1995).

5. Tu, M. "A 512-State PAM TCM code for HDSL2," *ANSI Contribution T1E1.4/97-300* (September 1997).

6. "Working Draft for HDSL2 Standard," *ANSI Contribution T1E1.4/99-006* (January 1999).

VDSL

Background

History

The origins of very high-speed digital subscriber line (VDSL) technology are rooted in ADSL. As we mentioned in Chapter 4, Bellcore originally envisioned ADSL for video on demand (VoD) applications. VDSL was seen as the natural evolution from ADSL, but with higher asymmetric speeds than ADSL—downstream rates up to 52 Mbps and upstream rates up to 6.4 Mbps.

Starting in the early 1990s, there were two primary drivers for VDSL standards:

■ Growing demand for high-bandwidth multimedia applications with integrated voice, data, and video
■ The availability of relatively cheap fiber in the ground (as telcos and other service providers began laying enormous amounts of fiber along utility lines, railroad tracks, and similar rights of way to upgrade their backbone bandwidth). Fiber close to the premises became available in the form of fiber digital loop carriers, also known as optical network units (ONUs). ONUs became especially popular in new residential developments and multidwelling units. Developers found that it was cost effective to lay fiber in conduits along with other utilities in a developing community. On the other hand, although backbone upgrades and new developments made it possible to get fiber close to the premises, it was still expensive to run fiber to every home or apartment unit.

The fundamental premise of VDSL was to offer an alternative to the fiber-to-the-home (FTTH) architecture for both narrowband and broadband services—to provide the means to develop a "full-service" access network. Since those initial efforts in the early 1990s, several standards bodies and special interest groups have started work on VDSL. Among them are ANSI, ETSI, ITU, the ADSL Forum, the ATM Forum, DAVIC, and the FSAN Initiative.

Although ADSL and VDSL are related technologies, there are important differences between the two (other than the obvious difference in speeds):

■ **Asymmetric and symmetric services.** The original idea of VDSL was to offer asymmetric service at higher speeds than ADSL. Since then, however, industry efforts began focusing on symmetric services as well. As a result, current specifications call for both asymmetric and symmetric operation. It is expected that asymmetric services will be provided primarily for residences and symmetric services mainly for businesses.

■ **Frequency ranges.** ADSL and VDSL are complementary in the frequency bands over which they operate. ADSL frequencies start above the POTS band and top off around 1.1 MHz. VDSL frequencies begin above the ADSL band and top off around 20 MHz.

■ **Transceiver power requirements.** Since VDSL is expected to support greater power concentrations in harsh environments (typically, in ONUs located in street cabinets), the requirements for VDSL transceivers are much more stringent.

■ **Upstream multiplexing.** VDSL delivers very high bandwidth to the premises; therefore, it is reasonable to expect that the bandwidth might be shared among multiple CPEs, or even multiple residences served from a single ONU in a daisy chain—in other words, a point-to-multipoint configuration. In this case, some form of upstream multiplexing would be required to arbitrate access onto the shared medium, i.e., the high-speed local loop.

At the time of this writing, VDSL technology is still in the definition stage. Although some pre-standard products exist, and draft standard texts have been proposed, questions on line coding and upstream multiplexing protocols still remain. Standards organizations and special interest groups are working to answer these questions as quickly as possible. The final version of the international standard, G.vdsl, will be developed by the ITU.

Basic Network Architecture

Asymmetric Services

The network architecture for asymmetric services is shown in Figure 6.1. This architectural diagram, based on the full-service access network (FSAN) initiative architecture, is important as a backdrop for the discussion of the technology. A detailed discussion of the FSAN architecture is reserved for a later chapter in this book.

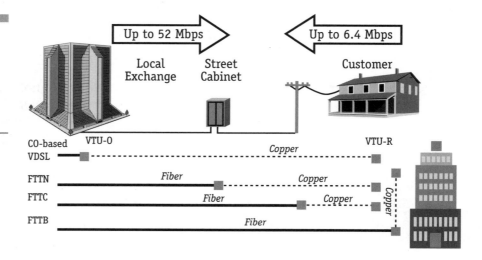

Figure 6.1
Basic network architecture for VDSL asymmetric services (Source: ETSI)

The architecture assumes that an ONU, terminating the fiber run, is close to the premises. Copper wire pairs from the ONU are then used to carry the signals over the last few Kft to every home. In rare cases where the premises is within a few Kft of the serving central office, the ONU can be eliminated altogether.

As Figure 6.1 shows, the FSAN architecture incorporates various fiber access alternatives. The major differences come from the positioning of the ONU. Except in the case where fiber extends all the way to the premises (FTTH/FTTB), copper loops are used to provide the high-bandwidth transport over the last few Kft. In most cases, consideration has been given to the use of VDSL as the primary transport technology. However, it is expected that service providers will use ADSL and VDSL in their migration strategies toward the FSAN architecture.

Symmetric Services

The network architecture for symmetric services is shown in Figure 6.2. As the figure shows, this offers a very cost-effective alternative to multiple copper pairs and inverse multiplexing equipment.

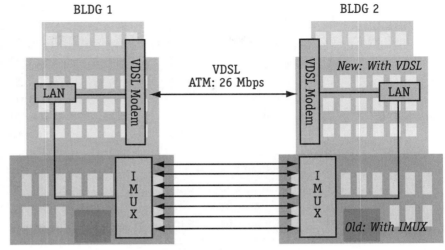

Figure 6.2
Basic network
architecture for
VDSL symmetric
services
(Source: DAVIC)

BLDG 1   BLDG 2

VDSL
ATM: 26 Mbps

New: With VDSL

LAN

Old: With IMUX

IMUX—Inverse Multiplexer

Applications

VDSL fulfills the vision of a full-service network because it can support narrowband applications and enhance certain ADSL/HDSL-based broadband applications, and can also enable new classes of multimedia applications. A few examples are summarized in Table 6.1:

Table 6.1
Full range of VDSL services

Narrowband services supported	Narrowband/broadband services enhanced	Broadband services enabled
• E-mail • Newsgroups • Chat rooms • Discussion lists	• Web surfing • Software distribution • Integrated voice and data access • Telecommuting • "Safe" Web browsing • Live multicast audio on demand (AoD) • E-commerce • Push services • Internet gaming • Remote home automation	• Live multicast VoD with multiple simultaneous video streams • High-quality Internet gaming • Multi-party videoconferencing • Campus LAN interconnections

Facilities that can benefit from VDSL systems include multi-dwelling units (hospitals, hotels, and apartment complexes) and campus environments (universities, government buildings, industrial parks, and planned community developments). A few examples are given below:

■ **Hotels.** Hotels have already been successful in offering video-on-demand services to their guests. In addition, business travelers want the convenience of high-speed Internet access from their hotel rooms. To provide these services, hotel management can install intelligent set-top boxes in every room that can be connected via existing copper pairs to ONU(s) located either in the basement of the hotel or on every floor. VDSL technology can then be used to transport the bits to and from the rooms and the ONU.

■ **Apartment complexes.** Apartment complex owners are realizing the value of adding multimedia services (including access to local content) and high-speed Internet access to their facilities in order to attract tenants and increase their revenue. In this scenario, fiber cabinets can be located in a central location such as a clubhouse or office administration building, allowing copper-pair loops to fan out to every apartment. As in the hotel scenario, intelligent set-top boxes can be located in each apartment to provide integrated voice, data, and video services over VDSL.

■ **Universities.** Universities can benefit from an infrastructure where students can use interactive video to communicate with their teachers from the dorms or with specialized training professionals in remote locations. They can also check out instructional videos from a centralized electronic library. In this scenario, fiber cabinets can be located close to the dorms and university buildings, with copper pairs that terminate in each room and classroom. Fiber can be used to aggregate the traffic onto a high-speed backbone.

■ **Government buildings.** Government buildings located close to each other can be interconnected using symmetric DSL services rather than by the traditional method of using multiple wire pairs and HDSL/HDSL2 with inverse multiplexers to aggregate the bandwidth. In this case, a pair of VDSL modems operating at symmetric speeds can eliminate the need for multiple wire pairs between buildings and several inverse multiplexers, thereby resulting in considerable cost savings.

▌ **Industrial parks.** Industrial parks close to a CO can be served by what is sometimes called "CO-based VDSL." Symmetric VDSL can be used to service industrial parks with a cost-effective alternative to multiple T1/E1 or T3/E3 connections.

Standards and Special-Interest Group Activities

ANSI, ETSI, and ITU

As with other DSL-related efforts, the ANSI, ETSI, and ITU standards bodies have been active in the development of VDSL standards.

ANSI study group T1E1.4[1] began the VDSL study project in March 1995. The efforts of this group resulted in the development of the VDSL System Requirements Document, which covers aspects such as reference configuration and description, data rates, interworking and service requirements, transmission and impairments, and operations, administration, management and provisioning (OAM&P).

In Europe, the ETSI TM6[2] began studying VDSL requirements in 1996 and adopted a VDSL standards project under the title High-Speed (metallic) Access Systems (HSAS). They compiled a list of objectives, problems, and requirements that have been published in a Functional Requirements Document. Among the ETSI study group's preliminary findings were needs for an active network termination (NT) and payloads in multiples of 2.3 Mbps.

The work of the ANSI and ETSI study groups was taken up by the ITU Study Group 15[3] to develop work plans toward the internationalization of the VDSL standard around G.vdsl. The following important VDSL requirements have been identified by these standards organizations:

▌ Robustness in the presence of common impairments (e.g., bridged taps, crosstalk, impulse noise, and radio-frequency ingress).
▌ Spectral compatibility with existing DSLs.
▌ Low power consumption: less than 1.5 W per modem. This requirement is driven by the fact that the most likely deployment scenario for VDSL modems will be in ONU equipment, similar to DLC deployments for ADSL. ONU cabinets are subject to limited space, harsh environmental conditions, and difficulty (or impossibility) of

implementing forced air cooling. Therefore, the power budget is extremely limited.

■ Capable of restricting the transmit power spectral density (PSD) in the amateur radio bands to –80 dBm/Hz to avoid interference.

ADSL Forum

One of the working groups within the ADSL Forum is the VDSL study group[4], which was formed in late 1997. This group reviews the progress in standards bodies and writes technical reports on upper-layer issues, such as:

■ Layer 2 and 3 protocol transport (e.g., PPP over ATM)
■ Customer interfaces (e.g., ATM25)
■ Customer wiring /CPE configurations
■ Network interfaces
■ OAM&P.

The ATM Forum

The ATM Forum[5] has defined a 51.84 Mbps interface for private network UNIs and a corresponding transmission technology. It is also reviewing architectures, distribution, and delivery of ATM all the way to the premises over transports such as ADSL and VDSL.

DAVIC

The Digital Audio-Visual Council (DAVIC)[6] was one of the first organizations to take a position on VDSL starting around December 1995. DAVIC is a non-profit association established in 1994, with a membership of over 157 companies from more than 25 countries representing multiple sectors of the audio-visual industry—manufacturing (computer, consumer electronics, and telecommunications equipment), service (broadcasting, telecommunications, and CATV), government agencies, and research organizations.

DAVIC has published specifications to create an industry standard for end-to-end interoperability of broadcast and interactive digital audiovisual information and multimedia communication. The specification consists of multiple parts, including definition of a delivery system architecture that encompasses VDSL:

- Part 1: Description of Digital Audio-Visual Functionality
- Part 2: System Reference Models and Scenarios
- Part 3: Service Provider System Architecture
- Part 4: Delivery System Architecture and Interfaces
- Part 5: Service Consumer System Architecture
- Part 6: Management Architecture and Protocols
- Part 7: High- and Mid-Layer Protocols
- Part 8: Lower-Layer Protocols and Physical Interfaces
- Part 9: Information Representation
- Part 10: Basic Security Tools
- Part 11: Usage Information Protocols
- Part 12: System Dynamics, Scenarios, and Protocol Requirements
- Part 13: Conformance and Interoperability
- Part 14: Contours: Technology Domain

Full-Service Access Network (FSAN) Initiative

In July 1995, several telcos and leading network providers organized themselves to expedite the introduction of full-service networks. This organization, known as the Full Service Access Network (FSAN) initiative[7], is a group of major telcos representing several million access lines worldwide working together with their strategic equipment suppliers to agree upon a common access system for the provision of broadband and narrowband services. This common access system is documented in an FSAN requirements specification shared with relevant standardization bodies and industry groups such as ANSI, ETSI, ITU, IETF, DAVIC, and the ATM and ADSL Forums.

The key motivation for the FSAN initiative was that enough similarities existed among participating telcos' requirements for future access networks to suggest that significant benefits could be achieved by adopting a common set of specifications. The volume of access lines represented by the telcos in the FSAN membership offered the opportunity to dramatically cut the cost of deploying new technologies and the full range of integrated services.

FSAN activities have progressed in four phases:

- **Phase 1 (July 1995 to June 1996).** During this phase, FSAN members identified the technical and economic barriers to broadband access network introduction. An ATM passive optical network (APON) was identified as the most promising approach to achieving large-scale, full-service access network deployment that could meet the evolving service needs of network users. It was shown that this APON approach could support a wide range of "FTTx" (fiber-to-the-building, -cabinet, -curb, and -home) access network architectures.

■ **Phase 2 (July 1996 to February 1997).** The work effort during this phase concentrated on devising a common set of requirement specifications. As consensus was achieved within the working teams, the results were assembled into a consistent set of requirement specifications and presented in March 1997 at the VIII International Workshop on Optical/Hybrid Access Networks.

■ **Phase 3 (March 1997 to December 1998).** During this phase, it was recognized that further work was necessary to achieve the required level of detail in three different areas: optical access network, customer network, and OAM&P.

■ **Phase 4 (The current phase at the time of this writing).** The focus of this phase is to complete the specification and drive toward field trials.

VDSL Alliance and VDSL Coalition

The VDSL Alliance and VDSL Coalition are special-interest industry groups that have organized around promoting the two key choices for VDSL line codes:

■ The VDSL Alliance[8] is composed of a number of companies working together to develop a specification based on discrete multi-tone (DMT).

■ The VDSL Coalition[9] is composed of a number of companies working together to develop a specification based on CAP/QAM.

Both groups have developed draft specifications for review by the relevant standards organizations.

Technology Discussion

As with other DSL technologies, VDSL has upstream and downstream channels. These channels occupy the frequency spectrum well above the voiceband and ISDN to allow service providers to overlay VDSL on top of existing POTS or ISDN service. Like ADSL, VDSL offers two paths to match the propagation delay and impulse noise protection requirements of various applications:

1. A slow path that is robust in the presence of impulse noise that lasts up to 500 (s with a one-way propagation delay of less than 20 ms
2. A fast path that is sensitive to impulse noise with a one-way propagation delay of less than 1.2 ms.

If both paths are established, the link operates in dual-latency mode; otherwise, it operates in single-latency mode (either slow or fast as established during the startup sequence). It is generally expected that asymmetric services will be delivered over ATM transport, and symmetric services will be delivered over either ATM or STM transport.

The method in which the upstream and downstream channels are modulated is the subject of line coding. Like any other evolving DSL technology, VDSL is not exempt from the line-code debates. Another area of debate in standards development has to do with the upstream multiplexing scheme to support point-to-multipoint configurations.

Modulation Schemes

The options for VDSL line codes can be divided into single-carrier and multi-carrier modulation schemes. Originally, four different line codes were proposed for VDSL (two single-carrier and two multi-carrier):

- **Carrierless amplitude phase (CAP) modulation.** This single-carrier line code is the same as described in Chapter 4 on ADSL and ADSL lite.
- **Simple line code (SLC).** This single-carrier line code is a version of four-level baseband signaling that filters the baseband and restores it at the receiver.

■ **Discrete multi-tone (DMT).** This multi-carrier line code is the same as described in Chapter 4 on ADSL and ADSL lite.

■ **Discrete wavelet multi-tone (DWMT).** This multi-carrier line code uses wavelet transforms to create and demodulate individual carriers.

Of these, SLC and DWMT have been ruled out. Today, CAP/QAM has been proposed as the single-carrier line coding representative, whereas DMT is representative of multi-carrier line codes.

Upstream Multiplexing

There are two choices for network termination—active or passive. These options are shown in Figure 6.3.

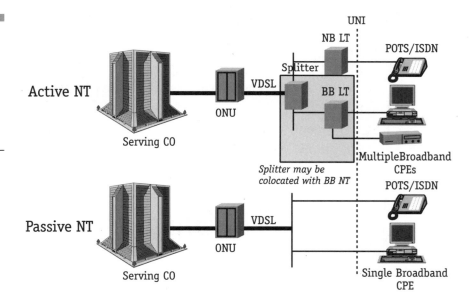

Figure 6.3

Options for upstream multiplexing with active and passive NTs (Source: ETSI)

If an active network termination is present, the mechanism to multiplex upstream cells or data channels becomes the responsibility of the premises network. This can be achieved by a star configuration where each CPE is connected to a switching or multiplexing hub that is typically incorporated into the VDSL unit, and is responsible for the arbitration of access to the shared upstream channel.

If a passive network termination is present, the mechanism to multiplex upstream cells or data channels must be built in. A collision detection mechanism similar to Ethernet might be used to resolve the issue that the upstream channels for each CPE must share a common wire. However, this strategy fails because it does not guarantee bandwidth. Therefore, the two choices for upstream multiplexing with passive NTs are:

- **Time division multiplexing.** This involves invoking a cell-grant protocol in which the downstream frames generated at the ONU or further up the network contain a few bits that grant access to a specific CPE during a specified period subsequent to receiving a frame. A granted CPE can then send one upstream cell during this period. Upon receiving the grant, the CPE transmitter must turn on, send a preamble to condition the ONU receiver, send the cell, and finally turn itself off. The disadvantage to this technique is that the ONU may grant access to a CPE even when that CPE does not have anything to send.
- **Frequency division multiplexing.** This involves dividing the upstream channel into frequency bands and assigning one band to each CPE. This method avoids any media access control with its associated overhead, although a multiplexer must be built into the ONU. The disadvantage is that it either (a) restricts the data rate available to any one CPE or (b) imposes a dynamic inverse multiplexing scheme that lets one CPE send more than its share for a period.

In addition, to support a full-duplex operation, both upstream and downstream channels must be carried simultaneously. As described in Chapter 3, there are two techniques for full duplex operation—frequency division multiplexing (FDM) and echo cancellation (EC). In FDM, the upstream and downstream channel frequencies are separated, whereas in EC the two channel frequencies overlap. Figure 6.4 shows an example of the FDM and EC techniques as they apply to DMT.

Echo cancellation is impractical, however, because (a) self-NEXT increases with frequency, so it becomes almost impossible to implement at the high frequencies over which VDSL operates; and (b) symmetric services are not feasible because the required overlapping bandwidth is too high. Therefore, two approaches have emerged to support the goals of arbitrating upstream usage and full-duplex operation:

Figure 6.4

Options for upstream/ downstream duplexing CDMT example

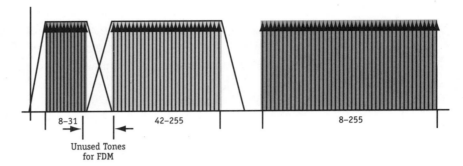

Frequency Division Multiplexing (FDM) with Separate Upstream and Downstream Channels

Echo Cancellation (EC) with Overlapped Upstream and Downstream Channels

■ **Time division duplexing (TDD).** With TDD, subchannels in the frequency spectrum are used for either upstream or downstream transmission. In other words, a subchannel is capable of supporting either direction of transmission; but it does not *simultaneously* carry upstream and downstream transmission. Another name for TDD is "ping-pong" transmission.

■ **Frequency division duplexing (FDD).** With FDD, subchannels in the frequency spectrum are exclusively used for upstream or downstream. The subchannels are separated by guardbands.

VDSL Proposals

Given the choices for line coding and upstream channel multiplexing, it is no wonder that segments of the industry have organized to propose choices. Two special-interest groups have proposed draft standards texts:

■ The VDSL Alliance proposed DMT line coding and time division duplexing.

■ The VDSL Coalition proposed CAP/QAM line coding and frequency division duplexing.

At the time of this writing, neither proposal has gained definitive approval by the standards bodies. The issues are still being discussed, with final approval expected sometime in 1999. In the following section, the details of each proposal are presented.

VDSL Alliance Proposal for DMT Line Coding and Time Division Duplexing (TDD)

This proposal is also known as synchronized DMT or SDMT[10]. In this proposal, all lines within the same binder group are loop timed, i.e., they are locked onto a common network clock. This means that the signal transmissions of all lines within the binder are synchronized such that they "ping" or "pong" at the same time, which avoids NEXT. Another feature of the proposal is that it allows the same frequency band (i.e., a DMT bin or subchannel) to be used for either upstream or downstream transmissions during different time periods.

The mechanism to arbitrate upstream channel usage is implemented through the use of grant bits in superframes (a superframe is nothing but a collection of consecutive DMT symbols, each carrying bits). In the VDSL Alliance proposal, a VDSL superframe consists of 20 DMT symbols, with each DMT symbol transmitted in 25 µs. Therefore, the total amount of transmission time for a superframe is 500 µs. Figure 6.5[11] shows how asymmetric and symmetric transmission is achieved.

■ For asymmetric transmission in the ratio of 8:1, 16 downstream symbols are transmitted first, followed by a single guardtime symbol, then by two upstream symbols, and finally by one more guardtime symbol. The guard times are necessary to account for channel propagation delay and echo response time.

■ For symmetric transmission in the ratio of 1:1, nine downstream symbols are transmitted first, followed by a single guardtime symbol, then by another nine upstream symbols, and finally one more guardtime symbol.

The VDSL Alliance has indicated the following benefits in this (SDMT) proposal:

Benefits of DMT Line Coding

■ Robustness to line impairments such as RFI. DMT monitors the line conditions and computes the bit-carrying capacity of each subchannel based on its S/N ratio. If a subchannel experiences RFI, then it may not be used at all in favor of other subchannels.

Figure 6.5
SDMT superframes
for asymmetric
and symmetric
transmission
(Source: DAVIC)

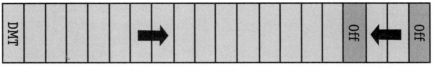

Asymmetric Mode 16:1:2:1

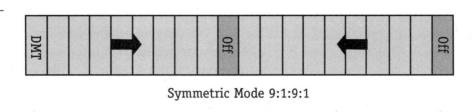

Symmetric Mode 9:1:9:1

20 DMT Symbols = 500 μs, Each DMET Symbol is 25 μs

- Impulse mitigation. DMT spreads the impulse energy over all frequencies (although not quite uniformly), and an impulse peak energy reduction of around 10 dB is the result.
- Bit swapping. With this technique, the same line rate can be maintained in the presence of changing line conditions. When a receiver detects degradation in one or more subchannels, it computes a modified bit distribution among the subchannels that maintain the line rate. This change is communicated back to the transmitter, which implements the new bit loading.
- Spectral compatibility with ADSL can be achieved by disabling subchannels below 1.104 MHz. When ADSL service is not present in the same binder, subchannels below 1.104 MHz can be utilized for higher data rates.
- Spectral compatibility with POTS, ISDN, and HDSL
- Interference with amateur radio can be avoided through the use of a spectral mask.

Benefits of TDD

- Flexibility in the ratio of upstream and downstream data rates, supporting both asymmetric and symmetric services
- Reduced complexity as result of the fact that, in TDD mode, a modem can either transmit or receive at any particular time, imply-

ing a single Fourier Transform (FT) at any given time. Therefore, the hardware requirements are minimized to computing only a single FT.

VDSL Coalition Proposal for CAP/QAM Line Coding and Frequency Division Duplexing (FDD)

In this proposal[11], there are four possible carriers denoted as 1D, 2D, 1U, and 2U—two for downstream transmission and two for upstream. These carriers are transmitted in separate frequency bands, as shown in Figure 6.6. A particular band allocation is defined by (a) the number of applied carriers, and (b) the values of the carrier band-separating frequencies f_1–f_8. The boundary values are f_1=0.12 MHz and f_8=20 MHz. Both the number of applied carriers and the values of f_1–f_8 depend on a transmission profile. If the transmission profiles are changed, a variety of services can be supported, i.e., asymmetric/symmetric and low rate/high rate.

Figure 6.6

FDD proposal for asymmetric and symmetric transmission (Source: BT Labs)

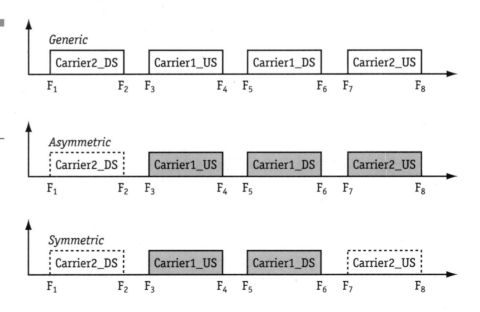

Examples of spectral allocations of standard asymmetric and symmetric transmission profiles are shown in Tables 6.2 and 6.3, respectively[11].

Table 6.2

Example of spectral allocation of standard asymmetric transmission profiles (Source: ANSI Contribution T1E1.4/98-045R1, June 1998)

Profile	Carrier	Carrier frequency f_c, (MHz)	Lowest frequency f_{LOW}, (MHz)		Highest frequency f_{HIGH}, (MHz)		Maximum PSD (dBm/HZ)
AS-04NAG-A	2D	10.1925	2.42	F_5	17.97	F_6	-60
AS-14NAG-A	1U	1.4850	0.84	F_3	2.13	F_4	-51
AS-24NAG-A							
AM-04NAG-A	2D	5.9775	2.10	F_5	9.88	F_6	-57
AM-14NAG-A	1U	1.2825	0.815	F_3	1.79	F_4	-51
AM-24NAG-A							
AL-04NAG-A	2D	4.0500	2.11	F_5	5.998	F_6	-54
AL-14NAG-A	1U	1.2825	0.815	F_3	1.79	F_4	-51
AL-24NAG-A							

Table 6.3

Example of spectral allocation of standard symmetric transmission profiles (Source: ANSI Contribution T1E1.4/98-045R1, June 1998)

Profile	Carrier	Carrier frequency f_c, (MHz)	Lowest frequency f_{LOW}, (MHz)		Highest frequency f_{HIGH}, (MHz)		Maximum PSD (dBm/HZ)
SS-02NAG-A	2D	6.2775	2.40	F_5	10.17	F_6	-57
SS-12NAG-A	1U	1.4850	0.84	F_3	2.13	F_4	-51
	2U	14.9175	11.03	F_7	18.81	F_8	-60
SM-02NAG-A	2D	4.3875	2.44	F_5	6.33	F_6	-54
SM-12NAG-A	1U	1.4850	0.84	F_3	2.13	F_4	-51
	2U	12.2850	10.3	F_7	14.229	F_8	-57
SM-24NAG-A	2D	4.3875	2.44	F_5	6.33	F_6	-54
	1U	1.4850	0.84	F_3	2.13	F_4	-51

The VDSL Coalition has indicated the following benefits in this proposal:

Benefits of CAP/QAM Line Coding

■ Well-understood, mature, and cost-effective technology
■ RFI that can be mitigated through the use of adaptive notched filters
■ Spectral compatibility with ADSL achieved by placing VDSL in upstream and downstream transmit spectra above 1.1 MHz. As an option, a VDSL PSD boost in the upstream direction could be used

to combat ADSL downstream power leakage above 1.1 MHz. When ADSL service is not present in the same binder, frequencies above 300 kHz can be utilized for higher data rates.

- Spectral compatibility with POTS, ISDN, and HDSL using flexible band allocation
- Interference with amateur radio can be avoided through the use of notched filters
- Full interoperability between CAP/QAM transceivers (since the two line coding techniques are so closely related). The receiver can automatically determine the line code during initialization without requiring a special training sequence[12].

Benefits of FDD

- Inherently avoids self-NEXT
- No need for central synchronization, which could be problematic for certain competitive deployment scenarios where multiple parties share the same binder group
- Can easily mix different services, i.e., symmetric/asymmetric and high rate/low rate.

Other Options

In addition to the above proposals, there is one more possibility—DMT line coding with FDD (the choice of CAP/QAM line coding with TDD has been ruled out). This is the basic idea behind a proposal known as Zipper[13], as shown in Figure 6.7. Zipper has the following features:

- DMT line coding extended to 2,048 subchannels
- Filterless FDD VDSL transceiver with maximum telco flexibility
- Support for any frequency plan from dc to 11 MHz, with a granularity of 5 kHz
- Each DMT subchannel is set and controlled by the telco in terms of the direction of transmission (upstream or downstream) and transmit power (including power boost or "shutoff").

This increased flexibility provides for excellent spectral compatibility. However, the downside is that it obviously requires synchronization between the sending and receiving modems and, ideally, between all lines in the same binder. It also requires extra computational capability and precision in the transceiver. For this reason, Zipper has been proposed as a future upgrade from SDMT systems.

Figure 6.7
Zipper
(Source: VDSL)

ANSI Requirements Documents

The ANSI and ETSI requirements documents form the basis for internationalization efforts of VDSL technology by the ITU. In particular, the ANSI document[14] provides detailed requirement specifications for the following topics:

- Reference configuration and description
- Data rates, interworkings, and service requirements
- Transmission and impairments
- Operations, administration, maintenance, and provisioning (OAM&P).

The remainder of this technology discussion is concerned with the ANSI requirements document.

VDSL Reference Model

The reference models are shown in Figures 6.8 and 6.9. In the models, there are eight interfaces for specification, as shown by the eight vertical lines. Figure 6.9 illustrates the generic interfaces for the copper access section.

Referring to the diagrams, the γ-O interface is between the application and the transport protocol specific (TPS) processing at the service provider end of the line. The TPS-TC (transmission convergence) processing is outside the scope of the ANSI requirements document. The TPS-TC interface to the physical media-specific (PMS) TPS is the α interface at the service provider side. The α interface is a logical and application independent interface. It is sometimes called "HAPI" for hypothetical application independent interface. At the α interface, provision is made for both "fast" and "slow" delay paths in both directions.* An additional logical interface between the PMS-TC and the PMD core modem is known as the δ-O interface.

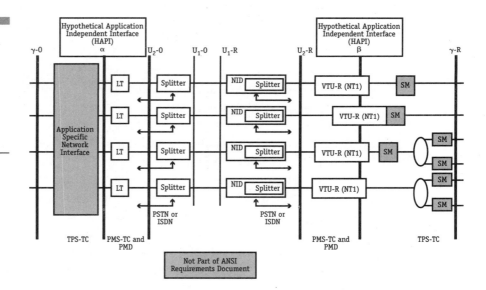

Figure 6.8

VDSL reference model (Source: ANSI Contribution T1E1.4/98-265, August 1998)

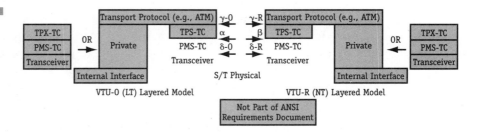

Figure 6.9

Generic reference model with hypothetical interfaces (Source: ANSI Contribution T1E1.4/98-265, August 1998)

---

* High-latency channels are designated "S" (for slow) and low-latency channels are designated "F" channels (for fast). However, the F channels may in fact have modest interleaving if the application definition of low latency allows. Although provision is made for dual-latency paths in both directions (i.e., at both the VTU-O and VTU-R), support for a single-latency path is required.

The γ-R interface is between the application and the TPS-TC processing at the customer end of the line. The application implementation is called a service module (SM) at the customer side of the line. This TPS-TC (transmission convergence) processing is also outside the scope of the ANSI requirements document. The TPS-TC interface to the physical media specific (PMS) TPS is the β interface at the customer side of the line. As with the a interface, provision is made at the β interface for both "fast" and "slow" delay paths in both directions. An additional logical interface between the PMS-TC and the PMD core modem is known as the δ-O interface.

In the reference models, the VDSL modem at the ONU is called the VDSL transmission unit–optical or VTU-O. The VDSL modem at the premises is called the VDSL transmission unit–remote or VTU-R.

The VTU-O (sometimes known as line termination or LT) converts digital data to and from the continuous-time physical-layer VDSL signals at the U_2-O interface. The VTU-R (sometimes known as network termination 1 or NT1) converts digital data to or from the continuous-time physical layer VDSL signals at the U_2-R interface. It performs the termination of the VDSL modulation scheme, link control and maintenance functions, and provides a logical β interface to the customer's equipment. The reference model does not imply specific ownership of the NT1 equipment by customer or network operator. The NT2 converts between NT1 signals and those of the customer-premises network.

The service splitters separate VDSL signals from other services, which can be either PSTN/POTS signals or ISDN signals. Since PSTN/POTS or ISDN signals can occupy the same physical media as the VDSL signal through the use of the service splitters, the U_1 reference point refers to copper-pair media carrying composite signals, while the U_2 reference point specifies the VDSL modem ports.

The access termination point (ATP) network interface device (NID) specifies the protection and distribution cable termination. The service module (SM) is an applications device that accepts the VDSL bitstream and implements the application. Examples of SMs are set-top box interfaces and personal computer interfaces.

At both the α and β interfaces, one or two possibly asymmetric bidirectional channels are supported. Associated with each data flow is implicit or explicit byte synchronization, which is maintained across the VDSL link. The modem provides the master clock for the downstream channels, which may be expressed as bit or byte frequency. Clock-rate adaptation is the responsibility of the application-dependent TPS-TC layer (e.g., by idle-cell insertion/deletion in the case of ATM). Since these are logical interfaces, data may in fact be transferred in any format, including bits and bytes at a constant rate, but may also be bursty when related to a clock signal. However, the average rate is assumed to be constant, and the depth of any buffering required is implementation dependent. Both the α and β interface flows are determined by OAM parameters at the VTU-O.

VDSL Data Rates

Tables 6.4 and 6.5 show the ANSI requirements for asymmetric and symmetric data rates, respectively.

Table 6.4

ANSI requirements for VDSL asymmetric data rates (Source: ANSI Contribution T1E1.4/98-043R3, June 1998)

Loop Length	Downstream	Upstream	Range Kft (km)
Short	52 Mbps	6.4 Mbps	1 (.3)
	34/38.2 Mbps	4.3 Mbps	
Medium	26 Mbps	3.2 Mbps	3 (1)
	19 Mbps	2.3 Mbps	
Long	13 Mbps	1.6 Mbps	4.5 (1.5)
	6.5 Mbps	0.8/1.6 Mbps	6 (2)

Table 6.5

ANSI requirements for VDSL symmetric data rates (Source: ANSI Contribution T1E1.4/98-043R3, June 1998)

Loop Length	Downstream	Upstream	Range Kft (km)
Short	34 Mbps	34 Mbps	
	26 Mbps	26 Mbps	1 (.3)
Medium	19 Mbps	19 Mbps	
	13 Mbps	13 Mbps	3 (1)
Long	6.5 Mbps	6.5 Mbps	4.5 (1.5)
	4.3 Mbps	4.3 Mbps	
	2.3 Mbps	2.3 Mbps	

VDSL Activation and States

The following activation requirements are defined in the ANSI document:

- **Cold-start process.** This process starts when power is first applied to the transceiver after intrusive maintenance, or if there has been significant change in the line characteristics (e.g., due to thermal effects). Intrusive maintenance can also apply when the transmission parameters (e.g., data rate, noise margin, spectral masks, and class of service) are altered. The requirement is that the cold-start process should be complete in less than 10 seconds, and if not, an indication must be provided by the VTU-O to the operator of the inability to start.

- **Normal-start process.** This process applies when either the VTU-O or the VTU-R activates from a power-down state. Power down is reached when a transceiver has its AC removed on purpose (typically when forced by the customer). This process applies only if there has been little or no change in the line characteristics. It also applies when there is accidental AC removal or failure at the customer premises, provided the transceiver can store all the necessary data and parameters to avoid the cold-start process. The requirement is that the normal-start process should be complete within five seconds (typically, two seconds), and if it is not, the cold-start process should begin.

- **Warm-start process.** This process applies to transceivers when they have reached synchronization and steady-state transmission and have subsequently responded to a deactivation request, which means entry is through the idle state. This is the usual method of activating the VDSL transmission system upon receipt of a first incoming or outgoing broadband call request. This process can be initiated after both the VTU-O and VTU-R transceivers are in power-saving mode. The requirement is that the warm-start process should be complete within 100 ms; if it is not, either the normal-start process and/or the resume-on-error process should begin if synchronization loss has occurred.

- **Resume-on-error process.** This process applies to transceivers that lose synchronization during transmission, e.g., due to a large impulse hit or an interruption longer than a "micro-interruption" of 10 ms. This applies only if there has been no change in line characteristics, and when the clock-frequencies recovery circuits can still predict sample timing. The requirement is that the resume-on-error process should complete within 300 milliseconds, and if it does not, the normal start process should begin.

ANSI further specifies five mandatory and one optional state in the requirements document:

- **Steady-state transmission state.** This is entered through successful completion of any of the cold-start, normal-start, warm-start, or resume-on-error processes. This means that full clock and frame synchronization has been achieved, and filter adaptations have been performed. Steady-state transmission may be exited upon power loss or entering the idle state.

- **Idle state.** A VDSL transceiver may enter this state to save ONU power and reduce unwanted RF emissions. Before it enters this state, there must be confirmation to the UNI and the network that the VDSL transmission has terminated. This state is exited when (a) a call request occurs through the warm-resume process, (b) synchronization loss is detected, or (c) power loss is detected.

- **Power-down state.** This is the state when there is full removal of power.

- **Power-off state.** This is the state prior to installation or first application of power.

- **Dynamic power savings state.** This optional state is intended to reduce the overall power consumption of the VTU-O transceiver and to reduce the crosstalk level and RFI radiation of the VDSL system. It could be used when ATM or other application links are active, but not consuming the full VDSL bandwidth. It alternates with the steady-state transmission state. No loss of application data should be tolerated when the VDSL transceiver moves between steady state and dynamic power-savings states. This state implies "hot resume."

Current State of the Technology

Deployment Issues and Possible Solutions

At the time of this writing, VDSL is still in definition. However, the following are some important deployment issues currently under investigation, even as the standard is being developed.

Spectral Compatibility and Binder Group Management

This issue has been discussed in detail in the chapter on HDSL and HDSL2 (Chapter 5). For VDSL, the impact of crosstalk in a binder group is quite critical because of the frequencies over which it operates, and the fact that there could be other high-frequency VDSL transmissions in the same binder (i.e., self-crosstalk). Studies have shown that both NEXT and FEXT increase with frequency. This issue has provoked the most discussion in the industry on the choice of line coding and upstream multiplexing.

Interference with Amateur Radio

Issue

VDSL invades the frequency ranges of amateur radio, and every above-ground telephone wire is an antenna that both radiates and attracts energy in amateur radio bands. As Table 6.6 shows, amateur radio bands overlap the transmission band of VDSL (which utilizes up to about 20 MHz).

Table 6.6

Amateur radio bands

Low (MHz)	High (MHz)
1.81	2.00
3.5	4.0
7.0	7.1
10.1	10.15
14.0	14.35
18.068	18.168
21.0	21.45
24.89	24.99
28.0	29.7

Possible Solutions

As discussed in Chapter 5 on HDSL, PSD masks are an effective way to minimize the impact of interference with any other transmission. This is the rationale behind the standards organizations requirement that the transmit PSD in the amateur radio bands be restricted to –80 dBm/Hz[14].

Interference with Home Networking

Issue

One of the options for home networking is to use the residential phone wiring that is already in place—an option proposed by the HomePNA. Unfortunately, this can cause problems because:

- The proposed frequency spectrums of VDSL and HomePNA technologies significantly overlap each other.
- Situations are likely where VDSL and HomePNA output energy onto adjacent wire pairs within the same binder group that serves a neighborhood.

As shown in Figure 6.10, a binder group fans out from an ONU in the neighborhood. Consider the situation where one wire pair from the binder terminates in a residence served through VDSL, and a second wire pair from the same binder terminates in a neighboring residence without VDSL service, but with HomePNA (i.e. the second residence happens to use a phone wire-based network for local networking). The second residence would not use a splitter device, since the telco service is still POTS and the data networking is completely "local" to the premises. However, the energy put out on the premises phone network can also bleed over onto the copper loop in the absence of an isolation function such as a splitter. Since the HomePNA frequencies are well above the ADSL spectrum, this is not a problem for ADSL, but it does present a problem for VDSL because of crosstalk:

- HomePNA interferes with VDSL as it bleeds over to the VDSL transmission in an adjacent pair. NEXT affects the downstream rate, and FEXT affects the upstream rate (although its impact is not as severe). This results in reduced capacity.
- VDSL in nearby homes interferes with HomePNA—depending upon the situation, the worst case can affect the whole network.

Figure 6.10
VDSL and HomePNA

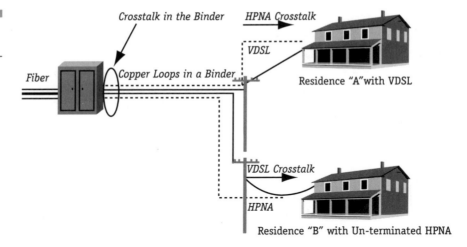

As we mentioned earlier, this issue is specific to VDSL. However, it could be a real problem for VDSL deployment given the likelihood of several million residences running HomePNA technology in the time-frame when VDSL becomes widely deployed.

Possible Solutions

One possible solution is to include splitters or other devices that provide isolation. However, this is not very practical, especially when "pure" HomePNA deployments (i.e., those residences that only want to do local networking over phone lines, while maintaining traditional POTS) are not required to install a splitter. The only viable long-term solution is to ensure spectral compatibility between the two technologies through a combination of spectrum allocation and PSD masks. Standards organizations are beginning to address the issue from this perspective.

A major difference between this issue and the previous one (interference with amateur radio) is that the amateur radio issue was understood by the VDSL standards development bodies well enough in advance to incorporate it into their requirements. However, HomePNA is relatively new, and the appropriate solution is still being worked through.

Active versus Passive Network Termination (NT) Devices

This issue was discussed in detail in Chapter 4 on ADSL and ADSL lite. For VDSL, the choice of active versus passive NTs has an impact on CPE architectures that support a full-service access network.

Future Directions

■ **Premises architectures.** Although it is not directly related to VDSL, the advent of ADSL/VDSL technologies has prompted a lot of industry activity around the notion of extending the high-speed access provided by ADSL/VDSL through a home distribution network. The basic idea is to share the high-speed, "always-on" connection among several networked information appliances (PCs, TVs, alarm systems, lighting controls, and so on). Unfortunately, just as an assortment of broadband technologies exists, several home networking technologies have also begun to emerge based on different transmission media—telephone wire, power line, and wireless. The "residential gateway" concept was developed by a group of service providers and equipment manufacturers* to provide access with a device that hides the complexity of both the access technology and the home distribution network technology from the customer[15]. Figure 6.11 illustrates this concept.

A flexible residential gateway design would allow a technician (or the consumer) to use software independently to upgrade or swap the home networking technology, just as it would allow a service provider to upgrade or swap the broadband access technology. Therefore, the concept has now gained value as performing both a physical isolation function (between service provider and customer wiring) and a logical isolation function (between the service provider's network and the customer's home network). Implied in this "improved" concept is the notion that the residential gateway will have routing and firewall capabilities. Gateway architectures are discussed in greater detail later in this book.

---

* This group has subsequently disbanded, but efforts concerned with the residential gateway concept are continuing. The Open Service Gateway Initiative is another industry group (**www.osgi.org**) that has taken on the charter of residential gateways.

■ **Operations, administration, maintenance, and provisioning (OAM&P).** As in the case of ADSL, OAM&P and the impact of VDSL on core network architectures are important topics for discussion and are ongoing areas of focus for the ADSL Forum.

Figure 6.11
Residential gateway concept

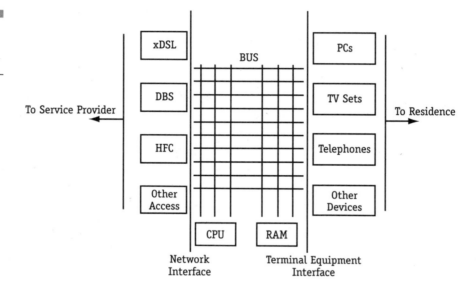

Summary

Around the world, deregulation of the telecommunications industry has resulted in competition among telephone carriers, cable TV service providers, utilities, and others to provide a "full-service" access network to the consumer—one that cost effectively integrates voice, data, and video. Cable companies have proposed the hybrid fiber/coax (HFC) architecture as the basis for such an access network. In the race to compete, telephone companies have looked toward DSL technologies to provide a comprehensive and cost-effective answer. For telcos, an attractive topology is the combination of using fiber cable to feed neighborhood ONUs and existing or new copper to connect the "last mile." This topology, called fiber-to-the-neighborhood, encompasses fiber-to-the-curb with short drops and fiber-to-the-basement, serving tall buildings with vertical drops. The enabling technology for FTTN is VDSL, which can transmit very high data rates over short reaches of twisted-pair copper telephone lines; however, the actual data rate depends on loop length.

VDSL is a natural complement to and the logical upgrade path from ADSL. As loop lengths shrink (either from natural proximity to a CO or from increased deployment of fiber-based access nodes), VDSL enables more channels and capacity for services that require high data rates. In other words, not only does VDSL support and enhance existing narrowband and broadband services; it enables new services that offer rich multimedia content.

While many companies and people in the industry share this vision, it is important to note that, in contrast to other DSL technologies, VDSL is still in the formative stages of standards development. Therefore, questions about issues such as line coding and upstream multiplexing protocols remain. Standards organizations such as the ITU and special-interest groups such as the ADSL Forum are working to answer these questions as quickly as possible to drive consensus toward the final version of the physical-layer standard, known as G.vdsl, and for access architectures around G.vdsl.

References

1. Information is available at the official Website at **www.tl.org/tlel/el4home.htm**.
2. Information is available at the official Website at **www.etsi.org**.
3. Information is available at the official Website at **www.itu.int**.
4. Information is available at the official Website at **www.adsl.com**.
5. Information is available at the official Website at **www.atmforum.com**.
6. Information is available at the official Website at **www.davic.org**.
7. Information is available at the official Website at **btlabsl.labs.bt.com/profsoc/access**.
8. Information is available at the official Website at **www.vdslalliance.com**.
9. Information is available at the official Website at **www.vdsl.org**.
10. Jacobsen, K., (Ed.), "VDSL Alliance SDMT VDSL Draft Standard Proposal," *ANSI Contribution T1E1.4/98-265* (August 1998).

11. Oksman, V. Ed., "VDSL Draft Specification," *ANSI Contribution T1E1.4/98-045R1* (June 1998).

12. Garth, L. Yang, J. and Werner, J. J. "A Dual-Mode Receiver for CAP and QAM," *ANSI Contribution T1E1.4/98-274* (August 1998).

13. Isaksson, M. et al., "Zipper—A Flexible Duplex Model for VDSL," *ANSI Contribution T1E1.4/97-016* (February 1997).

14. Cioffi, J. M. "VDSL Systems Requirements Document," *ANSI Contribution T1E1.4/98-043R3* (June 1998).

15. Information is available at **http://www.interactivehq.org/councils/html2/feigel/rg.htm**.

Fundamentals of Higher-Layer Technologies

Overview

In previous chapters we discussed various DSL technologies that dealt with the physical layer, i.e., Layer 1 in the OSI protocol stack. In this chapter, the important technologies used on top of DSL technologies are reviewed. We will discuss key technologies and their roles, or values, to DSL access as part of the end-to-end architecture. There will also be a discussion of how services are affected by these technologies from the user's perspective. The key higher-layer technologies discussed in this chapter are ATM, TCP/IP, point-to-point protocol (PPP), PPTP/L2TP (tunneling protocols), audio compression (G.7xx series of ITU standards), and video compression (JPEG, H.261, MPEG-1, MPEG-2, and MPEG-4). These technologies provide the essential functions that enable various broadband applications to coexist in a DSL access network. An example of a broadband application that incorporates many of these technologies is videoconferencing.

Asynchronous Transfer Mode (ATM)

In the telecommunications world, the acronym "ATM" does not stand for automatic teller machines. Instead, it stands for asynchronous transfer mode, a revolutionary communications technology developed in the 1980s to take advantage of advances in the fields of computers and communications. The first research on ATM and its related technologies was published in 1983 by two research centers, CNET and AT&T Bell Labs (now part of Lucent Technologies). In 1984, the Alcatel research center in Antwerp (Belgium) began further to develop the ATM concept. Around 1988, several telcos began to study the possibility of delivering integrated voice, data, and video services over what they termed a broadband Integrated Services Digital Network or BISDN. BISDN was seen as a new opportunity to introduce enhanced services, not just as a step up from ISDN.

ATM was proposed to the ITU as the technology to deliver these enhanced broadband services and was adopted in 1989 over the alternative synchronous transfer mode (STM) proposal. Unlike STM, which is based on time-division multiplexing, ATM does not waste slots that go empty when a sender has no data to send in a given time period.

Instead, ATM is a form of statistical multiplexing based on several key principles, including fast packet switching. As noted in Chapter 1, ATM gets its name from the fact that packetized data are *asynchronously* sent from the sender to the receiver in small units called *cells*, whose size is small and uniform in order to multiplex connections among multiple pairs of senders and receivers efficiently.

ATM was quickly embraced by both the computer and communication industries, and in 1991, the ATM Forum was formed to accelerate the development in the marketplace of interoperable products. Today, the Forum boasts a significant membership of vendors from many different countries. Although the ATM Forum is not a true standards body, many of the interoperability specifications it has released have become de facto standards due to the depth and breadth of industry support behind them. In this chapter, a broad overview of ATM is presented. (This is going to be a challenge, since volumes of books have been written on ATM!) For additional details on ATM, see references 1, 2, and 3.

Principles of ATM

ATM is based on several key principles:

- **Fast packet switching.** The advantage of packet switching over circuit switching is that packet switching does not require dedicated physical circuits. ATM extends the concept of packet switching even further—the packet sizes are chosen to be uniform *and* small. The resulting units that carry information are referred to as cells. The advantage of cell switching (also known as cell relay or fast packet switching) is that, since the cell size is always the same, the transmission delay is predictable and the buffering in the intermediate link queues is greatly simplified. Further, since the cell size is small, the delay characteristic for voice playback is quite short.*

- **Low cell overhead processing.** Not only is the cell size small (48 bytes for the data), the ATM cell header is only five bytes. The ATM cell is shown in Figure 7.1.

---

* In fact, in the original discussions, Europeans wanted to keep the cell size to only 32 bytes to support voice applications. However, too small a cell size adds to network overhead for data applications. Therefore, the United States and other countries proposed 64 bytes. Finally, a compromise was reached to keep the cell size to 48 bytes.

Figure 7.1

The ATM cell

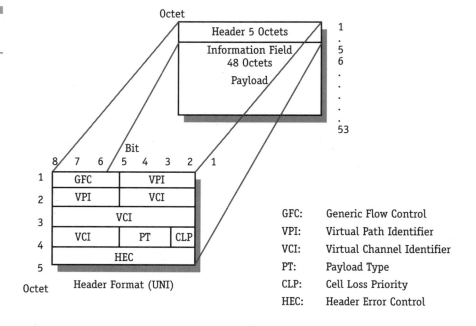

Header Format (UNI)

GFC: Generic Flow Control
VPI: Virtual Path Identifier
VCI: Virtual Channel Identifier
PT: Payload Type
CLP: Cell Loss Priority
HEC: Header Error Control

Low cell overhead is key to achieving the high bit rates that ATM is capable of supporting. Therefore, the total cell size is 53 bytes, small enough that processing can be implemented in high-speed, low-cost silicon. Furthermore, the cell-forwarding decision (i.e., to which output port the cell needs to be shipped on its way to its destination) is a simple lookup of two integer values—a virtual path identifier (VPI) and a virtual channel identifier (VCI). The meaning of VPI and VCI will be explained in the context of the ATM protocol model. For now, it is enough to understand that looking at a pair of integers (representing the VPI/VCI values) and making a decision on where to route the cell can be done quickly and cost effectively in silicon. The forwarding decision is based on mapping the incoming VPI/VCI value pair on an input port to an outgoing VPI/VCI value pair on an output port.* In

---

* The VPI/VCI values are only of "local" significance, i.e., within the context of an ATM switch. A cell arriving at an input port with a particular VPI/VCI value pair in its header is usually modified to a different VPI/VCI value pair just prior to its egress on the output port (this also implies that the header checksum must be recalculated). The next ATM switch downstream also switches the cell from its input port to its output port based on the incoming VPI/VCI value in the header. This process continues down the line, until the cell finally exits the network on the output port of the egress ATM switch.

other words, ATM switching can be done in the hardware, which enables bit rates such as 622 Mbps to be supported quite easily. The cell forwarding process is shown in Figure 7.2.

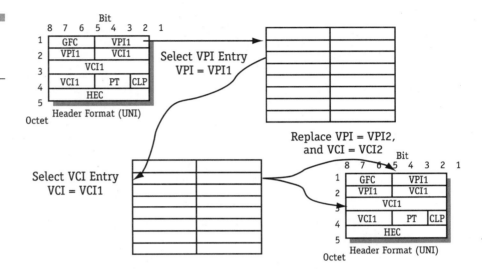

Figure 7.2

The ATM cell forwarding process

- **Sequential cell delivery.** In an ATM network, cells are delivered in exactly the same sequence as they entered the network. This principle follows from the previous requirement to make forwarding decisions as simple as mapping the incoming VPI/VCI value pair on an input port to an outgoing VPI/VCI value pair on an output port. This means that the route must be established when the connection is initially set up, since the mapping cannot be altered in the middle. In turn, this implies that packets must be in sequence, as there is no mechanism to deliver packets on alternate paths. In this sense, ATM resembles circuit switching since the route is established at the beginning and resources are reserved along the route. However, the underlying principle of ATM is still packet switching. This apparent contradiction can be resolved as follows: in circuit switching, the established route is a set of physical links. In ATM, the established route consists of virtual paths and virtual connections within virtual paths. It is possible (and often the case) for multiple virtual paths to exist across a single physical link. Therefore, it is possible to statistically multiplex cells belonging to multiple connections across a single physical link. These cells can be carried across either separate virtual paths (over the same physical link) or multiple virtual connections within a single virtual path (again, over the same physical link). This is illustrated in Figure 7.3.

Figure 7.3

Links, virtual paths, and virtual channels

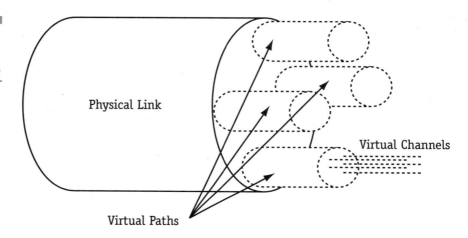

Physical Link

Virtual Channels

Virtual Paths

- **Statistical multiplexing.** Since ATM is based on packet switching, it is flexible in supporting both streaming (e.g., voice and video) and bursty (e.g., data) traffic. In ATM, regardless of the originating traffic characteristics (such as bit rates), the traffic is broken up into uniform units, i.e., cells. Figure 7.4 shows how traffic of different speeds, namely 64 Kbps, 2 Mbps, and 34 Mbps, is chopped into equal-sized packets or cells by a "chopper" or a "cell slicer." Therefore, within the ATM network, only cells are processed as they are rapidly transported from one end of the connection to the other. Upon egress, the constituent cells are regrouped back into the original traffic. This is referred to as *ATM adaptation*, which we will discuss later. The ability of ATM cells to carry any type of traffic is a significant advantage over circuit switching because it efficiently manages and simplifies the use of shared network resources.

- **Admission control and quality of service.** Although ATM is a unifying technology that can transport any kind of traffic, this does not imply that the transport of cells from multiple sources is treated with equal priority. On the contrary, ATM has a strict procedure for new connections being added to the ATM network, called connection

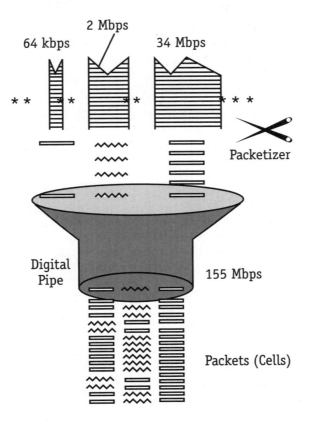

Figure 7.4

Traffic chopped into uniform cells

admission control (CAC). Admission control ensures that, when a new connection is to be established, the network has enough resources to meet the desired QoS parameters of the source. In other words, there is a "traffic contract" between the source and the network. The source specifies the parameters of the traffic it is going to send (e.g., real time or non-real time, constant bit rate or variable bit rate, and so on), and requests the ATM network to meet certain QoS parameters (e.g., network latency and allowable cell error ratio). Admission control in the ATM network verifies whether the new connection (if accepted) would violate the QoS parameters of the connections in place. If the answer is no, the network reserves* resources and lets the connection proceed. If the answer is yes (i.e., all available resources have already been reserved), the connection is

---

* In this manner, ATM employs some of the characteristics of circuit switching, while retaining the benefits of packet switching.

rejected, and the sender has to retry until the network is ready to accept the call. This is the ATM equivalent of a "busy signal." This policy ensures that once a connection is accepted the sender can reasonably expect the network to honor its QoS parameters. What happens, however, if the network conditions change so that, even without new connections, the network becomes heavily loaded? For example, consider the situation when a link goes down, thus causing traffic to be rerouted onto alternate paths and resulting in a heavily utilized alternate link. A flow control mechanism to regulate incoming traffic until the network becomes less congested seems an obvious choice, but too complex a flow control mechanism would unnecessarily slow down cell processing during normal operation. This is the basis for the next principle of ATM.

■ **Simplified flow control.** In ATM, there is no flow control mechanism in the traditional sense. Complex mechanisms would be difficult to implement at high bit rates. Instead, there is a simple congestion control mechanism—every cell has a cell loss priority (CLP) bit in the five-byte header. Normally, this bit is set to zero, meaning that the network transports the cell as normal. However, when the bit is on, meaning it is set to one, the cell may be discarded.* The ATM ingress switch that detects a source violating its traffic contract can start tagging the CLP bits in cells from that source. A source may also choose voluntarily to tag certain cells with CLP=1. In either event, tagging the cells by turning on the CLP bit does not imply that cells are immediately discarded; it only means they may be discarded if such action is warranted. When a true congestion situation occurs, any ATM switch along the path can discard tagged cells until the congestion condition downstream clears up. Note that endpoints are not notified when cells are discarded. They must use their higher-level protocol error recovery mechanisms to resend any lost data. In the early days of ATM, there were no feedback mechanisms for either the network or the receiver to the source to regulate traffic in real time. However, real-time regulation is necessary to support certain classes of applications (e.g., file transfers). Therefore, ATM now incorporates a feedback mechanism when the source sends resource

---

* All cells, regardless of whether CLP equals 0 or 1, must be delivered in sequential order. In other words, an ATM switch may not prioritize delivering cells with CLP = 0 over cells with CLP = 1, thus violating the sequential delivery principle.

management (RM) cells that are looped back by the receiver. The looped back cells contain information for the source to adjust its offered traffic rate according to real-time conditions.

Advantages (and Disadvantages) of ATM

The advantages of ATM are:

- **Flexibility.** ATM can transport any kind of information, regardless of how it originated or what the information characteristics are. This results in one simple universal network for voice, data, and video.
- **Efficiency** in the use of available network resources, and the consequent reduction of transport costs due to statistical multiplexing.
- **Reduction** of operation, administrative, and maintenance costs due to simplification of network processing.
- **Provision** for traffic contract, i.e., service level agreement and enforcement through admission control.
- **Scalability.** ATM can scale up to high bandwidths across a variety of physical media, and speeds can range from 1.544 Mbps over copper to 622.08 Mbps over single-mode fiber. Aside from a change in the physical media dependent layer, the rest of the ATM processing functionally remains the same. Table 7.1 summarizes some of the physical media:

Table 7.1

Examples of physical media supported by ATM (Source: Adapted from the ATM Forum)

Bit Rate	Physical Media	Distance
1.544 Kbps	Twisted pair	3,000 ft (900 m)
25.6 Mbps	Unshielded twisted pair (UTP)—Cat 3	330 ft (100 m)
51.84 Mbps	UTP—Cat 3	330 ft (100 m)
	Multimode fiber	1.2 miles (2 km)
	Single-mode fiber	9 miles (15 km)
155.52 Mbps	UTP—Cat 5	330 ft (100 m)
	Multimode fiber	1.2 miles (2 km)
	Single-mode fiber	15 Km
622.08 Mbps	Multimode fiber	990 ft (300 m)
	Single-mode fiber	9 miles (15 km)

Despite the wide acceptance and benefits of ATM, it is also important to recognize that ATM is a compromise in some ways, and therefore, it has a few disadvantages:

■ ATM does not handle voice as efficiently as isochronous networks do. Voice over ATM requires special handling to keep the delays to a minimum.

■ ATM does not handle data as efficiently as traditional frame relay does. ATM detractors are quick to point out the ATM "cell tax," i.e., the overhead associated with chopping data into small cells, each of which takes up header information.

■ ATM does not handle video as easily as isochronous networks do. Streaming video that originates as large information frames has to be chopped up to be transported across an ATM network and then reassembled.

However, despite its drawbacks, ATM remains a viable technology. The point to remember is that, although ATM does not perform any single function as well as dedicated voice, data, or video networking technologies do, it enables the delivery of these services sufficiently well in an integrated way. The benefits of having such an integrating technology far outweigh its drawbacks.

ATM Standards

A number of standards bodies have developed standards for ATM—the familiar ones are the ITU, ANSI, and ETSI. In addition, the ATM Forum, although not a true standards body, has played a dominant role in shaping the specifications that have become de facto standards. ATM networks can be both private and public with possible interconnections between them, as illustrated in Figure 7.5. Therefore, the following standard ATM network interfaces have been defined:

■ **User network interface (UNI).** UNI can be either private or public.

■ **Network node interface (NNI).** NNI can also be either private or public. The term private network node interface (P-NNI), is used for the interface between two private ATM networks. Interconnection between two public ATM networks is called *broadband inter-carrier interface* or *B-ICI*.

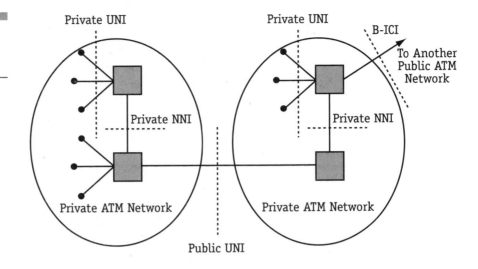

Figure 7.5
Standard ATM network interfaces

Table 7.2 shows some of the relevant specifications developed by the ATM Forum (note that there are a few that refer specifically to residential broadband). For a complete list, please refer to the ATM Forum's Website.

Table 7.2
Partial list of ATM forum approved specifications as of March 1999 (Source: ATM Forum)

Technical Working Group	Approved Specifications	Approved Date
Control Signaling	Addressing Addendum for UNI Signaling 4.0	February 1999
ILMI (Integrated Local Management Interface)	ILMI 4.0	September 1996
LAN Emulation /MPOA	LAN Emulation over ATM 1.0	January 1995
	LAN Emulation Client Management Specification	September 1995
	LANE 1.0 Addendum	December 1995
	LANE Servers Management Spec v1.0	March 1996
	LANE v2.0 LUNI Interface	July 1997
	LAN Emulation Client Management Specification Version 2.0	October 1998

continued on next page

Table 7.2

continued

Technical Working Group	Approved Specifications	Approved Date
	LAN Emulation over ATM Version 2, LNNI Specification	February 1999
	Multi-Protocol Over ATM Specification v1.0	July 1997
	Multi-Protocol Over ATM Version 1.0 MIB	July 1998
Network Management	Customer Network Management (CNM) for ATM Public Network Service	October 1994
	M4 Interface Requirements and Logical MIB	October 1994
	M4 Interface Requirements and Logical MIB: ATM Network Element View	October 1998
	CMIP Specification for the M4 Interface	September 1995
	M4 Public Network View	March 1996
	M4 "NE View"	January 1997
	Circuit Emulation Service Interworking Requirements, Logical and CMIP MIB	January 1997
	M4 Network View CMIP MIB Spec v1.0	January 1997
	M4 Network View Requirements & Logical MIB Addendum	
	ATM Remote Monitoring SNMP MIB	July 1997
	SNMP M4 Network Element View MIB	July 1998
	Network Management M4 Security Requirements and Logical MIB	January 1999
Physical Layer	Issued as Part of UNI 3.1: 44.736 DS3 Mbps Physical Layer, 100 Mbps Multimode Fiber Interface Physical Layer, 155.52 Mbps SONET STS-3c Physical Layer, 155.52 Mbps Physical Layer	
	ATM Physical Medium Dependent Interface Specification for 155 Mbps over Twisted Pair Cable	September 1994
	DS1 Physical Layer Specification	September 1994
	Utopia	March 1994
	Mid-range Physical Layer Specification for Category 3 UTP	September 1994
	6,312 Kbps UNI Specification	June 1995
	E3 UNI	August 1995
	Utopia Level 2	June 1995

continued on next page

Table 7.2 continued	Technical Working Group	Approved Specifications	Approved Date
		Physical Interface Specification for 25.6 Mbps over Twisted Pair	November 1995
		A Cell-based Transmission Convergence Sublayer for Clear Channel Interfaces	January 1996
		622.08 Mbps Physical Layer	January 1996
		155.52 Mbps Physical Layer Specification for Category 3 UTP (See also UNI 3.1, af-uni-0010.002)	November 1995
		120 Ohm Addendum to ATM PMD Interface Spec for 155 Mbps over TP	January 1996
		DS3 Physical Layer Interface Spec	March 1996
		155 Mbps over MMF Short Wave Length Lasers, Addendum to UNI 3.1	July 1996
		WIRE (PMD to TC layers)	July 1996
		E-1 Physical Layer Interface Specification	September 1996
		155 Mbps over Plastic Optical Fiber (POF) Version 1.0	May 1997
		155 Mbps Plastic Optical Fiber and Hard Polymer Clad Fiber PMD Specification Version 1.1	January 1999
		Inverse ATM MUX Version 1.0	July 1997
		Inverse Multiplexing for ATM (IMA) Specification Version 1.19	March 199
		Physical Layer High Density Glass Optical Fiber Annex	February 1999
	Routing and PNNI Addressing	Augmented Routing (PAR) Version 1.0	January 1999
		ATM Forum Addressing: User Guide Version 1.0	January 1999
		ATM Forum Addressing: Reference Guide	February 1999
	Residential Broadband	Residential Broadband Architectural Framework	July 1998
		RBB Physical Interfaces Specification	January 1999
	Service Aspects and Applications	Frame UNI	September 1995
		Circuit Emulation	September 1995
		Native ATM Services: Semantic Description	February 1996

continued on next page

Table 7.2
continued

Technical Working Group	Approved Specifications	Approved Date
	Audio/Visual Multimedia Services: Video on Demand v1.0	January 1996
	Audio/Visual Multimedia Services: Video on Demand v1.1	March 1997
	ATM Names Service	November 1996
	FUNI 2.0	July 1997
	Native ATM Services DLPI Addendum Version 1.0	February 1998
	API Semantics for Native ATM Services	February 1999
	FUNI Extensions for Multimedia	February 1999
Security	ATM Security Framework Version 1.0	February 1998
	ATM Security Specification Version 1.0	February 1999
Signaling	UNI Signaling 4.0	July 1996
	Signaling ABR Addendum	January 1997
Traffic Management	Traffic Management 4.0	April 1996
	Traffic Management ABR Addendum	January 1997
Voice & Telephony over ATM	Circuit Emulation Service 2.0	January 1997
	Voice and Telephony over ATM to the Desktop	May 1997
	Voice and Telephony over ATM to the Desktop	February 1999
	(DBCES) Dynamic Bandwidth Utilization in 64 KbpsTime Slot Trunking Over ATM, Using CES	July 1997
	ATM Trunking Using AAL1 for Narrowband Services v1.0	July 1997
	ATM Trunking Using AAL2 for Narrowband Services	February 1999
User-Network Interface (UNI)	ATM User-Network Interface Specification V2.0	June 1992
	ATM User-Network Interface Specification V3.0	September 1993
	ATM User-Network Interface Specification V3.1	1994

ATM Protocol Model

As Figure 7.6 shows, the ATM protocol model can be thought of as a "layered hourglass" model that incorporates Layer 1 and a portion of Layer 2 of the OSI reference model.

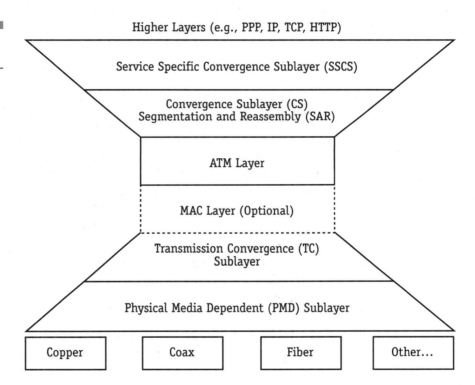

Figure 7.6

ATM protocol model

The ATM hourglass model from the bottom up is as follows:

- **Physical media-dependent (PMD) sublayer.** This layer is responsible for handling the specifics of the physical medium, whether it is copper, coax, or fiber. It is also responsible for the bit transmission capability (such as bit timing/alignment and line coding). The purpose of the PMD sublayer is to provide a logical bit interface to the layer above it, the transmission convergence sublayer.
- **Transmission convergence sublayer.** This layer is responsible for carrying ATM cells on the physical medium, adapting it to the bit transmission capability of the PMD sublayer. The TC sublayer pro-

vides framing and cell header error check (HEC) generation/verification. Another function of the TC layer is to insert and extract idle cells in order to adapt the cell rate to the physical medium. Since the TC layer is independent of the physical medium, the same TC can operate over multiple media.

■ **MAC layer.** The MAC layer is optional (and usually nonexistent) in most systems. Its main purpose is to provide a mechanism to support legacy shared-medium physical architectures. As previously described in Chapter 1, cable TV's tree-and-branch network is an example of such an architecture. The MAC layer can be used to isolate the ATM layer (which still behaves as if it is running over a point-to-point network) from the shared medium. This allows the upper layers from the ATM layer up to the application layer to function without any change.

■ **ATM layer.** This layer sits above the TC layer or the MAC layer, if one is present. The ATM layer is where functions such as cell header generation and extraction (but not header error check—recall that this is done at the TC layer), VPI/VCI translation, and cell multiplexing/demultiplexing occur. In other words, this is where all the action associated with cell manipulation occurs. As we mentioned earlier, the ATM cell header consists of five bytes (see Figure 7.1). The header functions are briefly described below:

  ▪ *Generic flow control (GFC).* GFC is envisaged to provide contention resolution and simple flow control for shared medium-access arrangements at the customer premises equipment (CPE), i.e., to support the function of the optional MAC layer.

  ▪ *Virtual path identifier (VPI).* The VPI is used to identify a virtual path that exists over a physical medium. The VPI field is either eight bits (for cells at UNI) or 12 bits (for cells at NNI). As we mentioned earlier, a virtual path contains several virtual channels because it is sometimes efficient to switch on only the VPI, rather than the VPI/VCI.*

---

* This technique was influenced by the idea behind cross-connect switches in the telephony world. Consider, for example, a telephone switch in the middle of the continental United States that processes long distance calls. It is likely that the switch will experience a lot of East Coast-to-West Coast calls that transit through it, or vice versa. It is inefficient to design the switch to process entire telephone numbers in order to switch the calls; it is sufficient for the switch to understand the area codes on the West Coast in order to send the calls in the right direction. Once the call reaches a switch on the West Coast, the switch at that point can make further call routing decisions based on more than just the area code.

■ *Virtual channel identifier (VCI).* The VCI is used to identify the virtual channel within a virtual path. The VCI field occupies 16 bits in the header.

■ *Payload type (PT).* Payload type is a three-bit field, encoded as shown in Table 7.3.

Table 7.3 Payload type encoding	Payload Type Value	Meaning	
	000	User Data Cell	Congestion *not* experienced, and *not* the *last* cell for AAL-5 protocol data unit (PDU)
	001	User Data Cell	Congestion *not* experienced and the *last* cell for AAL-5 PDU
	010	User Data Cell	Congestion experienced, and not the last cell for AAL-5 protocol data unit (PDU)
	011	User Data Cell	Congestion experienced, and the last cell for AAL-5 protocol data unit (PDU)
	100	OAM Cell	Segment F5 flow related
	101	OAM Cell	End-to-end F5 flow related
	110	Traffic Management Cell	Resource management cell (used to provide flow control feedback to the source)
	111	Reserved	

The reference to AAL 5 in the above table will be discussed shortly in the context of the ATM adaptation layer (AAL).

■ *Cell loss priority (CLP).* The function of the CLP bit was discussed earlier. If CLP is a zero, this indicates a "high-priority" cell that should not be discarded. Alternatively, if CLP is a one, this indicates a "low-priority" cell that may be discarded when there is network congestion.

■ *Header error control (HEC).* The HEC is a CRC calculation on the first four bytes of the header field* for error detection and correction. The HEC CRC sequence is verified at each intermediate point along the path to reduce cell loss probability and misrouting due to cell header errors. Note that since the VPI/VCI value pair is often altered when it transits through an ATM switch (refer to Figure 7.2), the HEC must be recalculated.

---

* No error control is performed by the ATM network for the actual payload.

■ **Segmentation and reassembly (SAR) sublayer.** This sublayer sits above the ATM layer. It is actually one of the sublayers that form the "common part" of the ATM adaptation layer. As the name implies, the purpose of AAL is to adapt traffic (offered by applications from the upper layers above) to the ATM layer.* The functionality of AAL can be divided into (a) those tasks that are common, regardless of the traffic type, and (b) those tasks that are specific to the traffic type. One obvious common function is the need to chop traffic into cells. This is the role of the SAR sublayer. SAR breaks up the offered traffic in 48 bytes of data† (and adds "padding" bytes, if necessary, to make up a full cell). On the receiving end, the process is reversed. There is an important advantage in separating out SAR as a specific sublayer—this makes the SAR portion of AAL application layer independent.

■ **Convergence sublayer (CS).** This sublayer sits above the SAR layer and is the second layer that belongs to the "common part" of the ATM adaptation layer. In this sublayer, the term "common" refers more to the processing that is done for a particular "class of service" (a term applied to quantify the traffic type). However, the CS performs unique functions for different classes of services. Originally, four classes of services named A, B, C, and D were considered, which resulted in equivalent AAL types numbered 1 through 4. Later, AAL 3 and AAL 4 were combined to a single AAL type labeled AAL 3/4. In addition, a new category called AAL 5 was defined. The mapping of AAL types to service classes is summarized in Table 7.4. Of all the AAL types, AAL 5 is the most popular and has been widely adopted by the data communications industry. Furthermore, AAL 5 is used not only to transport data, but also to transport ATM signaling.

■ **Service-specific convergence sublayer (SSCS).** This sublayer sits above the common part of AAL, i.e., above the convergence sublayer. Its purpose is to provide service-dependent functions such as reliable data transfer. Note that the ATM layer itself does not ensure reliable data transfer.

---

* It is important to note that ATM adaptation occurs *only* at the endpoints, and never in the switches. In other words, switches are concerned with the PMD, TC, and ATM layers. The optional MAC layer may exist in a switch to accommodate endpoints that require a MAC.

† Note that the header is generated at the layer below, i.e., the ATM layer, and not by the SAR.

Table 7.4

AAL types

	AAL Service Class			
	A	**B**	**C**	**D**
AAL Types	AAL 1 or AAL 5	AAL 2 or AAL 5	AAL 3/4 or AAL 5	AAL 3/4 or AAL 5
Bit Rate	Constant Bit Rate (CBR)		Variable Bit Rate (VBR)	
Timing Requirement	Required			Not Required
Connection Mode	Connection Oriented		Connectionless	

Although it is useful to map the ATM protocol model to the OSI seven-layer model, it is important to note that the OSI model was designed for data communications. When the ITU looked at the design of BISDN to carry integrated voice, video, and data, it determined that the OSI model was inadequate. Therefore, the ITU defined a protocol reference model, which, like the OSI model, also supports the idea of multiple layers, though it adds the new concept of planes. There are three distinct planes: the user, control, and management planes, sometimes referred to as the U, C, and M planes, respectively. The U plane deals with user data flow, the C plane is concerned with control protocols (i.e., signaling protocols for connection establishment and teardown), and the M plane is concerned with management protocols for operations, administration, and maintenance (OAM) functions. The BISDN ATM protocol reference model is shown in Figure 7.7.

Figure 7.7

BISDN ATM protocol reference model

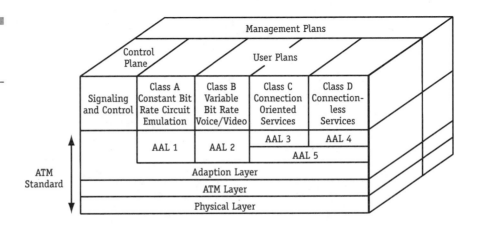

ATM Addressing

In previous discussions, we have seen that cells are routed from one place to another via ATM switches that map the incoming VPI/VCI value pair on an input port to an outgoing VPI/VCI value pair on an output port. During data transfer, an ATM address is not required. However, it is required to identify endpoints for the purpose of setting up and tearing down connections. Furthermore, the addressing must be flexible enough to allow interconnection of both private and public ATM networks as shown in Figure 7.5. ATM addresses are 20 bytes long and have three defined formats. A portion of the address called the authority and format identifier (AFI) specifies the exact format and addressing authority.

- **ITU E.164.** This format is like a telephone number and is recommended by the ITU. E.164 addresses are used in public ATM networks and are assigned by public network operators.
- **Data country code (DCC).** This carries LAN addresses in the format specified by the IEEE 802 recommendation. The DCC values are allocated and assigned to each country's ISO National Member Body.
- **International code designator (ICD).** This format is specified by the ISO then allocated and assigned by the ISO 6523 registration authority.

In addition to individual addresses, ATM supports group addresses. This allows "anycasting," which means a source can make "anycast" calls using any well-known ATM address. A connection can then be completed to any of the end systems that provide the desired service.

ATM Signaling

In ATM terminology, a virtual connection (VC) is the end-to-end connection established between two ATM addressable endpoints. Virtual connections can be classified into permanent virtual connections (PVCs) or switched virtual connections (SVCs):

- **Permanent virtual connections (PVCs).** The term "permanent virtual" seems like an oxymoron. How can something be permanent and virtual at the same time? In the world of ATM, however, it makes perfect sense. Yes, the connection is still virtual between two endpoints (i.e., sharing a physical link with other virtual connections); however, it is permanent in the sense that, once set up, it is never torn down until there is a link or node failure. PVCs are automatically established upon (re)initialization from the source endpoint, i.e., the entire path (VPI/VCI value pairs at each node along the way) is pre-established and reserved a priori. Therefore, as soon as the physical layers are up, PVCs can be established. In contrast, SVCs require a signaling procedure to establish the connections. Another difference is that PVCs are established in the management plane; SVCs are established in the control plane.

■ **Switched virtual connections (SVCs).** As mentioned above, SVCs require a signaling procedure, which includes distinct connection establishment, data transfer, and connection teardown phases. Connection establishment requires reservation of the resources within the network to support the new connection, otherwise admission control will reject the connection request. Once established, SVCs support the data transfer phase in the same manner as PVCs do. Note that the signaling procedure (e.g., for connection establishment) requires the exchange of messages in the network. How are these messages sent? Over a signaling VC, of course! A signaling VC is actually a PVC, since it must be available at all times for ATM endpoints to set up and tear down VCs.

The UNI signaling protocol is based on the ITU standard Q.2931, which is the broadband extension of the Q.931 protocol developed for ISDN. (Recall that ATM was the technology adopted for the implementation of BISDN.) Q.2931 consists of a service-specific convergence sublayer (SSCS) part and a common part. The common parts are the AAL 5 convergence sublayer and the SAR sublayer. The SSCS part consists of a service-specific connection-oriented protocol (SSCOP) and a service-specific connection function (SSCF). SSCOP provides the reliable delivery mechanism absent from the lower ATM layers. As we shall see later in this book, SSCOP messages play an important role in the discussion of G.lite and power management.

ATM Operations, Administration, and Maintenance (OAM)

In an attempt to provide a framework for standardized management information and formats until official ITU standards are developed, the ATM Forum developed the Interim Local Management Interface (or ILMI). Later, the term "interim" in ILMI was replaced with "integrated," so ILMI now stands for integrated local management interface. ILMI uses the simple network management protocol (SNMP) to provide status, configuration, and control information for an ATM interface.

In ATM, operations, administration and maintenance protocols are defined to perform functions such as performance monitoring, fault detection and isolation, and system protection. Five functional layers have been defined: F1 for section level, F2 for line level, F3 for path level, F4 for virtual path level, and F5 for virtual channel level. Of these, the first three are performed by the link protocols and are specific to media. For example, F1, F2, and F3 exactly map to the SONET/SDH functions. The two higher layers, F4 and F5, are ATM loopback tests. F4/F5 OAM "flows," i.e., ATM cells carrying management protocol information, are useful in fault detection and isolation.

Transmission Control Protocol/Internet Protocol (TCP/IP)

The term TCP/IP is derived from two protocols that are an integral part of the protocol suite—the transmission control protocol (TCP) and the Internet protocol (IP). The growth of the Internet has made TCP/IP one of the most well-known networking protocol suites, and IP has often been termed the "lingua franca" of the Internet. In this section, we will briefly (or as briefly as we can for, like ATM, the topic of TCP/IP has spawned multiple books!) review the history of TCP/IP, the standardization process, and some of the key protocols in the suite. For additional details on TCP/IP, see references 4, 5, and 6.

The origin of TCP/IP can be traced to the 1960s when the U.S. government's Advanced Research Projects Agency (ARPA), developed a network called ARPANET, which eventually evolved into the Internet we know today. In 1971, an agency called the Defense Advanced Research Projects Agency (DARPA) took over the management of ARPANET. During this time, the need developed for a set of networking protocols that would interlink multiple computers, regardless of their operating systems or underlying hardware. DARPA's sponsored research led to the development of protocols that became the basis for TCP/IP. By 1978, TCP/IP was sufficiently stable for public demonstration. Through the 1980s, TCP/IP continued to be enhanced. In the 1990s, with the growing importance of the Internet, TCP/IP became a dominant networking protocol with implementations in various devices such as PCs, PDAs, and intelligent appliances.

TCP/IP Standards

The protocols of the TCP/IP suite have been continually enhanced through a unique process called Request for Comments (RFCs), which is managed by the Internet Engineering Task Force (IETF). RFCs are used to describe the internal workings of the Internet. Some describe network services or protocols and their implementations, whereas others summarize policies. TCP/IP standards are always published as RFCs, although not all RFCs specify standards.

TCP/IP standards are not developed by a committee, but rather by consensus. Anyone can submit a document for publication as an RFC. Documents are reviewed by a technical expert, a task force, or the RFC editor, and then assigned a status. The status specifies whether a document is being considered for a standard. There are five status assignments of RFCs as shown in Table 7.5.

Table 7.5
Status assignments of RFCs

Status	Description
Required	Must be implemented on all TCP/IP-based hosts and gateways.
Recommended	All TCP/IP-based hosts and gateways are encouraged to implement the RFC specifications. Recommended RFCs are usually implemented.
Elective	Implementation is optional. Its application has been agreed upon, but is not a requirement.
Limited Use	Not intended for general use.
Not Recommended	Not recommended for implementation.

If a document is being considered as a standard, it goes through the stages of development, testing, and acceptance known as the Internet standards process. These stages are formally labeled *maturity levels*. Table 7.6 lists the three maturity levels for Internet standards.

Table 7.6

Maturity levels for Internet standards

Maturity Level	Description
Proposed Standard	A proposed standard specification is generally stable, has resolved known design choices, is believed to be well understood, has received significant community review, and appears to enjoy enough community interest to be considered valuable.
Draft Standard	A draft standard must be well understood and known to be quite stable, both in its semantics and as a basis for developing an implementation.
Internet Standard	The Internet standard specification (which may simply be referred to as a standard) is characterized by a high degree of technical maturity and by a generally held belief that the specified protocol or service provides significant benefit to the Internet community.

When a document is published, it is assigned an RFC number. The original RFC is never updated. If changes are required, a new RFC is published with a new number. Therefore, it is important to verify that one has the most recent RFC on a particular topic.

RFC text can be reviewed in several ways. The simplest way to obtain any RFC or a full and up-to-date, indexed listing of all RFCs published is to access the IETF on the World Wide Web[7]. RFCs can also be downloaded from various sources such as **nis.nsf.net**, **nisc.jvnc.net**, **venera.isi.edu**, **wuarchive.wustl.edu**, **src.doc.ic.ac.uk**, **ftp.concert.net**, **internic.net**, or **nic.ddn.mil**, or searched at sites such as those found in the index referenced in 8.

TCP/IP Protocol Architecture

The TCP/IP protocol suite maps to a four-layer conceptual model known as the DARPA model, which was named after the U.S. government agency that initially developed it. The four layers of the DARPA model are: application, transport, Internet, and network interface. Each layer in the model corresponds to one or more layers of the seven-layer open systems interconnection (OSI) model. Figure 7.8 shows the TCP/IP protocol suite.

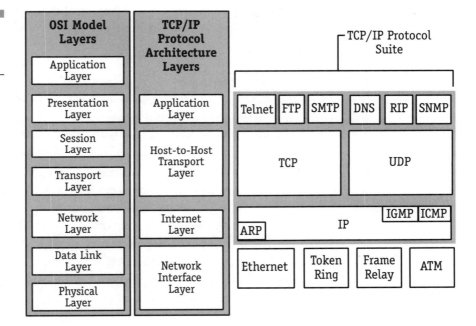

Figure 7.8
TCP/IP protocol suite

Network-interface Layer

The network-interface layer (also called the network access layer) is responsible for putting TCP/IP packets on the network medium and removing TCP/IP packets off the network medium. TCP/IP was designed to be independent of the network access method, frame format, and medium so that it could be used to connect differing network types. This included LAN technologies, such as Ethernet or token ring, and WAN technologies, such as X.25 or frame relay. Independence from any specific network technology gives TCP/IP the ability to be adapted to new underlying technologies such as asynchronous transfer mode (ATM).

Internet Layer

The Internet layer, analogous to the network layer of the OSI model (layer 3), is responsible for addressing, packaging, and routing functions. Note that it does not take advantage of any sequencing and acknowledgment services that may be present in the lower layers. An unreliable network interface layer is assumed, and reliable communications through session establishment and the sequencing and acknowledgment of packets is the responsibility of the transport layer.

The core protocols of the Internet layer are IP, ARP, ICMP, and IGMP:

■ The **Internet protocol (IP)** is a routable protocol responsible for IP addressing and the fragmentation and reassembly of packets. It is a connectionless, unreliable datagram protocol primarily responsible for addressing and routing packets between hosts. Connectionless means that a session is not established before exchanging data. Unreliable means that delivery is not guaranteed, although IP will always make a best-effort attempt to deliver a packet. An IP packet might be lost, delivered out of sequence, duplicated, or delayed. IP does not attempt to recover from these types of errors. The acknowledgment of packets delivered and the recovery of lost packets is the responsibility of a higher-layer protocol, such as TCP. IP is defined in RFC 791[9].

An IP packet consists of an IP header and an IP payload. Table 7.7 describes the key fields in the IP header.

Table 7.7

Key fields in the IP header

IP Header Field	Function
Source IP Address	The IP address of the original source of the IP datagram
Destination IP Address	The IP address of the final destination of the IP datagram
Identification	Used to identify a specific IP datagram and to identify all fragments of a specific IP datagram if fragmentation occurs
Protocol	Informs IP at the destination host whether to pass the packet up to TCP, UDP, ICMP, or other protocols
Checksum	A simple mathematical computation used to verify the integrity of the IP header
Time to Live (TTL)	Designates the number of networks on which the datagram is allowed to travel before being discarded by a router. The TTL is set by the sending host and is used to prevent packets from endlessly circulating on an IP internetwork. When forwarding an IP packet, routers are required to decrease the TTL by at least one.

If a router receives an IP packet that is too large for the receiving network, IP will fragment the original packet into smaller packets that will fit on the downstream network. When the packets arrive at their final destination, the IP at the destination host reassembles the fragments into the original payload. This process is referred to as fragmentation and reassembly. Fragmentation can occur in environments that have a mix of networking technologies, such as Ethernet and token ring.

- The **address resolution protocol (ARP)** is responsible for the resolution of the Internet-layer address with the network interface-layer address, such as a hardware address. When IP packets are sent on shared-access, broadcast-based networking technologies such as Ethernet or token ring, the media access control (MAC) address corresponding to a forwarding IP address must be resolved. ARP uses MAC-level broadcasts to resolve a known forwarding IP address to its MAC address.* ARP is defined in RFC 826[10].

- The **Internet control message protocol (ICMP)** is responsible for providing diagnostic functions and reporting errors or conditions regarding the delivery of IP packets. For example, if IP is unable to deliver a packet to the destination host, ICMP will send a "destination unreachable" message to the source host. ICMP is defined in RFC 792[11]. Table 7.8 shows the most common ICMP messages.

	ICMP Message	Function
Table 7.8 *Common ICMP messages*	Echo Request	Simple troubleshooting message used to check IP connectivity to a desired host
	Echo Reply	Response to an ICMP echo request
	Redirect	Sent by a router to inform a sending host of a better route to a destination IP address
	Source Quench	Sent by a router to inform a sending host that its IP datagrams are being dropped due to congestion at the router. The sending host then lowers its transmission rate. Source quench is an elective ICMP message and is not commonly implemented.
	Destination Unreachable	Sent by a router or the destination host to inform the sending host that the datagram cannot be delivered

* A complementary protocol that resolves a known MAC address to an IP address is called reverse ARP (RARP).

■ The **Internet group management protocol (IGMP)** is responsible for the management of IP multicast groups. An IP multicast group, also known as a host group, is a set of hosts that listen for IP traffic destined for a specific multicast IP address. Multicast IP traffic is sent to a single multicast address, but processed by multiple IP hosts. A given host listens in on a specific IP multicast address and receives all packets sent to that IP address, along with other hosts on the network that choose to listen to the same IP address. IGMP is defined in RFC 1112.[12]

Some additional aspects of IP multicasting are:

■ Host group membership is dynamic, meaning hosts can join and leave the group at any time.

■ A host group can be of any size.

■ Members of a host group can span IP routers across multiple networks. This situation requires IP multicast support on the IP routers and the ability of hosts to register their group membership with local routers. Host registration is accomplished using IGMP.

■ A host can send traffic to an IP multicast address without belonging to the corresponding host group.

For a host to receive IP multicasts, an application must inform IP that it will be receiving multicasts at a specified destination IP multicast address. If the network technology supports hardware-based multicasting, then the network interface is told to pass up packets for a specific multicast address. For example, with Ethernet, the network interface card can be programmed to respond to a multicast MAC address corresponding to the desired IP multicast address.

Transport Layer

The Transport layer (also known as the host-to-host transport layer) is responsible for providing the application layer with session and datagram communication services. It encompasses the responsibilities of the OSI transport layer and some of the responsibilities of the OSI session layer.

The core protocols of this layer are the transmission control protocol (TCP) and the user datagram protocol (UDP):

■ **Transmission control protocol (TCP)** provides a one-to-one, connection-oriented, reliable communications service. TCP is responsible for the establishment of a connection, the sequencing and acknowledgment of sent packets, and the recovery of packets lost during transmission. Connection-oriented means that a connection must be established before hosts can exchange data. Reliability is achieved by assigning a sequence number to each segment transmitted. An acknowledgment is used to verify that the other host has received the data. For each segment sent, the receiving host must return an acknowledgment within a specified period for bytes received. If an acknowledgment is not received, the data is retransmitted. TCP is defined in reference[13].

TCP uses byte-stream communications, where data within the TCP segment is treated as a sequence of bytes with no record or field boundaries. Table 7.9 describes the key fields in the TCP header.

Table 7.9

Key fields in the TCP header

Field	Function
Source Port	TCP port of sending host
Destination Port	TCP port of destination host
Sequence Number	The sequence number of the first byte of data in the TCP segment
Acknowledgment Number	The sequence number of the byte the sender expects to receive next from the other side of the connection
Window	The current size of a TCP buffer on the host sending this TCP segment to store incoming segments
TCP Checksum	Verifies the integrity of the TCP header and the TCP data

TCP Ports

A TCP port provides a specific location for delivery of TCP segments. Port numbers below 1,024 are well known and are assigned by the Internet Assigned Numbers Authority (IANA). Table 7.10 lists a few of the well-known TCP ports. A complete list of assigned TCP ports is available in the source listed in reference 14.

Table 7.10

Some well-known TCP ports

TCP Port Number(s)	Description
20, 21	FTP (data channel, control channel)
23	Telnet
80	Hypertext transfer protocol (HTTP) used for the World Wide Web

TCP Three-way Handshake

A TCP connection is initialized through a three-way handshake whose purpose is to synchronize the sequence number and acknowledgment numbers on both sides of the connection and exchange TCP window sizes and other TCP options, such as the maximum segment size. The following steps outline the process:

- The client sends a TCP segment to the server with an initial sequence number for the connection and a window size indicating the buffer size on the client to store incoming segments from the server.
- The server sends back a TCP segment containing its chosen initial sequence number, an acknowledgment of the client's sequence number, and a window size indicating a buffer size on the server to store incoming segments from the client.
- The client sends a TCP segment to the server containing an acknowledgment of the server's sequence number.

TCP uses a similar handshake process to end a connection. This guarantees that both hosts have finished transmitting and that all data were received.

- **User datagram protocol (UDP)** provides a one-to-one or one-to-many, connectionless, unreliable communications service. This means that neither the arrival of datagrams nor their sequencing is guaranteed. UDP does not recover lost data through retransmission. Rather, it is used when the amount of data to be transferred is small (such as the data that would fit into a single packet), when the overhead of establishing a TCP connection is not desired, or when the applications or upper-layer protocols provide reliable delivery.

Simple network management protocol (SNMP) uses UDP. UDP is defined in reference 15. Table 7.11 describes the key fields in the UDP header.

Table 7.11

Key fields in the UDP header

Field	Function
Source Port	UDP port of sending host
Destination Port	UDP port of destination host
UDP Checksum	Verifies the integrity of the UDP header and the UDP data
Acknowledgment Number	The sequence number of the byte the sender expects to receive next from the other side of the connection

UDP Ports

To use UDP, an application must supply the IP address and UDP port number of the destination application. A port provides a location for sending messages and functions as a multiplexed message queue, meaning that it can receive multiple messages at a time. A unique number identifies each port. It is also important to note that UDP ports are distinct and separate from TCP ports even though some of them use the same number. Table 7.12 lists well-known UDP ports. For a complete list of assigned UDP ports, see the reference listed in 14.

Table 7.12

Some well-known UDP ports

UDP Port Number	Description
53	Domain name system (DNS) name queries
69	Trivial file transfer protocol (TFTP)
137	NetBIOS name service
138	NetBIOS datagram service
161	Simple network management protocol (SNMP)

Application Layer

The application layer provides applications with the ability to access the services of the other layers and defines the protocols that applications use to exchange data. There are many application-layer protocols and new ones are constantly being developed.

The most widely known application-layer protocols are those used for the exchange of user information:

- The hypertext transfer protocol (HTTP) is used to transfer files that make up the pages of the World Wide Web.
- The file transfer protocol (FTP) is used for interactive file transfer.
- The simple mail transfer protocol (SMTP) is used for the transfer of mail messages and attachments.
- Telnet, a terminal emulation protocol, is used for remote log-in to network hosts.
- In addition, the following application layer protocols help facilitate the use and management of TCP/IP networks:
  - The domain name system (DNS) is used to resolve a host name to an IP address.
  - The routing information protocol (RIP) is a protocol that routers use to exchange routing information on an IP internetwork.
  - The simple network management protocol (SNMP) is used between the network management console and network devices (routers, bridges, and intelligent hubs) to collect and exchange network management information.

PPP and Tunneling

The origin of the point-to-point protocol (PPP)[16] is actually rooted in another protocol called serial line interface protocol (SLIP). SLIP was developed in the 1980s as a way of allowing hosts and router devices to communicate with each other over low-speed serial lines (1,200–9,600 b/s). It was a simple packet-framing protocol for IP on a serial line. Despite its simplicity, however, SLIP suffered from some serious deficiencies:

- **Addressing.** Both computers in a SLIP link needed to know each other's IP addresses for routing purposes.
- **Type identification.** SLIP had no type field, therefore, only one protocol (namely IP) could run over a SLIP connection.
- **No error detection/correction.**
- **No options for compression and encryption.**

In the late 1980s, PPP was devised as a method to address the deficiencies of SLIP. Like TCP/IP, PPP is not a single protocol; it actually refers to a suite of protocols based on the services the individual control protocols provide:

- Link control protocol (LCP)
  - Low-layer media-dependent encapsulation
  - Error detection
  - Negotiation of low-layer framing options
- Authentication control protocol (ACP)
  - Mechanisms for user authentication
- Network control protocol (NCP)
  - Mechanism to establish the network protocol (e.g., IP, IPX, Banyan VINES)
- Compression control protocol (CCP)
  - Mechanism to establish the compression methods
- Encryption control protocol (ECP)
  - Mechanism to establish the encryption methods.

As with other Internet-related protocols, the PPP protocol suite was developed and enhanced through the process of RFCs within the framework of the IETF. Some of these RFCs are generally applicable to all implementations (e.g., PPP framing), whereas others are specific to a particular technology (e.g., STAC compression). PPP was widely implemented in a variety of products including standalone routers and remote access servers developed by networking companies such as 3Com, Cisco, Bay Networks, and many others. In addition, Microsoft incorporated PPP into its operating system as part of its dial-up networking (DUN). In the 1990s, companies such as Cisco and Microsoft played an active role in enhancing PPP.

Building upon its extensibility, PPP was enhanced in 1994 with multilink PPP[17] to allow multiple physical links to be logically treated as a single link, thereby aggregating the bandwidth. Later, the suite was enhanced with tunneling protocols such as the point-to-point tunneling protocol (PPTP) and the layer 2 tunneling protocol (L2TP). These protocols allowed a PPP connection to be tunneled through an intermediate network such as the Internet.

For additional details on PPP, there are references such as those listed in 18 and 19.

PPP and the OSI Layer

As Figure 7.9 shows, the PPP protocol suite is considered to be part of the link layer and the network layer.

Figure 7.9

PPP and the

OSI layer

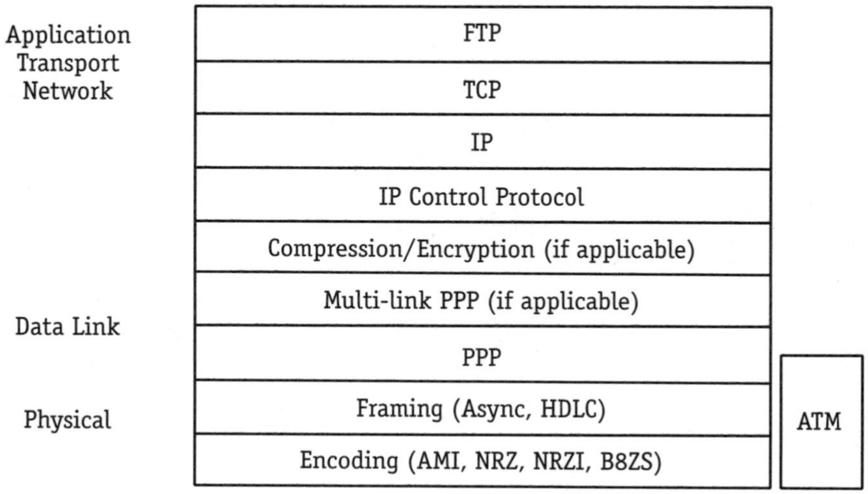

Application Transport Network	FTP
	TCP
	IP
	IP Control Protocol
	Compression/Encryption (if applicable)
Data Link	Multi-link PPP (if applicable)
	PPP
Physical	Framing (Async, HDLC)
	Encoding (AMI, NRZ, NRZI, B8ZS)

(ATM spans Data Link and Physical)

In the PPP protocol suite, the link-control protocol is concerned with establishing access over a generic "link." The link can, in fact, be anything that connects two unique endpoints. As Figure 7.10 shows, it could be a leased line, a connection between two ISDN terminal adapters, or a connection across the PSTN between two modems.

Through the efforts of the IETF, the following are some of the relevant Requests for Comments that specify PPP* over different kinds of links:

■ PPP over frame relay
■ PPP over SONET/SDH

---

* Strictly speaking, the RFCs specify how the link control protocol (LCP) operates over specific links. Other protocols of the PPP protocol suite (e.g., network control protocols and compression control protocols) ride on top of LCP (and independently of LCP). Therefore, they operate in exactly the same manner regardless of the link.

- PPP over ISDN
- PPP over X.25
- PPP over AAL5[20]. The idea that PPP protocols can ride on top of an ATM PVC or SVC is utilized as part of the PPP over ATM over ADSL architecture, as we shall see later in the book.

Figure 7.10
Examples of PPP links

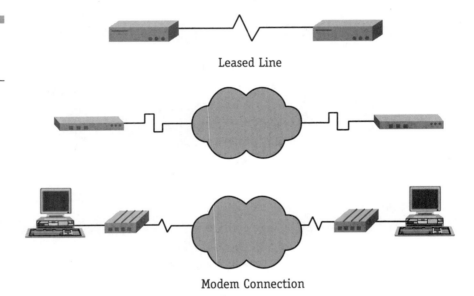

Leased Line

Modem Connection

The network control protocol (NCP) is concerned with establishing the protocol that can run over the link. For example, Internet protocol control protocol (IPCP) establishes IP packet flows across the link. Similarly, IPXCP establishes IPX packet flows across the link. It is quite possible for more than one type of protocol to coexist over the same link. In this way, more than one type of protocol can be supported over a link using separate network control protocols. This is an important advantage over SLIP, which supports only IP.

Control protocols are at the heart of PPP's flexibility. Different control protocols can be used to support not only multiple network protocols, but also features such as compression. The following are some examples of the options:

- Authentication control protocols
  - Password authentication protocol (PAP)

- ▌ Challenge handshake authentication protocol (CHAP)
■ Compression control protocols
    - ▌ Gandalf FZA compression protocol
    - ▌ Predictor compression protocol
    - ▌ Magnalink variable resource compression
    - ▌ STAC
    - ▌ V.42 bis
    - ▌ Microsoft point-to-point compression (MPPC)
■ Encryption control protocols
    - ▌ DES encryption (DESE)
    - ▌ Microsoft point-to-point encryption (MPPE)
■ Network control protocols
    - ▌ XNS IDP control protocol (XNSCP)
    - ▌ Banyan Vines control protocol (BVCP)
    - ▌ Internetwork packet exchange control protocol (IPXCP)
    - ▌ Appletalk control protocol (ATCP)
    - ▌ IP control protocol (IPCP).

All this flexibility, of course, comes at a price. The first issue is that each of the control protocols in the PPP protocol suite has its own header. Therefore, depending upon the actual set of control protocols required for a particular configuration, the overhead could be significant, which also adds to the packet processing. The second issue is that PPP's flexibility has led to a number of options. When two endpoints first try to establish a link using PPP, each endpoint does not know the options supported by the other. Therefore, each endpoint has to negotiate the applicable options to determine the options supported in common. This negotiation procedure can sometimes take a lot of processing and initialization time.

PPP Initialization

There are distinct phases of negotiation in a PPP dial-up session and each phase must complete successfully before data transfer can begin. These phases are shown in Figure 7.11.

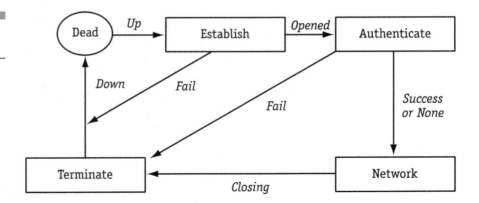

Figure 7.11
PPP initialization

Phase 1: PPP Link Establishment

PPP uses the link control protocol to establish, maintain, and end the physical connection. During the initial LCP phase, optional features (e.g., compression) are selected through a process of negotiation. Negotiation works on the following lines:

- **Configure request.** This implies that the initiator of the connection requests a set of optional features and their values.
- **Configure ACK.** This implies that the endpoint that would accept the connection agrees with the set of optional features. Agreement implies that not only are the proposed optional features satisfactory, but also the proposed values for the optional features (if any) are acceptable.
- **Configure NAK.** This implies that the endpoint that would accept the connection disagrees with the proposed value of an optional feature. The NAK reply can optionally propose an alternate value for the optional feature. The initiator can then resend the alternate value in a new "configure request" message.
- **Configure REJECT.** This implies that the endpoint that would accept the connection disagrees with the proposed optional feature (i.e., it does not support the requested feature). The initiator can then choose to ignore the feature and continue, or terminate the connection at this stage.

Note that during the link establishment phase, authentication protocols are selected (if either end desires authentication), but they are not actually implemented until the connection authentication phase. Similarly, during LCP, a decision is made on whether the two peers will negotiate the use of compression and/or encryption. The actual choice of compression/encryption algorithms and other details occurs after successful authentication.

Phase 2: User Authentication (Optional, but Usually Invoked)

Strictly speaking, the authentication phase is optional. It is invoked depending upon what was negotiated during link establishment. If, and only if, both endpoints agree to bypass authentication, the authentication phase is skipped. In practice, however, this phase is almost always invoked. During this phase, the initiator of the connection (typically, a client PC) presents the user's credentials to the endpoint that would accept the connection (typically, a remote access server). Most implementations of PPP provide one or more of the following authentication methods.*

- **Password authentication protocol (PAP).** PAP is a simple, cleartext authentication scheme. The NAS requests the user name and password, and PAP returns them in clear text (unencrypted). Obviously, this authentication scheme is not secure because a third party could capture the user's name and password and use it to get subsequent access to the remote access server and all the resources provided by the server. PAP provides no protection against replay attacks or remote client impersonation once the user's password has been compromised.
- **Challenge-handshake authentication protocol (CHAP).** CHAP is an encrypted authentication mechanism that avoids transmission of the actual password on the connection. The remote access server sends a challenge (consisting of a session ID and an arbitrary challenge string) to the client PC. The client PC must use a one-way

---

* Effective authentication methods provide protection against replay attacks and remote client impersonation. A *replay attack* occurs when a third party monitors a successful connection and uses captured packets to play back the remote client's response so that it can gain an authenticated connection. *Remote client impersonation* occurs when a third party takes over an authenticated connection. The intruder waits until the connection has been authenticated and then traps the conversation parameters, disconnects the authenticated user, and takes control of the authenticated connection.

*hashing algorithm** to return a response based on the user's password to the server. CHAP is an improvement over PAP in that the clear-text password is not sent over the link. Instead, the password is used to create an encrypted hash from the original challenge. The server knows the user's clear-text password and can therefore replicate the hash operation and compare† the result to the client's response. CHAP protects against replay attacks by using an arbitrary challenge string for each authentication attempt. CHAP protects against remote client impersonation by randomly sending repeated challenges to the remote client throughout the duration of the connection.

▌ **Microsoft challenge-handshake authentication protocol (MS-CHAP).** MS-CHAP is an encrypted authentication mechanism very similar to CHAP. As in CHAP, the remote access server sends a challenge. The difference is that the client PC returns a hash of the hash of the password, which provides an additional level of security. It allows the server (or an external centralized authentication server) to store hashed passwords instead of clear-text passwords, so that even if the server's password file is compromised, the passwords are not. MS-CHAP also provides additional error codes, including a password-expired code and additional encrypted client/server messages that permit users to change their passwords.

Phase 2a: PPP Callback Control (Optional)

Microsoft's implementation of PPP includes an optional callback control phase. This uses the callback control protocol (CBCP) immediately after the authentication phase. If callback is configured, then, after authentication, both the client PC and the remote access server disconnect. The remote access server then calls the remote client back at a specified phone number. This provides an additional level of security to

---

† One-way hashing algorithms are based on *one-way functions*, an essential mathematical tool used in cryptology. One-way functions work on the following premise. If y=f(x), then given x, it is easy to compute y. However, given y, it is computationally impossible (without expensive computers and several hours of number crunching) to recover x. Therefore, if x is a number that represents a password, only the hash value, i.e., y, is sent across the link. An intruder would not be able to recover x, the password, even though y can be observed.

† The comparison may be done either on the remote access server itself, or against a central authentication database server, such as one maintained by a Microsoft Windows NT primary domain controller (PDC), or on a remote authentication dial-in user service (RADIUS).

dial-up networking. The remote access server will only allow connections from client PCs originating from specific phone numbers.

Phase 3: Network-layer Protocol(s)

Once the previous phases have been completed, PPP invokes the various network control protocols (NCPs) selected during the link establishment phase (Phase 1) to configure protocols used by the client PC. For example, Internet protocol control protocol (IPCP) is used to configure the link to carry IP datagrams. Among other things, IPCP can be used to assign a dynamic address to the dial-in client PC.

PPP Data Transfer

Once the phases of PPP initialization have completed, PPP begins to forward data to and from the two peers. Each transmitted data packet is wrapped in a PPP header, which is removed by the receiving system. If data compression was negotiated between the endpoints terminating the PPP layer, then data are compressed by the sender prior to transmission and decompressed by the receiver upon reception (in either direction). Similarly, data are encrypted/decrypted depending upon what was negotiated during initialization. Figure 7.12 shows an example of PPP framing and headers for compressed data sent across the link.

Figure 7.12
PPP framing and headers

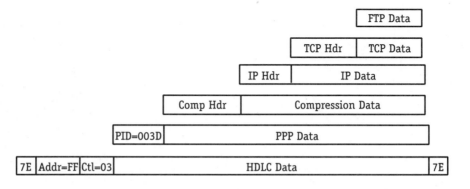

It is important to note that during the data transfer phase, the control protocols may continue to send control messages. For example, CHAP randomly sends repeated challenges throughout the duration of the connection as a security measure. LCP may send loopback messages (ECHO_REQUEST/ECHO_REPLY) to verify the status of the link. Some implementations may use link quality monitor (LQM) messages to determine when, and how often, the link is dropping data.

PPP and Tunneling

The basic idea of tunneling is simple—establish a tunnel across an intermediate network to extend the PPP link across the network. In other words, the endpoints that terminate PPP are outside the network, and the PPP link includes a "tunnel" through the network. This is shown in Figure 7.13.

Figure 7.13

Extending the PPP link through tunneling

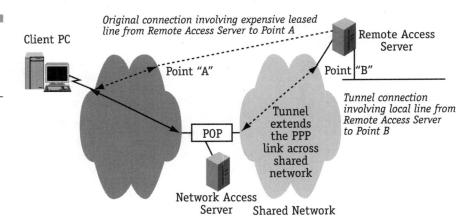

As with several technology innovations in the 1990s, the Internet was the driver for extending the PPP protocol suite with tunneling. With the growing popularity and availability of the Internet, many in the data communications industry felt that it would be useful to increase the usefulness of the Internet to transport PPP data across it. With the global availability of the Internet, PPP endpoints could be anywhere in the world. Furthermore, once a tunnel is established, it would be possible for more than one pair of endpoints to share a common tunnel. This would allow corporations to establish virtual private remote networks and avoid the cost of expensive leased lines as shown in Figure 7.14.

Figure 7.14
*Virtual private
remote networks*

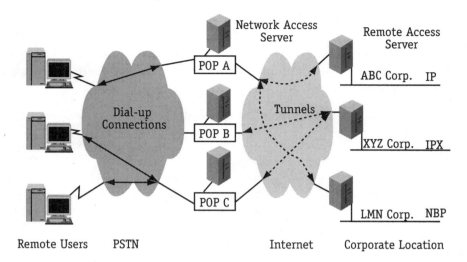

Generally, there are three computers involved in every deployment using tunnels:

- **A tunneling protocol client.** This is usually a client PC.
- **A network access server.** This is an intermediate computer that controls and maintains access to the tunnel.
- **A tunneling protocol server.** This is usually a remote access server.

Comparing a traditional dial-up connection to a tunneled connection, the difference between the two is the intermediate computer, which is known as the network access server. The role of the network access server is described later.

There are some important requirements to consider before tunneling PPP through the Internet (or any other network):

- **End-system transparency.** No special software must be required at either end, i.e., the tunnel must be transparent to the endpoints.
- **Addressing transparency.** The endpoints must be able to manage their network address space independent of the intermediate network through which the tunnel passes. For example, the endpoints could be running an entirely different protocol, such as IPX, and therefore, use IPX addressing even though the tunnel passes through an IP network.

■ **Authentication.** Access to the tunnel (ingress and egress) must be validated to allow secure connection through the network.

■ **Accounting.** The network provider must maintain tunnel usage records. This only makes sense since the tunnel is a resource provided by a network provider such as an ISP.

These requirements were taken into consideration in the development of tunneling protocols.

Tunneling Protocols

The following tunneling protocols were developed by the various companies collaborating to propose draft standards to the IETF (through the process of submitting drafts for RFCs):

■ **Layer 2 forwarding (L2F)[21].** Cisco, Nortel, and Shiva originally proposed this "historic" standard.

■ **Point-to-point tunneling protocol (PPTP)[22].** This was originally proposed by Ascend, 3Com, ECI-Telematics, Microsoft, and USR.

■ **Layer 2 tunneling protocol (L2TP)[23].** This is a combination of the L2F and PPTP proposals.

Today, PPTP implementations exist within Microsoft operating systems, as well as in certain embedded networking devices, such as standalone remote access servers developed by networking companies. Some embedded devices support both PPTP and L2TP implementations.

Tunnel Establishment

The following uses L2TP as the example tunneling protocol to discuss tunnel establishment in detail (PPTP works in a similar manner). To understand the protocol, we first need to establish some terminology:

■ **L2TP access concentrator (LAC).** The LAC is the L2TP term for the network access server. It is a device attached to one or more PSTN or ISDN lines capable of PPP operation. Typically, the LAC is located at an ISP's point of presence (POP).

■ **L2TP network server (LNS).** The LNS is the L2TP term for the tunneling protocol server and is usually the same as the remote access server used in traditional dial-up connections. Its function is to terminate the calls arriving at the LACs. Typically, the LNS is located at a corporation's headquarters.

Tunneling establishment typically involves three phases, and each phase must complete successfully before the next one can begin:

- **PPP connection and communication.** An L2TP client from the client PC uses PPP to connect to an LAC using an analog modem, an ISDN terminal adapter, or an XDSL modem. This connection uses the PPP protocol to establish the connection. The LAC partially authenticates the user through PAP or CHAP. The LAC may maintain a database mapping user names (or domain names such as xyzcorp.com) to "services," where each service is a tunnel endpoint located at the LAC (the other end of the tunnel is, of course, the LNS associated for the service). Once the LAC has authenticated the user, and approved the connection (i.e., allow user access to the tunnel), the process can continue to the next phase.

- **L2TP control connection.** Using the connection to the LAC established by the PPP protocol, the L2TP protocol creates a control connection from the L2TP client to an LNS. In other words, if a tunnel already exists from the LAC to the LNS, the user connection is added onto the tunnel. If the tunnel does not exist, a new one is established prior to adding the user connection. The tunnel is established using a packet-oriented PPP link, e.g., PPP over X.25 virtual circuits or PPP over frame relay PVCs. Adding a user connection to the tunnel implies that the LAC assigns an unused slot on the tunnel representing a "call ID" or "connection ID" to the previously established PPP connection between itself and the client PC. The LAC also sends a connect indication to the LNS over the tunnel. At this point, the LNS can either accept or reject the incoming connection request. If it rejects the connection request, it can return a cause to the LAC, which, in turn, can relay the rejection cause back to the client PC. Assuming the LNS accepts the connection, the process can continue to the next phase.

- **L2TP data tunneling.** Finally, the L2TP protocol creates IP datagrams containing encrypted PPP packets (i.e., encapsulated PPP packets), which are then sent through the L2TP tunnel to the LNS.

The LNS disassembles the IP datagrams, decrypts the PPP packets, and then routes the decrypted packets to the private network. This is shown in Figure 7.15.

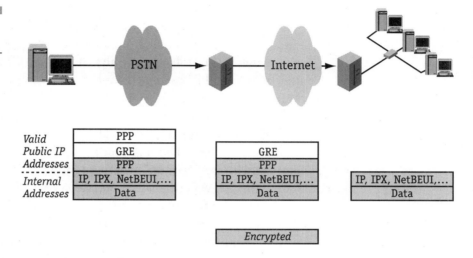

Figure 7.15

L2TP encapsulation

There are two kinds of messages associated with the L2TP protocol:

■ **Control messages.** Control messages are responsible for establishing, managing, and releasing sessions through the tunnel and for maintaining tunnel status.
■ **Payload messages.** Payload messages carry data through the tunnel. The header contains session information and optional acknowledgment and sequencing information.

Differences between Traditional Dial-up and Tunneling

Although the user experience of a traditional dial-up connection versus tunneling is the same, there are important differences between the two in the service models.

■ **Security/authentication.** In traditional dial-up, encryption is optional and authentication is a single-step process (between the client PC and the remote access server). In tunneling, the user data are always encrypted to protect against the fact that part of the link is across a shared (and most likely public) network such as the Internet. Furthermore, authentication is a two-step process. The first level of authentication is at the LAC to secure admission into the tunnel. The second level of authentication occurs at the LNS.

■ **Address allocation.** In traditional dial-up, the client PC's IP address is typically assigned by the ISP as part of IPCP. The client's IP address therefore defaults to whatever public IP address is dynamically assigned. With tunneling, the tunnel carries the address information transparently across the public IP address space of the Internet, so that the client PC's IP address can continue to remain as a privately assigned (and even static) IP address. In fact, the client PC's addressing need not even be IP based. It could, for example, use an IPX address. The IPX header information is then encapsulated along with user data in the IP datagrams that transit the Internet.

■ **Accounting.** In traditional dial-up, the accounting records are maintained only at the remote access server side. With tunneling, both the LAC and LNS can independently maintain accounting data (packet counts, connection attempts, start and stop times, etc.). This works well since the two may be owned by separate organizations. For example, the ISP may own the LAC, whereas the corporation may own the LNS. In this way, the ISP and the corporation may choose to keep independent, separate records or to correlate the two sets for tracking and auditing.

Audio Compression

In the early days of telephony, the need to achieve acceptable voice quality across long distances, while minimizing required bandwidth, drove the need for advances in speech compression. The invention of 64-Kbps pulse-code modulated (PCM) samples using μ-law coding* (in North America) and α-law coding (in Europe) was one of the factors that made possible the evolution of the PSTN to the digital age. Pulse code modulation enabled telephone network operators to offer uniform

---

* μ-law coding is a form of speech compression widely used in telecommunications because it improves the S/N ratio, without increasing the amount of data. Typically, μ-law compressed speech is carried in 8-bit samples.

high-quality voice service even across great distances. Often considered a reference point, 64K PCM voice is referred to as "toll-quality" speech.

By the 1980s, the PSTN had entered the era of the cellular phone and other aspects of the digital age. New applications such as digital cellular, voice messaging, videophones, multimedia documents, and Internet telephony required efficient digital speech coders.

G.7xx Series of ITU Standards

Initially, vendors developed their own proprietary schemes for speech compression, some of which are still around in standalone devices such as digital answering machines. However, the growing need for multiple devices to interact with each other prompted standards organizations to develop international methods to encode and compress audio. The ITU has developed a series of relevant standards as shown in Table 7.13. The source code for implementing some of these standards is available in the public domain.

Table 7.13
ITU telephony
speech coding
standards

Standard	Bit Rate	Frame Size/ Look Ahead	Year Finalized
G.711 PCM	64 Kbps	0.125 ms/0	1972
G.726 (G.721, .723*), GG.727 ADPCM	16, 24, 32, 40 Kbps	0.125 ms/0	1990
G.722 Wideband Coder	48, 56, 64 Kbps	0.125 ms/1.5 ms	1988
G.728 LD-CELP	16 Kbps	0.625 ms/0	1992, 1994
G.729 CS-ACELP	8 Kbps	10 ms/5 ms	1995
G.723.1 MPC-MLQ	5.3 & 6.4 Kbps	30 ms/7.5 ms	1995
G.729 CS-ACELP Annex A	8 Kbps	10 ms/5 ms	1996

*G.721 and G.723 were merged into G.726 in 1990.

The aim of efficient speech compression is to produce a compact representation so that, when it is reproduced, it is perceived by the listener to be "close" to the original. The two main measures of closeness are intelligibility and naturalness. Therefore, the goal of speech compression is to remove natural redundancy in speech samples, while maintaining the intelligibility and naturalness of the reproduction. Two of the important speech compression techniques used in the ITU standards are adaptive differential PCM (ADPCM) and code excited liner prediction (CELP). These techniques will be briefly discussed next, and for additional details, there are several references in the literature (see 24, 25, and 26).

ADPCM

Codecs based on PCM and ADPCM are waveform codecs, meaning that they attempt to produce a reconstructed signal whose waveform is as close as possible to the original. Waveform codecs do not care how the original signal was generated; they merely attempt to reproduce a close approximation to the original through sampling. The advantage of this approach is that waveform codecs are signal independent and can theoretically perform well even for non-speech signals. The disadvantage is that performance is dependent upon the sampling rate. At rates above 16 Kbps, waveform codecs produce high-quality speech, but performance drops at lower bit rates.

The difference between PCM and ADPCM codecs is that the latter attempt to predict the value of the next sample from the previous samples. In other words, ADPCM codecs take advantage of the fact that there are natural correlations present in speech samples due to the effects of the vocal tract and the vibrations of the vocal chords. If the samples are not too far apart in time and the predictions are effective, the error signal variance between the predicted samples and the actual speech samples will be small enough to be quite acceptable. This is the basis of differential pulse code modulation (DPCM) schemes. They quantify the difference between the original and predicted signals into discrete values that can be encoded as a bit stream. ADPCM codecs take the optimization one step further—the predictor and quantizer are adaptive so that they dynamically adjust their behavior according to the characteristics of the input speech signal. The process works as follows:

The input audio is filtered and digitized by an analog/digital (A/D) converter into a series of PCM samples. At any point in time, the PCM input sample X_n is compared to a PCM signal estimate, calculated from the previous sample, X_{n-1}. The resulting differential value is encoded with respect to the current step size of the quantizer and output as a four-bit ADPCM data sample (on which the decoder can subsequently operate). Simultaneously, the ADPCM data sample is fed back into the quantization step size determination logic to generate an estimate for the next input sample, i.e., X_{n+1}. When the ADPCM codec is reset, the quantization step size is initially set to a minimum value, and the estimated signal value is set to zero. As the processing proceeds, these values dynamically adapt, as discussed earlier. The output is a four-bit sample, of which the most significant bit (used as the sign bit) denotes the direction of the original signal—whether ascending or descending. Since ADPCM operation is independent of the signal, nearly any sound and any speech, irrespective of language, can be processed and reproduced with good quality. The optimizations in ADPCM results in a low bit rate, while maintaining voice quality, i.e., down to 32 Kbps instead of 64 Kbps for PCM. In other words, ADPCM can be used effectively to double the voice channel capacity. This is important for carrying voice across expensive links such as undersea cables.

CELP

Codecs based on code excited liner prediction (CELP) work on an entirely different principle than waveform codecs. As noted earlier, the advantage of the waveform codec is that it is signal independent, but its performance depends on the sample rate. For low bit rates, however, waveform codecs do not provide the necessary quality of speech reproduction so a different approach is required. This is where CELP comes in. CELP-based codecs do care how the signal was generated, or more specifically, how the vocal tract is excited by air forced into it through the vocal chords. The resulting speech sounds can be broken into different classes, depending on their mode of excitation:

- Voiced sounds are produced when the vocal chords vibrate open and closed, thus interrupting the flow of air from the lungs to the vocal tract and producing quasi-periodic pulses of air as the excitation. The rate of the opening and closing gives the pitch of the sound.
- Unvoiced sounds result when the excitation is a noise-like turbulence produced by forcing air at high velocities through a constriction in the vocal tract, while an opening in the vocal chords called the glottis is held open. Such sounds show little long-term periodicity.

■ Plosive sounds result when a complete closure is made in the vocal tract, and air pressure is built up behind this closure and released suddenly.

■ Some sounds cannot be considered to fall into any one of the three classes above, but are a mixture.

CELP codecs use an extensive code book of excitation waveforms and a closed-loop search mechanism for identifying the best excitation in the set for every frame of input speech without rigidly classifying the input as voiced or unvoiced. The above adaptively leads to natural sounding speech at relatively low bit rates. Further, if the spectral envelope in the speech model is estimated in a backward manner, based on a recent past history of quantized speech, the result is the low-delay CELP (LD-CELP) system shown in Figure 7.16. The LD-CELP algorithm has been shown to provide high-quality coding of telephone speech at 16 Kbps.

Figure 7.16
LD-CELP
encoding/decoding

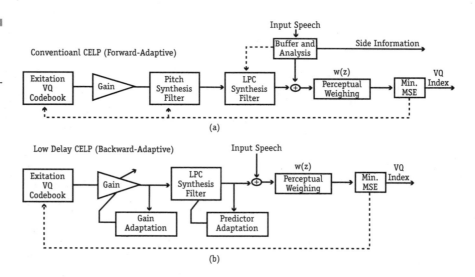

Video Compression

Direct transmission of digital video requires a high-bandwidth channel. Uncompressed NTSC TV signals, for example, require a bit rate of 100 Mbps, whereas other forms of uncompressed video have requirements as high as 400 Mbps. Clearly, this far exceeds any economically viable digital transport capability available in the copper-loop plant. Therefore, for cost-effective video communication and services, efficient coding techniques that reduce the transmission rate for a desired picture quality are essential. Advances in video compression came about in the late 1980s and early 1990s, especially through international standardization endeavors. Several standards bodies have been working on the algorithms to encode and compress video. All these algorithms are based on techniques that manipulate images, since a video is nothing but a sequence of frames, each representing an image. The most commonly used encoding process for images is based on the Joint Photographic Experts Group (JPEG) standard.

Video encoding can generally be divided into two broad categories: intra-frame and inter-frame coding. Intra-frame coding is used to remove spatial redundancy in single-frame images using JPEG or JPEG-like techniques. Inter-frame coding, also known as predictive coding, is used to remove temporal redundancy. It takes advantage of the strong correlation between consecutive video frames to achieve the highest possible compression ratio.

Joint Photographic Experts Group (JPEG) Standard

The JPEG[27] encoding process is shown in Figure 7.17.[28] The input image is divided into individual blocks of 8 × 8 pixels. The individual blocks are then processed in three steps:

- **Discrete cosine transform (DCT).** In the initial stage, the DCT of each block is calculated. DCT is similar to the discrete Fourier transform (used in voiceband and XDSL modem processing); it transforms an image from the spatial domain to the frequency domain. The input to DCT is an 8 x 8 array of integers; each integer is a value between zero and 255 representing the pixel's luminance. The output is an array of coefficients between –1,023 and +1,024 for the corresponding frequency representation.
- **Quantization.** Quantization is the key step necessary to achieve compression. It tries to group "similar" frequencies together so that they can be treated as a single entity. However, there is a tradeoff between quantization and image quality—a large quantization step size can

result in unacceptable image distortion. On the other hand, a small quantization step size results in lower compression ratios. The process takes advantage of the fact that the human eye has a natural high frequency roll-off; i.e., high frequencies are less important than low frequencies. Therefore, by using small quantization step sizes for the lower frequencies and high quantization step sizes for the higher frequencies, it is possible to achieve an optimal balance between compression and acceptable image quality. For example, the first frequency component may be multiplied by one to maintain full accuracy, whereas the last (i.e., 64th) frequency component may be multiplied by .001 since it is scarcely perceived by the human eye. The quantization step sizes are stored in a quantization matrix. After performing matrix multiplication between the DCT coefficients and the quantization matrix, the resulting output is a sparsely populated matrix (i.e., some of the matrix elements become zero).

■ **Encoding.** The output of the quantizer becomes the input to the binary encoder, which may use techniques such as Huffman encoding to remove redundancy. The result is a bit stream representing a compressed image, which can either be stored or transmitted.

Decoding a JPEG compressed image essentially involves an inverse of the steps above (i.e., reverse order).

Figure 7.17
JPEG encoding
process

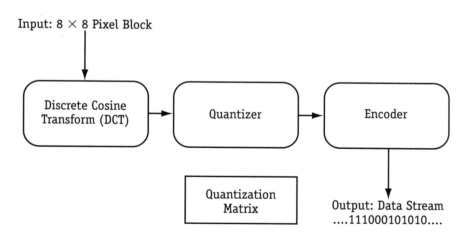

Input: 8 × 8 Pixel Block

Discrete Cosine Transform (DCT) → Quantizer → Encoder

Quantization Matrix

Output: Data Stream
....111000101010....

H.261 Developed by the ITU

The ITU[29] developed the H.261 standard[30] to enable videophone and videoconferencing services over ISDN. This standard specifies a real-time encoding/decoding system with a delay less than 150ms. The algorithm is capable of running over a defined set of transmission rates of p × 64Kbps (where p=1,2, ...30). The format for the input images is based on the common intermediate format (CIF) which is 360 pixels × 288 lines for luminance* and 180 pixels × 144 lines for chrominance.† The frames are non-interlaced, and the input rate is 29.97 frames per second for NTSC-compatible systems. For videophone applications where low bit rates are required, another format called quarter CIF (QIF) has also been defined in the standard. QIF is 180 pixels × 144 lines for the luminance and 90 pixels x 72 lines for chrominance.

The coding scheme for the H.261 standard can be summarized as follows. Each frame is divided into a distinct "macro block." Each macro block consists of one 16 × 16 luminance block and two 8 × 8 chrominance blocks. For each luminance block, the encoding algorithm finds a best-match block in the previous frame (also called motion estimation). For a typical video scene, it is highly likely that an object will sustain for some period. Motion estimation searches for a representation of the current macroblock from a previously coded picture. When a suitable representation is found, only the information that is *different* from that representation needs to be coded, thereby achieving compression. Although the standard does not specify how to find the best-match block, the most commonly used technique is to minimize the displaced block difference (DBD), resulting in a best-match motion vector. The encoder then decides whether intra-frame or inter-frame mode should be used. First, the motion-compensated predictive error (defined as the difference between the current block and the best-match block) is calculated. If this error is small, the difference is encoded. Otherwise, it is advantageous to encode the current block directly. In either case, a coding process similar to the JPEG standard is used to code the block.

Although the H.261 standard was originally designed for two-way videoconferencing, certain applications based on H.261 that utilize only one direction of video transmission may emerge in the future. For

---

* The brightness information of a video signal.
† The signal or portion of a composite signal that bears the color information.

example, educational applications are envisioned in which a lecture is encoded in real time and transmitted to remote classrooms and to students at home. A transmission rate of 1.5 Mbps could provide sufficient quality for coverage of the lecturer and the board or visuals from an overhead projector.

MPEG-1 Developed by ISO/IEC

MPEG stands for the Moving Pictures Expert Group[31]. It is a group working under the joint direction of the International Standards Organization (ISO) and the International Electrotechnical Commission (IEC). This group was formed around 1988 with the original motivation of developing a standard for storage (on devices such as CD-ROMs) of full-motion video at bit rates of 1 Mbps to 2 Mbps. MPEG standards actually cover three parts: MPEG audio, MPEG video, and MPEG system. In the context of this book, we shall focus mainly on the topic of MPEG video.

The recommended input picture size for the MPEG video standard is 360 × 240 pixels for luminance and 180 × 120 for chrominance. The frame rate is 29.97 frames per second for NTSC compatible systems, although these parameters can vary in the standard. The MPEG-1[32] coding scheme approved in 1992 by ISO/IEC is very similar to that of H.261. The major difference between the two is that MPEG-1 allows bidirectional motion estimation. In MPEG terminology, there are three kinds of frames, as shown in Figure 7.18:

■ **Intra-frames (I-frames).** This term is applied to frames coded as a still image with no past or future history used in the coding. The encoding scheme for I-frames is the same as that for JPEG. I-frames are considered "baseline" frames. There must be sufficient I-frames at regular intervals to support accurate decoding. Since the other types of frames (discussed below) work on motion estimation, using an incorrect I-frame for prediction increases the decoding error. Therefore, the periodic introduction of I-frames in the decoding stream ensures that there is a reference point.

■ **Predicted frames (P-frames).** This term is applied to frames that are coded based on prediction (using motion estimation) from the most recently constructed I-frame or P-frame. This technique works well in situations where the background stays the same through a sequence of frames as in a movie. If the predicted change is large (e.g., the background changes) then an I-frame must be used to reset the reference.

- **Bidirectional frames (B-frames).** This term is applied to frames that are coded based on prediction from the two closest frames, one in the past and one in the future. This process takes the difference between the two closest frames and averages that difference to construct a new frame. This increases the compression ratio (resulting in the smallest-sized frames), but adds computational complexity and delay.

Figure 7.18
I-, B-, and
P-frames in MPEG
compression

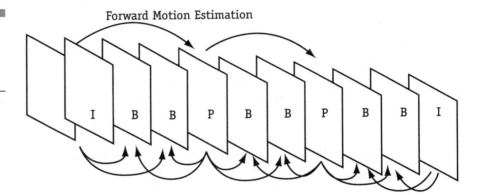

MPEG-1, has been used successfully to demonstrate that VCR-like video quality can be achieved at a bit-rate of 1.5 Mbps. Therefore, MPEG-1 enables XDSL technologies to deliver full-motion video over a copper-loop access network. However, although the standard can support a broad set of applications at bit rates below 1.5 Mbps, its quality and flexibility are not considered ideal for applications that demand higher bit rates. The MPEG-2 project was organized to meet the objective of better quality at bit rates between 4 Mbps to 9 Mbps.

MPEG-2 Developed by ISO/IEC

The motivation for MPEG-2 was to develop a high-quality video standard for broader and higher bit-rate applications including broadcasting, consumer electronics, and telecommunications. The MPEG-2[33] standard approved in 1994 by ISO/IEC is an extension of MPEG-1:

- MPEG-2 uses a 10 × 10 matrix for DCT, rather than the 8 × 8 matrix supported by both JPEG and MPEG-1. This provides an improvement of over 50 percent in the number of coefficients. Consequently, in MPEG-2, an I-frame is encoded in roughly 400,000 bits; in MPEG-1, an I-frame requires only 150,000 bits. For both B-frames and P-frames, MPEG-2 requires twice as many bits to encode as MPEG-1. The result of all of this is a much higher picture quality, again at the cost of greater computational complexity and delay. However, with advances in semiconductor technology such as digital signal processors (DSPs), the encoding/decoding algorithms can be efficiently implemented in low-cost silicon.
- MPEG-2 supports four resolution levels: (a) low, which is compatible with MPEG-1 images; (b) main, which is used for NTSC compatibility; and (c), two high-resolution levels for HDTV.

The MPEG-2 systems layer provides features necessary for multiplexing and synchronization of video, audio, and data streams:

- Video streams are broken into units called video access units. A video access unit corresponds to one of the image frames described above (I, P, or B). A collection of video access units is a video elementary stream, and several elementary streams can be combined and packetized to form packetized elementary streams (PES). PES can be stored or transmitted as they are, but are commonly converted into either program streams or transport streams. Program streams (PS) resemble the original MPEG-1 streams. They consist of variable-length packets and are intended for use in media where there is a very low probability of bit errors or data loss. Transport streams (TS) are fixed length. Each TS packet is 188 bytes long with 4 bytes of header information. The TS packets are intended for transport over media where bit errors or loss of information is likely.
- Synchronization information is built into the MPEG-2 system layer through time stamps. Two timestamps—the presentation videoconferencing time stamp (PTS) and the decoder time stamp (DTS)—are included in the PES packet header. These time stamps tell the

decoder when to display information to the end user and when to decode information in the decoder buffers, respectively. Obviously, this requires the clocks between the encoder and the decoder to be synchronized. There are several techniques to achieve synchronization, including those that leverage mechanisms built into an underlying ATM transport.

MPEG-2 plays an important role in the delivery of high-quality video over broadband networks. MPEG-2 is already used in many aspects of everyday life, for example:

- Direct broadcast satellite (DBS)
- Cable television (CATV)
- High-definition television (HDTV).

MPEG-4 Developed by ISO/IEC

The standards organizations began work effort on MPEG-4* in 1998. The MPEG-4 format is meant to become the universal language between broadcasting, movie, and multimedia applications. While former MPEG standards (MPEG-1 and MPEG-2) were only concerned with compression, MPEG-4[34] provides additional functionality to satisfy the needs of content publishers, service providers, and end users:

- For content providers, MPEG-4 enables reusable content production and copyright protection.
- For network service providers, MPEG-4 offers transparent information that can be interpreted and translated into the appropriate native signaling messages of each network. MPEG-4 also provides a generic QoS descriptor for different MPEG-4 media that can be translated to network QoS by network providers. This is important to enable services over broadband technologies such as XDSL.

---

* After MPEG-1 and MPEG-2, there was discussion of an MPEG-3 standard suitable for HDTV. However, the standards working group members later realized that they could simply extend MPEG-2 to cover this application. Therefore, MPEG-3 was skipped, and the next generation came to be known as MPEG-4.

■   For end-users, MPEG-4 brings higher levels of interaction with content, within the limits set by content publishers. It also brings multimedia to new networks, including those employing relatively low bit rates, and wireless communication.

MPEG-4 provides for standardization in four areas:

■   Representation of aural, visual, or audiovisual units called "media objects." These media objects can be of natural or synthetic origin, i.e., they can be recorded with a camera or microphone or computer generated.

■   Description of the composition of these objects to create compound media objects that form audiovisual scenes (A-V scenes).

■   Mechanisms to multiplex and synchronize the data associated with media objects, so that they can be transported over network channels providing a QoS appropriate for the nature of the specific media objects.

■   Interaction with the A-V scene generated at the receiver.

Videoconferencing

Throughout this book, we have used videoconferencing as an example of the possible applications enabled by broadband access networks such as XDSL. It is a natural fit for broadband access since it combines real-time audio and video processing. This requires quality of service (QoS) parameters, such as low latency and high bandwidth, to be effective.

Once it was the exclusive domain of business-to-business communication, but advances in technology have made videoconferencing almost affordable for consumers in the late 1990s. Technological advances and competitive forces are expected to make this application even more inexpensive and appealing to consumers in the years ahead. Anticipating the need for widespread audio/video communication, the ITU has, once again, been active in developing videoconferencing standards to enable interoperability among multiple vendors.

H.32x Series of Standards Developed by the ITU

H.32x[35,36] actually refers to a family of standards related to audio/video communications using videoconferencing equipment. These standards allow multiple vendors' videoconferencing equipment to interoperate using a common messaging format. Table 7.14 summarizes the H.32x series of standards[37]:

Table 7.14

A summary of the H.32x series of standards (Source: Ref. 37)

	H.320	H.321	H.322	H.323 V1/V2	H.324
Approval Date	1990	1995	1995	1996/1998	1996
Network	Narrowband switched digital ISDN	Broadband ISDN ATM LAN	Guaranteed bandwidth packet-switched networks	Non-guaranteed bandwidth packet-switched networks, (Ethernet)	PSTN or POTS, the analog phone system
Video	H.261, H.263	H.261, H.263	H.261, H.263	H.261, H.263	H.261, H.263
Audio	G.711, G.722, G.728	G.711, G.722, G.728	G.711, G.722, G.728	G.711, G.722, G.728, G.723, G.729	G.723
Multiplexing	H.221	H.221	H.221	H.225	H.223
Control	H.230, H.242	H.242	H.230, H.242	H.245	H.245
Multipoint	H.231, H.243	H.231, H.243	H.231, H.243	H.323	
Data	T.120	T.120	T.120	T.120	T.120
Communications Interface	I.400	AAL, I.363, I.400, TCP/IP AJM, I.361, TCP/IP PHY, I.400			V.34 Modem

Summary

In this chapter, we worked our way up the OSI layers to discuss a few key technologies above the physical layer. We began with a discussion of ATM—an important link-layer technology that provides the framework to support integrated voice, data, and video. We then discussed how the protocol of the Internet, TCP/IP, enables applications to transparently ride on top of various underlying link-layer technologies. Next, we looked at PPP and the related tunneling protocols that sit in the layers between ATM and TCP/IP. Above the transport layer (the domain of TCP/IP) are the session and presentation layers where specialized services such as encryption/decryption and compression/decompression are performed. We discussed some of the important standards-based audio and video encoding/compression technologies. Finally, we discussed videoconferencing as an example of a broadband-enabled application that is based on the standardization of several important underlying technologies.

References

1. Kumar, B. *Broadband Communications*, McGraw-Hill (May 1998).
2. Dutton, H., and Lenhard P. *Asychronous Transfer Mode (ATM) Technical Overview* (New York: Prentice-Hall, 1995).
3. Kwok, T. *ATM: The New Paradigm for Internet, Intranet, and Residential Broadband Services and Applications* (New York: Prentice-Hall, 1997).
4. Simoneau, P. *Hands-on TCP/IP*, McGraw-Hill (August 1997).
5. Stevens, W. R. *TCP/IP Illustrated, Volume 1: The Protocols*, Addison-Wesley (January 1994).
6. Black, U. D. *TCP/IP and Related Protocols*, McGraw-Hill (December 1997).
7. Information is available on the official Website at **www.ietf.org/ rfc.html**.
8. A searchable RFC index is available at **www.rfc-editor.org/rfc-search.html**.
9. Postel, J. "Internet Protocol," RFC 791 (September 1981).
10. Plummer, D. C. "Ethernet Address Resolution Protocol," RFC 826 (November 1982).
11. Postel, J. "Internet Control Message Protocol," RFC 792 (September 1981).

12. Deering, S. "Host Extensions for IP Multicasting," RFC 1112 (August 1989). See also RFC 2236.
13. Postel, J. "Transmission Control Protocol," RFC 793 (September 1981).
14. Reynolds, J., and Postel J. "Assigned Numbers," RFC 1700 (October 1994).
15. Postel, J. "User Datagram Protocol," RFC 768 (August 1980).
16. Simpson, W. ed., "The Point-to-Point Protocol," RFC 1661 (July 1994). See also RFC 2153.
17. Carr, D., Coradetti, T., Lloyd B., McGregor, G., and Sklower K. "The PPP Multilink Protocol," RFC 1990 (August 1996).
18. Carlson, J. *PPP Design and Debugging*, Addison-Wesley, 1997.
19. Sun, A. *Using and Managing PPP*, O'Reilly & Associates, 1999.
20. Gross, G. Kaycee, M., Li, A., Malis, A., and Stephens, J. "PPP over AAL5," RFC 2364 (July 1998).
21. Valencia, A., Littlewood, M., and Kolar T. "Cisco Layer Two Forwarding (Protocol)," RFC 2341 (May 1998).
22. Hamzeh, K., Pall, G. S., Verthein, W., Taarud, J., Little, W. A., and Zorn, G. "Point-to-Point Tunneling Protocol (PPTP)," draft-ietf-pppext-pptp-08.txt (February 1999).
23. Townsley, W. M., Valencia, A., Rubens, A., Pall, G. S., Zorn, G., and Palte, B. "Layer Two Tunneling Protocol: L2TP," draft-ietf-pppext-12tp-14.txt (February 1999).
24. Jayant, N. S.. and Noll, P. *Digital Coding of Waveforms,* New York: Prentice-Hall, 1985.
25. O'Shaughnessy, D. *Speech Communication: Human and Machine*, Addison-Wesley (1987).
26. Barnwell, T., Nayebi, K., and Richardson, C. H. *Speech Coding: A Computer Laboratory Textbook*, New York: John Wiley, 1996.
27. Information is available at the official Website at **www.jpeg.org/public/jpeghomepage.htm**.
28. Wallace, G. K. "The JPEG Still Picture Compression Standard," *Communication of the ACM* (April 1991).
29. Information is available at the official Website at www.itu.int.
30. "Recommendation H.261: Video Codec for Audio-Visual Services at p x 64 Kbps," *ITU Recommendation* (March 1993).
31. Information is available at the official Website at **drogo.cselt.stet.it/mpeg**.

32. Chiariglione, L. "MPEG-1: Coding of Moving Pictures and Associated Audio for Digital Storage Media at Up to About 1.5 Mbps," *ISO/IEC 11172* (June 1996).

33. Chiariglione, L. "MPEG-2: Generic Coding of Moving Pictures and Associated Audio Information," *ISO/IEC 13818* (July 1996).

34. Koenen, R. ed., "Overview of the MPEG-4 Standard," *ISO/IEC JTC1/SC29/WG11 N2725* (March 1999).

35. "Recommendation H.320: Narrowband Visual Telephone Systems and Terminal Equipment," *ITU Recommendation* (July 1997).

36. "Recommendation H.323: Packet-Based Multimedia communications Systems," *ITU Recommendation* (February 1998).

37. "A Primer on the H.323 Standard," Version 2.0 (DataBeam Corporation). Available from the company Website at **www.databeam.com/ h323/h323primer.html**.

Premises and Access Network Architectures

Overview

In this chapter, we will discuss how the various physical-layer DSL technologies described in previous chapters fit into real-world premises (customer location) and access network architectures. We will begin with a review of the key requirements for premises architectures developed by the ADSL Forum. Although these requirements were developed for residential customers, most of them will translate directly into the business environment (with the exception of any requirements for entertainment services). We will then study how some of the proposed architectures meet these requirements. We will then enter a discussion of the functional categories of access architectures, and the network elements that comprise them. In this context, we will review some commonly deployed DSL-based access architectures and discuss important architectural issues. The chapter concludes with a look at future directions for voice over DSL.

Premise Network Architecture Requirements

In Chapter 1, we discussed the reference model developed by the ADSL Forum for consistent definition of interfaces in a broadband architecture[1]. The Forum has subsequently refined the reference model for the development of CPE architecture requirements, as shown in Figure 8.1.

Figure 8.1

Refined reference model for end-to-end architecture (Source: ADSL)

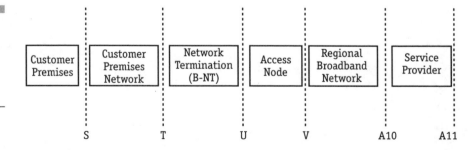

| Customer Premises | Customer Premises Network | Network Termination (B-NT) | Access Node | Regional Broadband Network | Service Provider |

S T U V A10 A11

Notes:
1) S, T, U, and V correspond to ITU practice. A10 and A11 are borrowed from DAVIC as there are no ITU equivalents.
2) Regional Broadband Network is referred to as Transport in the book text.

This model assumes that the proposed architectures are based on a PPP-over-ATM-over-ADSL protocol model at the U-interface as specified by the ADSL Forum[2]. For more information on the rationale for adopting a PPP-over-ATM-over ADSL protocol model, see the works referred to in 3 and 4. The ADSL Forum has developed a document targeted to residential customers desiring data access* that specifies the requirements to support a broadband services architecture consistent with this protocol model[3]. The requirements are based on the following market segmentation (the first two categories assume a single PC at the premises, whereas the latter two assume multiple PCs):

▮ **Internet access.** These are residential users who have a single PC to access the Internet. They may have accounts with multiple ISPs, but they do not plan on accessing multiple ISPs simultaneously.

▮ **Work at home.** These are residential users who use high-speed access for telecommuting. Such users typically have a single PC connected to a single destination (i.e., their corporate office). It is highly desirable that the home environment be indistinguishable from the office environment.

▮ **SOHO (office/home office).** These are small or branch offices with multiple interconnected PCs used for business purposes that will most likely be an "always-up" connection to a single destination (e.g., a corporate office or the Internet).

▮ **Multi-purpose residential users.** These are the technologically sophisticated residential users who typically have multiple PCs at home and access both ISPs and their corporate networks. Members of the household may require simultaneous access to different network service providers (NSPs).

The requirements are listed from three different perspectives: (a) CPE configuration, (b) U-interface, and (c) service definitions. They are labeled as mandatory (M), conditional (CR), or optional (O). These requirements are briefly described below.

---

* In 1999, the ADSL Forum took up a separate work effort to define the requirements and architectures to support derived voice channels over DSL. The goal is to build upon the baseline architectures developed for data access.

CPE Configuration Requirements

The following have been labeled as *conditional* requirements:

- **Simple administration and management.** Consumers cannot be expected to become experts at setting up and managing their premises networks, whereas corporate environments have experts who are hired to maintain local area networks.
- **Support for multiple PCs connected simultaneously.** According to several research organizations, the number of homes with multiple PCs is approaching a critical mass in 1999. In addition, a significant number of premises are expected to have multiple Internet capable devices. A premises network architecture must connect these PCs and devices within a home and allow them to share a common link to a network services provider.
- **Support for intrapremise networking.** From the previous requirements, it follows that the network architecture must also support PC/device communication within the premises without traversing the access network.

The following have been labeled as *optional* requirements:

- **Should not be ADSL specific.** The premises network architecture solution should not be ADSL-centric. It should easily adapt to other DSL variants and access technologies, such as cable and satellite.
- **Support for different in-home technologies.** There are a number of available technologies that can support the connection between a PC/device and a DSL modem. Examples include Ethernet, USB, IEEE 1394, ATM-25, HomePNA, power line, and distance wireless. These networking technologies differ in the bandwidth they support, the number and types of devices they can network together, and the physical media they operate over. It is also likely that many of these technologies will have to coexist in subscribers' homes. Therefore, the architecture should not preclude connection between in-home devices and the high-speed access network with any of these technologies.
- **Support for simultaneous connections to multiple NSPs.** It may be desirable for users to have simultaneous active sessions with different network service providers.

U-interface Requirements

The following have been labeled as *mandatory* requirements:

- **PPP support.** This requirement follows as a result of conformance to the ADSL Forum's protocol model at the U-interface[2]. As described in Chapter 7, PPP provides session, authentication, and IP configuration management.
- **Avoidance of additional crosstalk.** The premises network architecture should incorporate safeguards to prevent any electrical signals from leaking out of the premises onto the access copper pairs.

The following have been labeled as *conditional* requirements:

- **Providing dynamic rate adaptation feedback.** With physical layer technologies such as ADSL, the speed of the link may dynamically change due to variations in environmental or premises conditions. As described later in this chapter, this is especially true for G.lite modems operating in splitterless mode. When the data rate changes, it is necessary to inform the source of any rate reduction, i.e., when a PC/device is sending data, it must be notified to throttle back to the new rate. Therefore, the premises network architecture must provide flow control between the PCs/devices on the premises network and the DSL modem on the access network.
- **Supporting ATM switched virtual connections (SVCs) and permanent virtual connections (PVCs).** Service providers may choose to deploy PVCs, SVCs, or both. The premises network architecture must support the ability for the end-user PCs and devices to control SVC signaling in an SVC environment or to choose the proper PVC in a PVC environment.
- **Providing configuration information.** The premises network architecture should provide a mechanism to distribute necessary network configuration information to the end-user PCs and devices.

The following have been labeled as *optional* requirements:

- **Selecting the PVC based on QoS parameters.** For PVC-based architectures, the DSL modem must be able to select a PVC based on the QoS attributes required by the application.

Service Definition Requirements

Unlike the requirements based on CPE configuration or the U-interface, the requirements based on service definition depend upon the service itself. The following requirements have been identified:

- **Subscriber initiated sessions.** This means that the premises network architecture must enable the user on any PC/device to bring up or tear down a session to the desired NSP.
- **Dynamic NSP selection.** The premises network architecture must enable the user on a PC/device dynamically to pick the NSP with which to establish a session.
- **Compatibility with AAA services.** The premises network architecture must be compatible with the authentication, authorization, and accounting mechanisms utilized by the NSP.
- **Dynamic IP configuration.** The premises network architecture should enable the NSP to provide the end-user PC/device with the necessary IP configuration information on a per-session basis.
- **Network initiated sessions.** In addition to user initiated sessions, it is also necessary for the premises network architecture to support the ability for an NSP to initiate sessions to a customer's PC/device. This implies that each PC/device must be uniquely addressable by the outside world. If the PC/device is in power savings mode, there must be a mechanism to wake it up so it can respond to the incoming connection request.
- **Security.** The premises network architecture should be compatible with end-to-end security mechanisms such as IPSec (Internet protocol security) and should permit the use of firewalls at the access point.
- **Binding.** The premises network architecture should enable temporal connections of a PC/device to a particular virtual connection (VC).
- **Routing.** The premises network architecture should allow packet forwarding between different subnets within the premises and between the premises subnets and the access network.
- **Traffic prioritization.** The premises network architecture should enable different priorities of traffic (at least "high" versus "low").
- **Bandwidth reservation.** The premises network architecture should enable bandwidth reservation for applications that require it.

■ **Dual path latency.** The premises network architecture should reduce delay for certain multimedia applications such as streaming video.

Proposed Premises Network Architectures

A number of premises network architectures have been proposed and discussed at length within the ADSL Forum and elsewhere. Given the broad set of requirements above, it is safe to say that it is nearly impossible for a particular architecture to satisfy all of them. In this section, we present a detailed discussion of some of the popular architectures that have been proposed. We leave it up to the reader to judge the appropriateness of a particular architecture to a specific implementation that would enable the desired services. The architectures we will examine are:

■ Single host, internal DSL modem (e.g., ATU-R). This is a trivial case, but it provides the backdrop for discussion on the advanced cases.
■ Multiple hosts, internal DSL modem.
■ Single/multiple host(s), external DSL modem (e.g., ATU-R) over ATM-based premises distribution. This is a solution implemented at layer 2.
■ Single/multiple host(s), external DSL modem (e.g., ATU-R) over non-ATM based premises distribution. This is a solution implemented at layer 2.
■ Multiple hosts, any media premises distribution using routing. This is a solution implemented at layer 3.

Single Host, Internal DSL Modem

Rationale

The rationale for this architecture is that it is simple, and one of the early implementations to help jump-start the market for DSL deployment. The basic idea is to embed the DSL modem in the end-user PC/device—most common implementations consist of a PCI ADSL modem adapter preinstalled in one of the available slots in a PC. In some cases, the ADSL modem can support either G.dmt or G.lite, with a fallback to V.90 analog capability.

Protocol Model

The protocol model for the single host, single internal modem is shown in Figure 8.2. In this case, the S- and T-interfaces are internal to the host.

Figure 8.2

Protocol model for single host, internal DSL modem

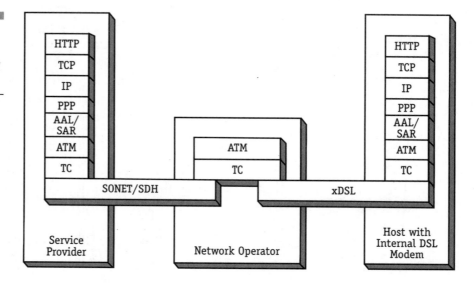

Benefits

This is likely to be the least-cost solution. It takes up no additional real estate, and mounting is simple from all perspectives—the user, the PC manufacturer, and the service provider.

■ The user experience is assumed to be similar to that of using a pre-installed V.90 modem. For example, in the Windows environment, the user would use dial-up networking (DUN) to connect over the ADSL link; the only perceived difference is the high speed.

■ The PC manufacturer preinstalls and configures the hardware and most of the software to enable access.

■ Service providers do not have to get consumers to "crack open the case" to install new hardware. In a situation where G.lite service is deployed in a splitterless mode, the provider does not have to send a technician to the site. The provider supplies the user with access parameters similar to the dial-up paradigm today, and the user configures these parameters using familiar applications.

Drawbacks

The ATM function is embedded in the PC, therefore, it requires either a hardware or software SAR. This, however, is not a major issue since either hardware or software solutions are now cost effective. Hardware SARs add a little extra cost to trade off against host CPU MIPs.

Open Issues

■ In the absence of an autoprovisioning system or SVC support, the user will have to configure the ATM stack with at least a VPI and VCI. For both PVCs and SVCs, the user will need to provide a user identifier and password for integration with AAA services at the NSP (however, this information can be stored locally in an encrypted file*). There have been discussions within the industry of utilizing fallback capability to V.90 analog, if available, to bootstrap to the new service by dialing a provider's Website and downloading the parameters necessary to provision for DSL.

■ Access to multiple providers in the absence of SVCs will also require some configuration at both ATM termination points.

■ To meet many of the requirements, advances in the capabilities of host operating systems or host operating system updates will be required. For example, it is difficult or impossible to support simultaneous sessions to multiple NSPs with some host operating systems. Currently, not all PC operating systems have PPP over ATM support, and even fewer embedded operating systems offer it today.† Currently, features such as multiple sessions to a single NSP may not be supported.

---

* This works for a PC, but not necessarily for embedded devices.
† This situation, however, is likely to change in the future as support for PPP over AAL5 is added to PC and embedded operating systems.

Multiple Hosts, Internal DSL Modem

Rationale

The rationale for this architecture is that, once DSL connectivity is established, there will be a need for other PCs/devices to share the DSL bandwidth. In this case, the PC/device (usually a PC) acts as a gateway, relaying traffic on behalf of other PCs/devices on the network (the gateway PC/device and the other PCs/devices are interconnected through a hub). The premises network could be any of the technologies based on Ethernet-like CSMA/CD, i.e., traditional 10/100 Base-T Ethernet, HomePNA, or IEEE 802.11 (wireless).

Protocol Model

The protocol model for a gateway PC is shown in Figure 8.3. The protocol model for a hub-attached PC/device is shown in Figure 8.4.

Figure 8.3

Protocol model for a gateway PC

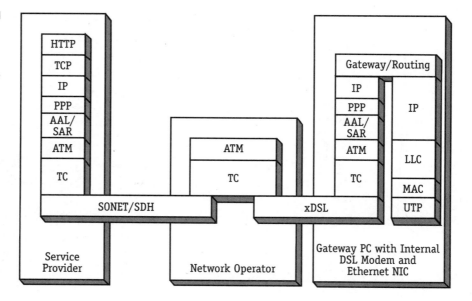

Benefits

- It is an easy step up to enable other PCs/devices to share the DSL bandwidth. New versions of PC operating systems are evolving to enable Internet connection sharing. Configuration wizards can walk the user through setting up both the gateway PC and any number of client PCs to share the DSL connection.

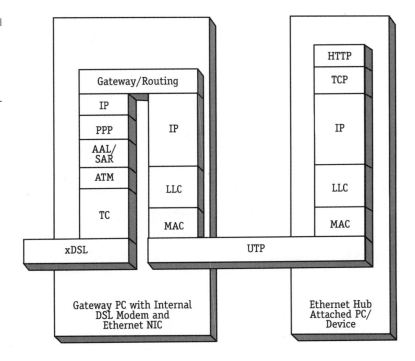

Figure 8.4

Protocol model for a hub-attached PC/device

Drawbacks

■ This architecture requires the gateway to be "always on," which may not be practical if the gateway is a PC.

■ If the gateway is a PC, it has to perform its traditional processing functions while relaying traffic for other PCs/devices on the network. This may put a strain on processing resources. It also brings up the question of the reliability of the PC when it is performing all these tasks. This is an important concern service providers.

■ Only the gateway PC can get the benefit of ATM connectivity (such as QoS). The other PCs/devices are still in a collision domain with QoS issues, unless mechanisms such as IEEE 802.1q are deployed.

Open Issues

■ The reliability of the gateway PC is a major concern. This will require robust hardware and software to ensure that various applications running on the gateway PC can continue to run without interfering with gateway functions. Given the abundance of applications that can run on a PC, it is impossible completely to "bullet-proof" the gateway PC, unless its functions are drastically limited to run only "tested" applications.

■ Leaving a PC on all the time to act as a gateway is not always practical. The industry has been making progress in its efforts to develop a PC that is "instantly available," but in today's market, they are still in the developmental stage.

Single/Multiple Host(s), External DSL Modem Over ATM-based Premises Distribution

Rationale

The rationale for this architecture is that the benefits of ATM can be extended all the way to the end-user PCs/devices. It also greatly simplifies the protocol model as shown below.

Protocol Model

The protocol model for the single host, ATM-attached external modem is shown in Figure 8.5.

Benefits

The circuit-switched nature of this model brings with it all the benefits of the existing telephony model, particularly if the host operating system and supporting infrastructure support end-to-end SVCs. Security, quality of service, and access policy models are particularly easy to provision and administer with this model using standard ATM mechanisms.

Limitations

■ ATM-25 network equipment is far less available than other premises distribution media (for example, Ethernet).

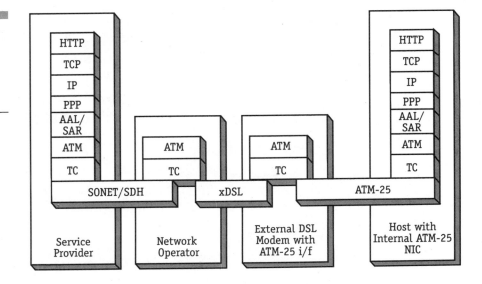

Figure 8.5
Protocol model for single host, ATM-attached DSL modem

- Support for ATM in prevalent host operating systems is spotty. Legacy systems such as Windows 3.1 and Windows 95 have no native support for ATM. Support on other platforms, as for Macintosh or UNIX, is even harder to find. For most of these legacy systems, the installation process will be fairly complex, requiring new network interface cards, drivers, and ATM software stacks. The comparatively small ATM-25 market puts this model at a disadvantage versus the much higher-volume, more broadly supported premises distribution models.

- Support for multiple hosts requires an ATM switch function at the premises, either separate or incorporated into the DSL modem. Currently, ATM switches in the core network are expensive, and it is unlikely that their prices will drop any time soon to levels within consumer reach. It might be possible to incorporate a low-cost ATM switch function into the hardware of the DSL modem, but the low-volume market for ATM makes this unlikely.

Open Issues

This model is well understood, and there are no major open issues other than the low-volume market for ATM-25. A minor issue is that host operating systems have to support PPP over AAL5 as specified in

RFC 2364[4] and (optionally) incorporate SVC support. Support for these features is starting to become available on the newer versions of some PC operating systems.

Single/Multiple Host(s), External DSL Modem Over Non-ATM Based Premises Distribution

Rationale

The overriding basis to consider non-ATM-based premises distribution is the lack of ubiquity of ATM-25-enabled end-user PCs/devices, and the associated cost of ATM-25 relative to alternative solutions. For the single-host case, the rationale for this architecture is primarily to provide a DSL migration path to the installed base of PC users that do not have a preinstalled DSL modem. The basic idea is to connect an end-user PC/device to an external modem that connects to the DSL line through some means other than ATM. Sometimes, the external modem is powered separately; otherwise, it may draw power from the PC. Initially, the means to connect the modem to the PC/device was through Ethernet; today, alternatives such as the universal serial bus (USB) or IEEE 1394 are appearing on the horizon.

For multiple hosts, the rationale is to share the DSL bandwidth. In this case, all the PCs/devices are connected via a common hub that is connected to the external modem. In other words, the external modem is shared much like a networked printer. The premises network could be any of the technologies based on Ethernet-like CSMA/CD, i.e., traditional 10/100 Base-T Ethernet, or HomePNA, or IEEE 802.11 (wireless).

Protocol Model

In the early days of ADSL trials, there were no standards to connect an external DSL modem to a host such as a PC. In an attempt to address this issue, different methods have emerged:

▪ Local tunnels can be based on the point-to-point tunneling protocol (PPTP) or layer 2 tunneling protocol (L2TP)[5]. Tunneling protocols (discussed in Chapter 7) have been proposed as a mechanism to connect external modems to hosts over Ethernet. The protocol diagram for a PC connected to an external modem via Ethernet using L2TP is shown in Figure 8.6.

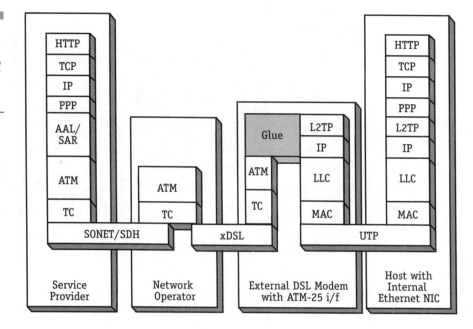

Figure 8.6

Protocol model for single-host, external DSL modem using L2TP

- PPP-over-Ethernet (PPPoE)[6] is specific to Ethernet connections. The protocol diagram for a PC connected to an external modem via Ethernet using PPPoE is shown in Figure 8.7.
- Broadband modem access protocol (BMAP)[7] has been proposed as a generic mechanism to connect external modems to hosts over Ethernet, USB, or IEEE 1394. The protocol diagram for a PC connected to an external modem via USB using BMAP is shown in Figure 8.8.

Benefits

Whether the protocol model uses local tunnels, PPPoE, or BMAP, the external modem architecture has the following advantages:

- It provides a migration path to DSL services for the installed base of PC users.
- It isolates the end-user PC/device from the type of service available from a local service provider e.g., ADSL versus HDSL versus VDSL.

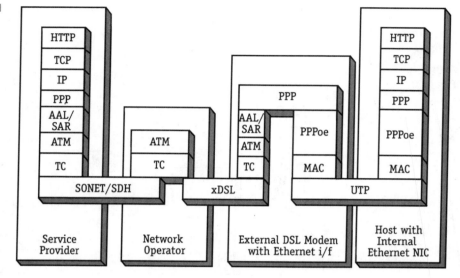

Figure 8.7
Protocol model for single host, external DSL modem using PPPoE

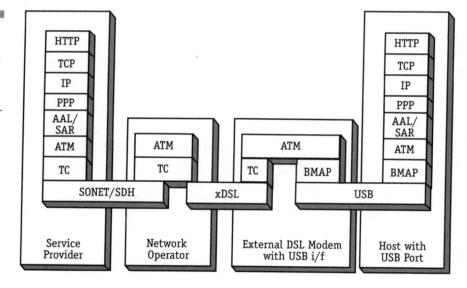

Figure 8.8
Protocol model for single host, external DSL modem using BMAP over USB

- New features and services can be added to the external modem through software downloads without impacting the host (PC/device).
- It supports multiple simultaneous sessions with multiple concurrent NSPs.

■ Tunneling support (especially PPTP) is already available in PC operating systems.

The following are the advantages of PPPoE:

■ PPPoE is simple. After a basic negotiation to discover the peer's MAC address, PPP frames are transported directly within an Ethernet frame, with a shim header to provide session multiplexing.
■ PPPoE can multiplex several PPP sessions over the same VC.

The following are the advantages of BMAP:

■ Unlike other Ethernet-based solutions, BMAP attempts to maintain full ATM connectivity with QoS to the end user PC/device.
■ It provides a uniform solution for Ethernet, USB, and IEEE 1394 connected modems.

Limitations

The following are the disadvantages of local tunnels (PPTP or L2TP):

■ The protocol stack at the end-user PC/device is complex and may not be practical for certain embedded devices.
■ The overhead cost of processing every packet to add/remove tunnel headers at either of the connections (PC and the DSL modem) is a consideration.
■ "Glue" code development is required at the DSL modem to terminate the tunneling control protocol and bridge PPP from a tunnel to a specific ATM VC.
■ Each end-user PC/device must have two IP addresses—one to terminate the PPP session at the far end, and the other to terminate the local tunnel.* This adds to the configuration complexity.

The following are the disadvantages of PPPoE:

■ PPPoE requires additional driver software (known as a "shim" in the RFC) to be installed in the operating system.

---

* The IP address to terminate the local tunnel can be a local (e.g., "network 10") address, since it is not seen outside of the premises network.

■ Unlike BMAP, PPPoE can support multiple PCs/devices sharing the DSL line through PPP session multiplexing. However, no mechanism exists today that can receive incoming calls over PPPoE. The PPPoE discovery mechanism is asymmetrical, and the discovery messages cannot be reversed to support incoming calls and routing to the appropriate PC/device.

■ PPPoE does not provide QoS support and may be limited to data-access applications only.

■ PPPoE works only with Ethernet modems.

The following are the disadvantages of BMAP:

■ Like PPPoE, BMAP requires additional driver software to be shipped with the BMAP-enabled modem and installed in the operating system.

■ The protocol stack at the PC/device is much more complex than a traditional dial-up—instead of PPP directly accessing the dial-up modem, this introduces Ethernet, ATM, and BMAP between PPP and the XDSL modem.

■ BMAP supports multiple PC/devices sharing the DSL line, but at a price—one entity (usually the PC) must act as a gateway and relay traffic from the others. This the same issue as a gateway PC with an embedded modem. Alternatively, the external modem may run multiple instances of BMAP, but this is still under study.

Open Issues

■ The industry has not yet settled on one definitive method to connect external modems over Ethernet. Today, local tunnels have been ruled out in favor of PPPoE or BMAP. In the long term, however, BMAP may get broader support since it works with other connectivity options as well, e.g., USB and its QoS support.

■ For PPPoE, the issues are QoS support and accepting incoming calls. There have been discussions about using the IEEE 802.1q standard to address the QoS question[8].

■ For BMAP, the issue of multiple instances of BMAP to support multiple hosts sharing a common DSL line is under study.

■ If multiple PCs/devices have to be supported, then the provisioning and SVC support issues are more complex than in the case of the single-host internal DSL modem.

Multiple Hosts, Any Media Premises Distribution Using Routing

Rationale

The primary rationale for this architecture is the growth of Internet-capable devices in the home, especially multiple PCs. With PC prices continuing to fall, the need to have Internet connection sharing becomes a driving factor. Routing is preferred in this environment because of the following considerations:

■ With a multiplicity of devices in the home, there is a need to have a measure of security at the T-interface point. Security can take various forms, including firewall capability to fend off hackers. Firewalls, in turn, imply processing the protocol stack up to the network layer (and above), which brings the firewall device into the domain of potentially performing routing functions as well.

■ As noted earlier, there are a variety of media choices for a premises distribution network—Ethernet, coax cable, HomePNA, power line, wireless, and IEEE 1394. It is unlikely that any one technology will prevail. Most experts now agree that homes will have multiple "networks," with each supporting a particular medium. The way to ensure connectivity between devices that reside on networks with disparate media is through routing.

■ Routing simplifies the end-user PC/device protocol stack as shown below.

The need to have a "networking appliance" or "residential gateway" that reliably performs router, firewall, and other functions has piqued the interest of the industry. In the simplest terms, the residential gateway (RG), also called an integrated access device (IAD), is nothing but a routing device. It has been optimized to address consumer needs for ease of use and has the flexibility to adapt to different broadband access media (DSL, cable, or satellite) and different premises distribution media (Ethernet, coax cable, HomePNA, power line, wireless, IEEE 1394, etc.).

Protocol Model

The protocol model for a host attached to a residential gateway is shown in Figure 8.9.

Figure 8.9

Protocol model for a host attached to a residential gateway

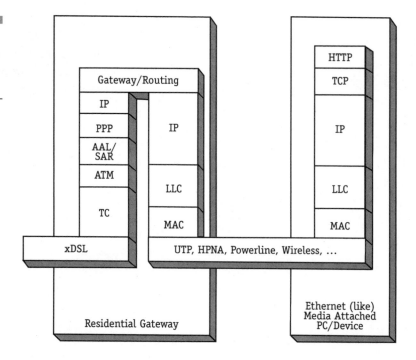

Benefits

Routing in a residential gateway device has the following benefits:

■ It requires no new software installation on the host devices at the customer premises. Most host operating systems provide support for TCP/IP. In addition, they also support the TCP/IP stack using the dynamic host configuration protocol (DHCP) to minimize configuration requirements.

■ The "skinny" protocol stack allows implementation even in low-cost embedded devices (such as hand-held devices).

■   It supports connectivity between devices on disparate media.

Limitations

The primary limitation of this architecture is that it typically does *not* extend ATM beyond the T-interface therefore, the benefit of ATM QoS all the way to the end-user PCs/devices is not available. In other words, this architecture makes the premises distribution network strictly an "IP world." There are those who would argue that this situation is not necessarily a bad thing. Still, it is important to point out that, unless the residential gateway can also offer direct ATM interfaces, the benefits of ATM to the end-user are reduced. However, since the U-interface does not change, the service provider gets all the benefits of the PPP-over-ATM architecture as described in the literature.[8]

Open Issues

■   Routing has a reputation for being complex, especially to configure and administer. There are concerns that this might not translate well into consumer environments. However, this reputation is somewhat unfair. While it is true that core network routers have to perform a variety of functions to support a variety of protocols, edge routers, such as in the architecture considered here, deal with a simple set of functions and usually one protocol, namely TCP/IP. It is possible greatly to simplify the functions of the routing device and make it easy to use and configure. If we restrict the router to a PPP-over-ATM network interface, there use requires no more configuration than a dial-up networking session, i.e., the PPP user name, the password, and the VPI/VCI. The challenge, however, remains for the industry to make this device as easy as possible to set up and administer.

■   The residential gateway has to be flexible in both hardware and software to make it "proof." At the same time, this flexibility should not exceed consumer price expectations.

■   The physical location of the residential gateway is not yet determined. Is it next to the network interface device, in a garage/basement/attic, next to a PC or TV, or embedded in a PC? It is likely that a multiplicity of implementations will emerge.

- Who owns and manages the residential gateway? Is it the service provider or the customer? What provisions are there for the service provider to get visibility into the device without overstepping privacy boundaries?
- There are currently no standards that define the architecture and interfaces of the residential gateway, although there is evidence that this is drawing increased industry attention. For example, the Open Service Gateway Initiative[9] is a group of companies that has specifically chartered itself to address this issue.

Functions of the Access Network

After a discussion of the premises network, let us now shift our attention to the access network. First, let us look at the important functions of an access network, as shown in Figure 8.10:

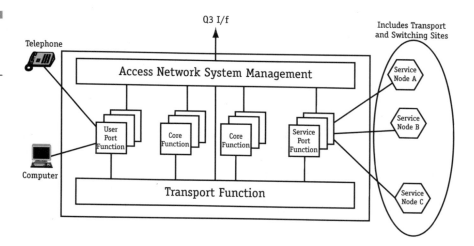

Figure 8.10

Access network functions

- User port interface functions
- Service port interface functions
- Core functions
- Transport functions
- Access network system management functions.

Each of these categories of functions is described below. It is helpful to keep these functional categories in mind as we describe various access architectures such as integrated digital loop carrier (IDLC), fiber-to-the-curb(FTTC), fiber-to-the-node (FTTN), fiber-to-the-home (FTTH), and competitive local exchange carrier (CLEC).

User Port Functions

User port functions (UPFs) adapt the specific user network interface (UNI) requirements to the core and access management functions. The access network may support several different access interfaces and UNIs that require specific functions according to the relevant interface specifications and access bearer capability requirements. Examples of UPFs are:

- Termination of the UNI
- Analog/digital conversion
- Signaling format conversion
- Activation/deactivation of the UNI
- Handling UNI bearer channels
- UNI testing

Service Port Functions

Service port functions (SPFs) adapt the requirements defined for a specific SNI (subscriber network interface) to common bearers for handling in the core function and select the relevant information for treatment in the access network system management function. Examples of SPFs are:

- Termination of the SNI functions
- Mapping of the bearer requirements and time-critical management and operational requirements into the core function
- Mapping of protocols, if required, for a particular SNI
- SNI testing.

Core Functions

Core functions (CFs) adapt the individual user or service port bearer requirements to the common transport bearers. This includes handling protocol bearers according to the required protocol adaptation and multiplexing for transport through the access network. Core functions can be distributed within the access network. Examples are:

■ Access bearer handling
■ Bearer channel concentration
■ Signaling information multiplexing
■ Circuit emulation for the ATM transport bearer.

Transport Functions

Transport functions (TFs) provide paths for common bearers between different locations in the access network and media adaptation for the relevant transmission media used. Examples of transport functions are:

■ Multiplexing
■ Cross-connection, including grooming and reconfiguration
■ Physical media functions.

Access Network System Management Functions

Access network system management functions (AN-SMFs) coordinate the provisioning, operation, and maintenance of the user port functions, service port functions, core functions, and transport functions within the access network. Furthermore, they coordinate operational functions with the service node via the SNI, and the user terminal via the UNI as defined in the relevant interface specifications.

Network Elements of a DSL-based Access Architecture

The major elements of the DSL access network are as follows:

■ Customer premises equipment (CPE)
■ Access loop
■ Colocation central office equipment
■ Transport
■ Switch/point-of-presence (POP) location.

These elements are illustrated in Figure 8.11.

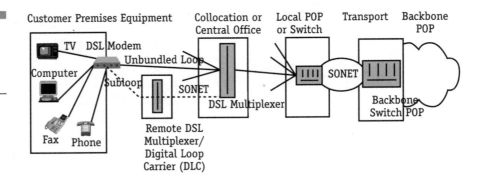

Figure 8.11
Elements of a DSL-based access architecture

Customer Premises Equipment (CPE)

Currently, CPE devices target specific markets or services, so they range from the simple to the advanced devices. The CPE is discussed in detail as part of the permise configuration in the previous section. The advanced versions of DSL CPE devices have a wide range of interfaces that enable a variety of devices such as telephones, computers, and fax machines to connect to the access network. Some typical CPE interfaces are shown in Figure 8.12.

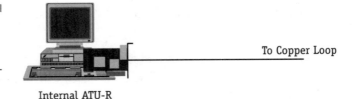

Figure 8.12
Typical CPE interfaces

Internal ATU-R

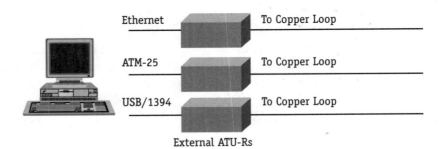

External ATU-Rs

Since CPE devices are based on services and functions, costs can range from as little as $100 up to $1,000. The requirements for CPE were discussed in detail earlier in this chapter.

Access Loop

The access loop has been the bottleneck for cost-effective broadband services. In the access network, there is dedicated connectivity from each customer to the colocation or CO equipment, whereas, in the rest of the network, facilities such as switching and transport are shared among multiple customers. Thus, the overall access cost is directly proportional to the cost of the loop. This fact is illustrated later in this chapter—bringing fiber closer to the customer reduces the cost per customer as long as the fiber is shared. However, once fiber reaches the last mile (where it is not shared), the overall cost of the network is directly proportional to the fiber deployment cost.

To deploy DSL successfully, the copper loop must meet the following general requirements:

- The loop should not have any loading coils.
- The loop length should be less than 18 Kft.
- Copper gauge mix should be limited.

Colocation or Central Office Equipment

Colocation or central office equipment is the physical location within the central office that terminates one end of the copper loop (the other end is terminated by the CPE). Several such copper loops (twisted pairs) are terminated on what is called a main distribution frame (MDF). In other words, the MDF terminates all loops from premises that are served by a particular central office. From the main distribution frame, wire pairs from the premises are "jumpered" over to new wire pairs to complete the connectivity to the appropriate equipment that provides the desired service. For example, with traditional POTS, pairs from the MDF are jumpered to new pairs that run up to a telephone switch such as a Lucent 5E or a Nortel DMS 100. If ADSL service is desired in addition to POTS, the jumpered wire pairs from the MDF first go to a splitter rack. From the splitter rack, two wire pairs are connected—one goes to the telephone switch just as before, and the second connects to a DSLAM (for ADSL). Intermediate (for example, on the MDF) test points are provided to allow technicians to troubleshoot the connection.

Prior to the Telecommunications Act of 1996 in the United States, the incumbent local exchange carriers (ILECs) owned the copper loops and other equipment in the central office. Since the passage of the Act, however, ILECs are required to provide colocation space in their central offices where alternate carriers called competitive local exchange carriers (CLECs) can install their own equipment. The ILECs are also required to unbundle network elements such as the copper loops and offer them to the CLECs. In other words, CLECs can lease the loops and then terminate them at their own colocated equipment in the ILEC's central office. Usually, the CLEC's equipment is isolated from the rest of the equipment in a metallic enclosure called a "cage." The CLEC must also negotiate interconnection agreements with the ILEC in every state on issues such as pricing, time intervals associated with building-out space, and the availability of colocation space and local loops. It can negotiate from a standard interconnection agreement offered by the ILEC, or it can opt to participate in an agreement previously negotiated by another CLEC.

The colocation cost for a CLEC consists of one-time and recurring costs. The one-time cost for a cage with power could run from $10,000 to $250,000, with the average being $50,000. The recurring charge is equal to about 20% of the one-time charge. Consequently, a CLEC interested in providing DSL coverage must assess its ability to recover the high cost of colocation.

Colocation or CO equipment must meet the following general requirements:

- A variety of network interfaces
- Support for a random mix of transceiver types
- Capacity of several hundred to about 1,000 lines/multiplexer
- Modular growth of line capacity
- Meeting network equipment building standards (NEBS) for central office installation
- Meeting appropriate environmental requirements for indoor/outdoor installation.

Transport

The basic function of transport network elements is to backhaul traffic from various colocation or CO equipment to a common switching point where it is switched or routed to the appropriate network. Fiber is usually deployed in this part of the network, so SONET is often the transport technology of choice for service providers. Figure 8.13 shows typical connectivity using SONET. SONET is based on the synchronous digital hierarchy (SDH), i.e., SONET equipment consists of add/drop multiplexers (ADMs) that provide bandwidth in multiples of 51.84 Mbps (called an OC-1). For example, OC-3 means 3×51.84 or 155.52 Mbps, and OC-12 means 622.08 Mbps. OC-3, OC-12, and even OC-48 links are now part of the SONET backbone in the transport network. Additional details on SONET/SDH can be found in the resources listed in references 10–13.

Figure 8.13
SONET in transport

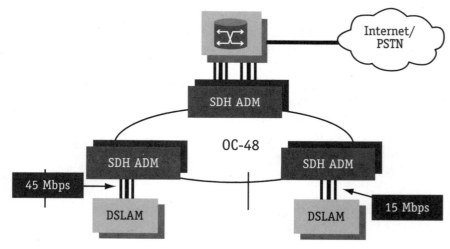

ADM = Add/Drop Multiplexer

With the growth of the Internet and deployments of DSL in the access network, traffic in the transport network has begun to push the limits of SONET. As a result, new technology that can take advantage of the wavelength used in fiber has become the new direction of the transport network. Recently, DWDM (dense wave division multiplexing)-based products, which promise to bring bandwidth relief to the transport portion of the access networks, have been announced. The proposed

DWDMring systems vary widely in their function, however. Some simply provide raw wavelengths, or "virtual fiber," and leave it up to an overlay network to provide formatting and management to make the bandwidth usable. Other systems provide integrated performance monitoring and may format the bandwidth into smaller units, such as OC-12 or OC-3.

Switch/Point-of-presence Location

The point of presence (POP) is the primary location for a network service provider and the demarcation point for the access network. The equipment that provides interconnections to other network and service providers resides here, and it is the first point in the network where end-user traffic is processed, i.e., switched for voice traffic, routed for data traffic, or sent through an ATM switch (independent of the traffic type). This location can also host value-added services for the customers. Figure 8.14 shows a typical switch site.

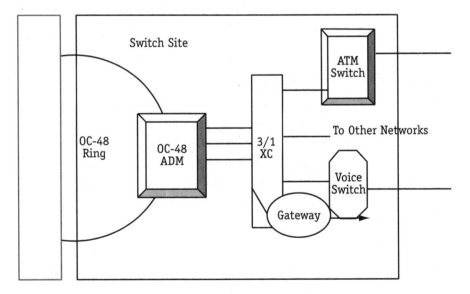

Figure 8.14

A typical switch site

The switch/POP location typically hosts equipment for a variety of services:

■ Voice switches for voice services
■ Crossconnects for interconnection

■ Routers and/or ATM switches for data traffic
■ Servers for hosting information (Internet servers, video servers, etc.)
■ Gateways to other networks
■ Specialized systems that provide specific functions, such as advanced services so a user can access the required services on an as-needed basis.

The service provider who operates and maintains the site usually owns all the equipment. Although switching equipment is not part of the access network, it is critical that the deployed switching platforms be designed for robust support of both narrowband and broadband services. Sometimes, providers deploy switching equipment designed to perform specific tasks (such as voice services alone). This increases the cost of providing and managing these services. CO equipment vendors are now developing platforms that can satisfy both data- and voice-traffic requirements, often within the same chassis. Software vendors are also developing solutions to simplify the provisioning of new services and integrate the management of these services into existing operations systems such as billing.

Common Access Architectures

In Chapter 1, we discussed the fundamental network topologies, star, bus, and ring. The star topology, which is the basis for architectures deployed by telephone companies around the world, has a dedicated cable from a central point to each node. It has the advantage of providing high-capacity connectivity to each subscriber if the need exists, without affecting other subscribers. Hence, bandwidth is not shared, which provides excellent security and manageability. The major disadvantage is that it requires more cables. Fortunately, the loop plant is a sunk cost that most providers have already recovered. A variation of the basic topology is the double star, where there is a mid-point entity (such as a digital loop carrier) that performs some level of concentration or multiplexing on the data before they reach the CO. All of the DSL architectures discussed in this section fall into either the star or double star category. Access topologies are illustrated in Figure 8.15.

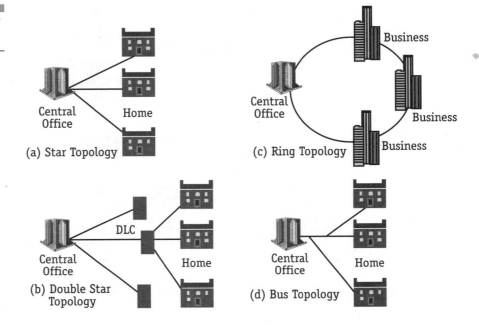

Figure 8.15
Access topologies

(a) Star Topology

(b) Double Star Topology

(c) Ring Topology

(d) Bus Topology

In the next couple of sections, we will discuss the architecture deployed by the network operator around the world. Some of the architecture are Integrated Digital Loop Carriers (IDLC) and other fiber-based architecture in distribution and the plant.

Integrated Digital Loop Carrier (IDLC)

Telephone companies, mostly United States, have deployed digital loop carrier (DLC) by utilizing fiber to extend the loop reach. As of 1998, about 50% of access lines were served by DLC. Figure 8.16 shows the growth of DLC in the access network.

Currently, as a means to bring digital technology closer to the subscribers, integrated digital loop carrier is the most equipment in the ILEC network. IDLC equipment typically uses SONET-compatible fiber technology on the trunk side to support various services.* The access rates are STS-1, DS-3, DS-1, fractional DS-1, digital data service up to 56 Kbps, ISDN, and POTS. The trunk transport rates are usually OC-3

---

* The older generation of IDLCs use traditional T-carrier over copper to connect the CO to the IDLC on the trunk side.

or OC-12. Most of the deployed equipment is based on Bellcore specifications (as contained in TR-303), and the architecture is designed to support both basic and enhanced telephony and data services. The hardware architecture and software features vary from vendor to vendor. In general, however, the following system features are supported by most:

- Remote software provisioning capability
- Compact and rugged shelf design to sustain outside plant operation
- Operations, administration, maintenance, and provisioning (OAM&P) features
- Flexible hardware architecture to accommodate future software upgrades
- Digital service to end-users, if needed.

Figure 8.16
DLC penetration rate

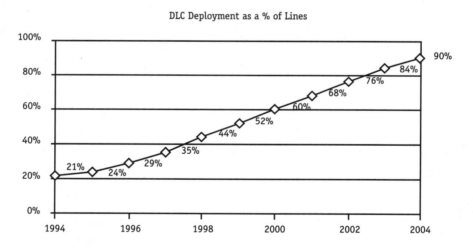

The IDLC architecture is depicted in Figure 8.17.

The advantage of IDLC is that it extends the reach from the central office. Instead of long loops running from the central office to a number of premises, a single fiber link (or two for redundancy) connects the CO to the IDLC. From the IDLC, copper loops fan out to serve the residences in a neighborhood. One can typically recognize IDLCs as the tall "green boxes" in one's neighborhood/subdivision. The disadvantages of the IDLC architecture are space and environmental constraints. Neighborhoods typically have aesthetic rules about the size and location of equipment. Also, since IDLCs are typically outdoors, they have to survive extreme ranges of temperatures. Due to these, line cards in the IDLCs have restrictions on the maximum board space and power they draw—in turn, this constrains the number of lines per card. The goal of DSL chip set vendors and equipment manufacturers is to maximize this number within the constraints of space and power.

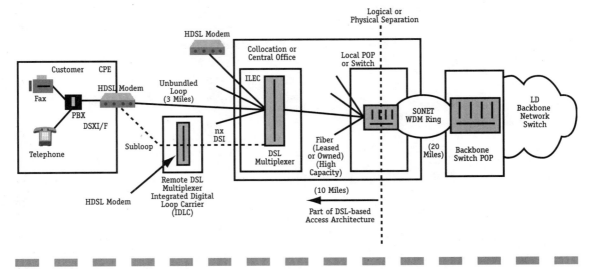

Figure 8.17 *IDLC architecture*

Fiber-to-the-node (FTTN) and Other Fiber-based Loop Architectures (FTTC, FTTH, and FTTB)

These architectures are similar to the IDLC architecture, with the exception that fiber is always used on the trunk side, and in a portion of the distribution plant. If fiber is deployed up to a node (which is equivalent to the pedestal in a neighborhood), it is called fiber-to-the-node (FTTN). If it is installed to the curbside of a home, it is called fiber-to-the-curb (FTTC). If it goes all the way to the home, it is called fiber-to-the-home (FTTH). Fiber used in the basement or wiring closet of an office building or complex is called fiber-to-the-building (FTTB). Additional details of fiber-based architectures may be found in references 14–18.

As mentioned in Chapter 6, fiber networks have either an active or passive network termination (NT):

■ If the termination is active, the NT contains electronics to perform the necessary multiplexing and optical-to-electrical conversion to transport all the services on a single fiber using different wavelengths. The advantage of the active NT-based architecture is the manageability it brings to the network. On the other hand, active architecture is more expensive to install.

■ If the termination is passive, a single fiber passes through the entire neighborhood with each subscriber assigned a logical wavelength to access the network. The NT contains a passive splitter that divides the signal carried over the distribution fiber to about eight subscribers. The advantage of a passive NT is its low cost because there are no active electronic components. The disadvantages are lack of privacy, complex power requirements, and reduced network management capability.

Fiber-to-the-node (FTTN) Architecture

Fiber-to-the-node architecture is typically used to deliver services over VDSL. Fiber runs all the way up to an optical network unit, which does the conversion necessary to carry signals from the fiber optic cable on one side to copper loops on the other. From the ONU, copper loops run to every premises in the neighborhood. The advantage of this architecture is that it requires no major modification to existing physical plant. It can be overlaid on the existing IDLC architecture although it might require upgrades to the line cards.

Fiber-to-the-curb (FTTC) Architecture

Instead of terminating the fiber at the ONU in the neighborhood, the fiber extends to the curbside from around eight to 24 homes can be connected via very short copper loops. The advantages here are increased capacity and reliability. The reliability increases as a result of the amount of fiber in the network and the short lengths of copper loops. However, FTTC can be very cost prohibitive when compared with the number of homes served.

Fiber-to-the-home (FTTH) Architecture

The extension of the FTTN architecture all the way to a residence is the fiber-to-the-home (FTTH) architecture. The optical-to-electrical conversion is performed when the signal reaches the home. Thus, it is necessary to deploy intelligent network components deeper in the network than in an FTTN network. Obviously, the economy of serving multiple subscribers with a single optical network unit is lost in the FTTH network.

At one extreme, each home could have a dedicated fiber connection to the colocation or central office equipment. In this case, the amount of upstream and downstream digital bandwidth available to each home would theoretically be very large, but implementation would be very expensive. This increased expense can be attributed mainly to the increased equipment, material, and labor costs incurred when installing an FTTH network. Costs can be driven down somewhat by splitting the fiber numerous times as it travels from the central office to the consumer. However, even when the fiber is split and shared among subscribers, FTTH is still much more expensive to deploy than FTTN.

Fiber-to-the-building/basement (FTTB) Architecture

Like FTTH, fiber-to-the-building/basement is an extension of the fiber-to-the-node architecture all the way to the premises—in this case, however, the premises is a multidwelling unit such as an apartment complex, a hotel, or an office park. The fiber is terminated at an ONU located in the basement or wiring closet of the building. The architecture takes advantage of the wiring inside the building to deliver broadband services (usually over VDSL) to the different units/offices. Obviously, this architecture is more expensive than FTTN because fiber has to extend all the way to the premises. However, it is cheaper than FTTH since the cost of the fiber can be distributed among multiple units in the building. Furthermore, if there are several buildings in close proximity (e.g., in a downtown location), the cost to run fiber to each building is not prohibitive. This architecture is also popular in parts of the world, such as Hong Kong and Singapore, that have dense residential populations living in apartment complexes.

CLEC Architecture

In the late 1990s, as a direct result of the 1996 Telecommunications Act, a new type of architecture emerged to help encourage competition for services in the local loop. In the last mile, the competitive local exchange carrier architecture is dependent upon the local ILEC infrastructure (since the CLEC leases the local loop from the ILEC). However, because CLECs can build the rest of the access network from the ground up, they have the flexibility to determine the market segments, services, and technologies they want pursue. In general, CLECs have targeted businesses and telecommuters, whose communications costs are paid for by corporations.

Figure 8.18

CLEC architecture

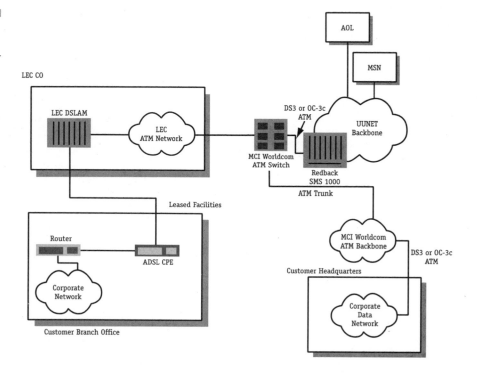

Figure 8.18 shows a typical generation CLEC architecture. It is quite flexible in terms of supporting traditional voice and enhanced broadband services as allowed within the regulatory environment. In this architecture, the CLEC deploys DSL-capable customer equipment that has interfaces to support data (and sometimes voice) services to the customer premises. Since the service is usually paid for by a corporation to enable its employees to telecommute, installation costs (including the cost of the CPE) and recurring monthly costs are not usually the primary concern—reliability is. This is where CLECs try to differentiate

themselves because they have been able to build out their backhaul infrastructure using state-of-the-art equipment geared to efficient transport of both data and voice.

The particular choice of DSL technology over the copper loop depends on the desired bandwidth and symmetrical versus asymmetrical service. It also depends on what can practically be run over the actual loop (recall that the CLEC is dependent upon the ILEC to provide the loop therefore, the loop quality has to be capable of supporting a particular technology). Regardless of the DSL technology, the loops terminate onto the appropriate line cards in a common DSLAM chassis. From the DSLAM, all the traffic is backhauled, typically over a SONET ring, to the CLEC's switching centers. In this way, a CLEC can limit the actual colocation space required, thus reducing expenses. The CLEC may either buildout the infrastructure to backhaul the traffic from the ILEC's central office site, or it may lease it from the ILEC. At the switching center, the backhauled traffic from several colocated DSLAMs in geographic proximity is aggregated and shipped to the end destinations. CLECs usually have value-added services at the switching center, such as local content Web hosting (on behalf of business customers), content caching, and virtual private network (VPN) services.

Although CLECs originally started by offering competitive data services, they are aggressively pursuing opportunities to enable voice services by utilizing the DSL bandwidth to extract digitally derived voice channels (i.e., containing packetized voice that can be shipped around as data).

Architectural Issues

There are several technological, economic, and regulatory issues related to DSL deployment. The physical-layer issues specific to each DSL technology were addressed in Chapters 4, 5, and 6 (for ADSL, HDSL, and VDSL, respectively). In this section, we will look at the architectural issues from operational, economic, and regulatory perspectives.

Operational

Loop Qualification

Loop qualification is the process of determining if a loop will support DSL transmission at a given rate. There are a number of factors that affect the ability of a loop to support DSL:

- The age, condition, and makeup of the loop, and the plant that it passes through
- The inclusion of DLC systems
- The desired transmission rate
- The DSL technology employed (e.g., G.dmt or G.lite)
- Services already carried on the loop (e.g., POTS)
- The loss introduced by the loop including the customer premises wiring
- Transmission impairments such as crosstalk, radio interference, and impulse noise.

Currently, DLC-based architectures present the greatest challenge for deployment. As noted earlier, DLC systems have several constraints in terms of space, power, and environment. At the time of this writing, line cards that meet these constraints are still being developed. Therefore, most deployments today require a direct loop between the CO and the premises (i.e., no intermediate DLC).

Interoperability

Physical-layer interoperability was discussed in Chapter 4 in the context of ADSL interoperability testing. However, physical-layer interoperability by itself is not enough. To deploy their solutions confidently, service providers must be assured that their equipment is interoperable with equipment in other parts of the network (and other interconnected networks) as part of the overall end-to-end architecture. This means that protocol stacks must be compatible with other.

For instance, if the service provider expects RFC 2364 support at the U-interface and the premises architecture does not allow it, there exists an incompatibility that cannot be resolved even though the physical ADSL link may be up. Therefore, one has to be careful when choosing premises architectures and specifying the capabilities of the CPE.

Unfortunately, true end-to-end interoperability can only be achieved through rigorous testing since all levels of the protocol stacks have to be interoperable with the appropriate partner stacks. In 1999, the ADSL Forum stepped up to the task of enabling this level of testing to happen in an open manner.

Provisioning

Provisioning involves several steps:

- Configuring the CO
- Configuring the CPE
- Verification of DSL link connectivity
- Configuration of the transport (e.g., ATM parameters)
- Verification of transport connectivity
- Configuration of the service (e.g., ID and password for Internet access)
- Verification of service availability.

It is obviously desirable to have these steps fully automated to make the process easy for the end-user. Unfortunately, automatic provisioning requires strict coordination between several network elements. These elements need to communicate each other in the correct order to complete the process accurately and seamlessly. This area of automatic provisioning is still the subject of study by the ADSL Forum and other industry groups. Various mechanisms (for example, the use of an "installation server") have been advocated to solve the issue, but no definitive answers have emerged. The following example of provisioning steps using an installation server will illustrate the current thinking on the subject:

- A customer places DSL service request at the installation server (the customer may access the server through traditional dial-up).
- The installation server collects information about the customer and generates an order number. The installation server may do some basic validation to ensure that service is available (and even initiate loop qualification to ensure that the service is, in fact, possible). In any event, assuming that the order can be processed, the installation server generates an order that triggers the network management function.

■ Network management then provisions the network with the appropriate parameters for the desired service. This might include setting up the ATM transport parameters as well as the maximum ADSL upstream and downstream parameters at the DSLAM.

■ When the CO equipment provisioning is complete, the network management returns a completed order status to the installation server. It also provides the installation server the parameters to be configured at the CPE to match the CO side.

■ The parameters are then downloaded from the installation server to the CPE.

■ The CPE is configured using the downloaded parameters. Once configuration is complete, the user can be instructed to begin the procedure of validating DSL service availability.

Test, Diagnosis, and Repair

Testing may be triggered when there is a fault report either as the result of customer notification or autonomously through a network alert. Management agents that continually monitor service-level parameters for threshold conditions issue the alerts. The trouble is identified as either an out-of-service condition or a degradation-of-service condition, and the trouble response priority is set accordingly. Appropriate test equipment and procedures should enable demarcation between all the network elements in the end-to-end architecture. This helps to diagnose the fault at a specific location and establish ownership (sometimes known as *sectionalization*). If the problem has been isolated to the access network, metallic line tests may be used to test the loop itself (after isolating it from the CO equipment and the CPE). Once the root cause of the problem has been identified, repair is assigned to the appropriate work crew. The trouble condition must be corrected in an appropriate time period in accordance with the service level agreement (SLA).

From an architectural perspective, there must be one or more management agents in each network element to allow the service provider's network management system to identify fault conditions. Whenever possible, the agents should communicate enough information to pinpoint the root cause. In addition, loopback testing at the various OSI layers plays an important role in sectionalization.

Performance Monitoring and Traffic Management

Performance monitoring and traffic management are related. Performance monitoring means watching the traffic levels to ensure that the QoS parameters do not exceed what has been specified in the service level agreement. If there is a violation of the traffic contract, appropriate action must be taken. This usually means sending a notification to the offending source(s) to slow down the offered data rate to get the network back into a decongested state. In extreme cases, data may be dropped altogether.

Traffic management tries to reduce the incidence of traffic contract violations by traffic shaping, i.e., modifying the offered rates on various links to ensure that appropriate traffic flow is maintained. In other words, this evens out the "peaks and valleys" in the traffic. If the traffic still exceeds allowable thresholds, feedback mechanisms have to be employed to notify the sender(s) to slow down.

As in the case of fault management, performance monitoring requires that the architecture support the ability of network management agents so they can report accurately the status back to the appropriate monitoring entities.

Scalability

Scalability is critical because the access network architecture should accommodate services growth without requiring massive infrastructure changes that could be capital intensive. It is for this reason that deployed equipment must be as "future-proof" as possible so as to allow system upgrades, either through software downloads or ease of hardware replacement (such as replacing the line card rather than the entire chassis).

Economic

An important aspect of DSL deployment is the cost of enabling and providing new services. Although the overall cost per bit transferred with DSL technologies is drastically lower than traditional T-carrier technology, the total cost to service customers can still be an issue. This is especially true if DSL services take away from existing revenue-generating services. The business case has to be compelling enough to entice the provider to deploy services, while fending off competitive threats from alternative access technologies such as cable.

The architectural model adopted has a direct impact on the cost. Fiber architectures can be capital intensive in the short term, but pave the way for advanced services in the future. In addition, the deployed equipment must be as proof as possible to adapt to a variety of services down the road.

Costs can be capital (one-time) or operational (recurring). Both capital and operational costs must be broken into baseline costs (independent of the service) and enhanced costs (dependent on the service). An example of advanced costs would be the capital and operational charges incurred when the provider offers entertainment services such as video on demand.

When developing the business case to deploy DSL, providers need to keep in mind the following points (based on a projection of estimated costs and revenues over a three- to five-year period in a given service area*). The following example is for baseline services

- Total number of subscriber loops in a service area
- Total number of qualified subscriber loops in a service area (derived from an estimate of the percentage that can be qualified)
- Total number of customers who will sign up for baseline service, assuming the loop has been qualified; this percentage estimate is known as the "take rate."
- Total baseline capital cost, which is the accumulated capital investment for the period for a service area
- Baseline capital cost per customer
- Total baseline operating cost per month
- Baseline operating cost per customer per month
- Expected baseline service revenue per customer per month.

Regulatory

In the United States, the regulatory issues that have an impact on the architecture come from the unique relationship that exists between the

---

* Typically, the service area will be metropolitan. Telephone companies like to think of service areas in terms of the local area transport areas (LATAs) they own and operate.

CLEC and the ILEC. On one hand, the CLEC and ILEC compete for the same end customer. On the other, the nature of DSL is such that the CLEC is dependent upon the ILEC to lease the local loop. To ensure fair pricing by the ILEC, the FCC* sets the lease price of the loop and other network elements based on a complex cost analysis of information provided by the ILEC and other sources. The lease price obviously varies by region.

However, although the government can regulate the lease price of the loop, it does not (and cannot) guarantee that a loop is always available to the CLEC. The ILEC can refuse colocation space to the CLEC if the ILEC can prove that it does not have enough space available in the central office. Furthermore, if the CLEC runs services that interfere with the ILEC's service, the ILEC can complain to the regulatory body to get the CLEC expelled from its network (the spectral compatibility issue can be significant in this case). Unfortunately, such issues are sometimes difficult to prove, and are often settled without litigation. There are still those in the industry who believe that, because of litigation barriers to entry, the Telecommunications Act of 1996 did not truly open the access network to competition.

G.lite Specific Issues

Thus far, we have reviewed architectural issues that are general to all forms of DSL. However, there are two issues specific to G.lite:

- Rate changes due to splitterless operation
- Maintaining a network presence during reinitialization.

Before we describe these issues, a quick review of the splitterless environment will set the context for the discussion. As we noted in Chapter 3, the primary function of a splitter, which can be located at both the CO side and the premises side, is to separate the DSL service from POTS. Despite the important role of splitters, in 1998, the industry had a strong motivation to eliminate the premises splitter. The

---

* FCC—The Federal Communications Commission is the regulatory body that oversees communications regulation.

installation of a splitter device requires a "truck roll," or slower deployments since it requires the network operator to send technicians to a site to install the equipment (and possibly new wiring) and verify the setup. In other words, the deployment rate is constrained by the availability and scheduling of qualified personnel. Therefore, to accelerate deployment, a splitterless operation is an option for G.lite. Figure 8.19 shows the various deployment options and the pros and cons of each.

Figure 8.19

G.lite deployment options

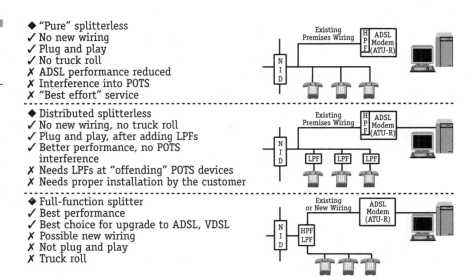

◆ "Pure" splitterless
✓ No new wiring
✓ Plug and play
✓ No truck roll
✗ ADSL performance reduced
✗ Interference into POTS
✗ "Best effort" service

◆ Distributed splitterless
✓ No new wiring, no truck roll
✓ Plug and play, after adding LPFs
✓ Better performance, no POTS interference
✗ Needs LPFs at "offending" POTS devices
✗ Needs proper installation by the customer

◆ Full-function splitter
✓ Best performance
✓ Best choice for upgrade to ADSL, VDSL
✗ Possible new wiring
✗ Not plug and play
✗ Truck roll

Rate Change Due to Splitterless Operation

Figure 8.19 shows that, in the pure splitterless operation, both POTS and ADSL services operate simultaneously. Consider the situation where a constant bit-rate (CBR) application (e.g., a videoconferencing application) runs over ADSL service that has negotiated QoS parameters based on the available data rate when none of the POTS devices are off hook. As we described in Chapter 4, when certain POTS devices go from an on-hook state to an off-hook state, the line impedance changes dramatically, and has an impact on the ADSL service. This usually results in a much lower data rate than when all POTS devices are on hook (i.e., no POTS device is active). Under these conditions, it is possible that the ADSL service cannot satisfy the QoS parameters previously negotiated by the application.

From this discussion, it follows that, in a pure splitterless environment, it is impossible to guarantee QoS parameters that require a specific line rate to be always available. In other words, in ATM terminology, the only classes of applications that can be supported under these conditions are unspecified bit-rate (UBR) or available bit-rate (ABR). Of course, installing distributed splitters or full-function splitters can eliminate this issue. Nevertheless, since pure splitterless operation is a plausible scenario, the rate change issue needs to be considered by both service providers and users.

To address the rate change issue, there have been discussions in the industry (e.g., the UAWG and ADSL Forum) on developing mechanisms for notifying the upper layers. The specific mechanisms are the subject of further study.

Maintaining a Network Presence during Reinitialization

There are two reasons why reinitialization could occur in the context of G.lite:

■ Reinitialization can be caused by line impedance changes during simultaneous POTS and ADSL service in a pure splitterless environment. The fast-retraining algorithm attempts to minimize the time of retraining through the use of previously stored profiles. However, in the worst case, fast retraining can revert to a full training sequence.

■ Reinitialization can also be caused when a DSL modem in low-power state is triggered to get back into normal state. One of the features of G.lite is the support for low-power states to satisfy regulatory requirements that a PC and associated devices powered by the PC (such as internal DSL modems) conserve power during periods of inactivity. Therefore, a modem can be put into a low-power state in accordance with the power management policy set by the user of the PC. Depending on the corresponding ADSL link state when a modem is in a low-power state,* the modem may experience re-initialization.

---

* The ADSL link states are loosely coupled to device power states. It is possible for a link state to be completely up even when the device itself is in low-power state. This is implementation dependent—in general, however, when a device is in the low-power state, the link state is unlikely to be fully operational. For example, an implementation could choose to tear down the PPP session and higher-layer applications, but keep the ATM circuits alive during low-power state.

Regardless of how re-initialization may occur, if it lasts longer than 10 seconds, it presents a problem to certain classes of applications that require a network presence to be maintained. Generally speaking, the problem is likely to result from a PC and its associated devices going into low-power state, since it is highly likely (and desired) that the state lasts longer than 10 seconds. The 10-second value comes from the fact that, in order to keep a user connection alive, there are various periodic keep-alive mechanisms at both the ATM and PPP layers that declare the link to be down if the receiving end does not respond within 10 seconds. Some of these mechanisms are illustrated in Figure 8.20.

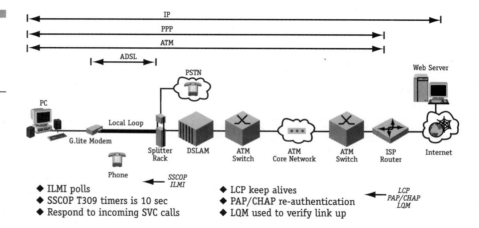

Figure 8.20

Keep-alive mechanisms for DSL-based access

From the perspective of maintaining network presence, applications can be classified as follows:

- **Client-to-server.** Internet access is an example of this application category. This class of application does not require the upper layers to be maintained when a PC and its associated ADSL modem go into low-power state. When the PC and its associated ADSL modem are restored to normal state, the upper layers (e.g., reestablishing ATM VCs and PPP sessions) can be reinitialized as if from a cold start. This will have no impact on use of the PC for Internet access from that point on.

- **Server-to-client.** Push technology (e.g., automated news feeds) is an example of this category. This class of application requires the upper layers to be "always available" even when the PC and its associated modem go into low-power state. In this scenario, the upper layers are not active, but they must be capable of being quickly restored, before the sending application gives up retrying.

■ **Peer-to-peer.** Internet telephony is an example of this category. This class of application requires the upper layers to be "always on," even when the PC and its associated modem go into low-power state. In this scenario, the upper layers must be active since the PC and its associated modem must be capable of instantly responding to calling applications that may not automatically retry.

Table 8.1 summarizes the differences between "always on" and "always available," although the exact implementation is vendor dependent.

Table 8.1	"Always On"	" Always Available"
"Always-on" versus "always-available"	PC is on	PC can be in low-power state
	ADSL link state is L0 or L1	ADSL link state is L3, L1, or L0.*
	Protocol stacks are up (ATM and PPP). The IP address is preserved.	Protocol stacks are down but capable of being restored on short notice when triggered (e.g., through a network-initiated request).
	Example applications supported: IP telephony, IP multicast, push technology	Example applications supported: On-demand BW and its related multimedia services

* The L2 link state is reserved, but not defined in G.Lite (G.992.2).

Future Directions

The topics discussed in this chapter have been considered primarily from the perspective of supporting data access. However, there is a growing interest in supporting voice services over DSL, referred to as VoDSL. Both ILECS and CLECs are motivated to support VoDSL:

■ ILECs are interested in conserving their copper loops, while satisfying the requirements for second and third telephone lines.

■ CLECs are looking for additional service differentiation and satisfying customer requirements for integrated access (i.e., voice and data).

Adding voice to DSL is currently the subject of active study within the ADSL Forum. The reference voice architecture under consideration by the ADSL Forum is an overlay on the reference data access architecture. Voice traffic is carried across the access network as digitally encoded voice samples using one of the several techniques described in Chapter 7 (e.g., G.711). Each "virtual telephone line" is known as a *digitally derived voice channel* or simply, a derived voice channel, which takes up a portion of the DSL bandwidth to carry the digital samples.

The architecture for derived voice channels typically consists of the following network elements:

- **Integrated access device (IAD).** As mentioned earlier, the term IAD is sometimes synonymously with the term residential gateway.* Regardless of the actual industry term, from the perspective of the voice architecture, the function of this device is to transport digitally encoded voice samples over the DSL link. The IAD multiplexes digital samples from several derived voice channels by first converting them into ATM cells prior to transport over the common physical DSL link. In other words, each derived voice channel maps onto a unique ATM virtual connection that transports cells containing digital samples that belong to that derived voice channel. The IAD may or may not perform the conversion from analog voice to digital samples. For voice transport over ATM (VtoA) architecture, the IAD is likely to perform the conversion to digital samples internally; the samples are then transported in ATM cells using either AAL1 or AAL2 adaptation. For a voice over IP (VoIP) architecture, the IAD does not have to do any internal conversion since the arriving data streams (originating from IP telephones) already have digital samples in IP packets. In this case, the IAD merely transports the IP packets containing voice samples as ATM cells using AAL5 adaptation (similar to how it handles data). The IAD provides interfaces into which POTS devices can be plugged.

- **DSLAM.** As in the case of the data access architecture, the DSLAM performs an aggregation function for ADSL traffic, which is trans-

---

* Usually, the term IAD refers to a device that integrates voice and data services to the premises, and the term residential gateway refers to a device that not only does this integration, but also incorporates other functionality (e.g., enabling video services).

parent to the DSLAM whether the traffic represents cells carrying voice samples or data. All the DSLAM sees are ATM cells at the U-interface. These cells are multiplexed onto a common upstream link that interfaces with an ATM switch. The DSLAM is thus common to both the data and voice architectures.

■ **ATM switch.** The ATM switch takes the cells and routes them to the appropriate destination based on the cell headers (i.e., the VPI/VCI values). The cells carrying data are transported, as before, through the ATM network to their destination. Usually this destination is a router that reassembles the data cells into packets for subsequent shipment across an IP network. The cells that carry voice are automatically identified as having a different destination (based on the VPI/VCI values in their headers) and are transported to another device called the interworking unit. The ATM switch is also common to both the data and voice architectures.

■ **Interworking unit (IWU).** The IWU performs a function inverse to the IAD's; it demultiplexes the incoming ATM cells carrying digital voice samples and reassembles them into individual voice channels. It also performs the necessary conversion to interface the individual voice channels to a Class 5 PSTN switch, much as a digital loop carrier does.

■ **Class 5 switch.** The Class 5 switch is already part of the PSTN infrastructure equipment in the central office. By connecting the Class 5 switch to the DSL-based broadband access network (via the IWU), the service provider can extend traditional voice services to derived voice channels. In other words, features such as caller ID and call waiting can be provided to any telephony device at the premises (connected via the IAD).

■ **Private branch exchange (PBX).** In some situations, it is desirable to connect a PBX to provide off-premises extensions (OPXs). This setup is of value to a telecommuter because the telephone at the residence (attached to a derived voice channel) behaves exactly like another corporate office extension. The telephone can be used simultaneously with data access over a common copper loop.

Summary

Although there are other options, the industry has generally agreed upon the PPP-over-ATM-over-ADSL end-to-end broadband service architecture advocated by the ADSL Forum. As service providers begin their deployments based on this architecture, DSL-based access networks need to conform to the U-interface specifications. This puts restrictions on the protocol model that is required at the U-interface. Premises networks, on the other hand, have to conform to the uniqueness of the customer premises environment. Therefore, an adaptation function occurs between the premises network on one side and the access network on the other. The adaptation function maps the protocol models at the T-interfaces to the uniform protocol model required at the U-interface.

In this context, we reviewed the key requirements for premises architectures developed by the ADSL Forum and examined how some of the proposed architectures meet these requirements. Next, we discussed the functions of access architectures, and the network elements that comprise these architectures. We used this as a backdrop to analyze some DSL-based access architectures and discuss them from operational, economic, and regulatory perspectives. We concluded the chapter with a discussion of some of important G.lite-specific issues and looked ahead to the future of voice over DSL.

References

1. Information is available at the official Website at **www.adsl.com**.
2. "Broadband Service Architecture for Access to Legacy Data Networks over ADSL," *ADSL Forum Technical Report 12*, Issue 1 (June 1998).
3. Payne, D., Wojewoda, J., Shenouda, H., Amin-Salehi, B., Shieh, P., and Tai, C. "References and Requirements for CPE Architectures for Data Access," *ADSL Forum Working Text 31*, Work in Progress (May 1999).
4. Gross, G., Kaycee, M., Li, A., Malis, A., and Stephens, J. "PPP over AAL5," RFC 2364 (July 1998).

5. Jeffery, M. et al., "Multiple Devices in the Home: Using Local Tunnels to Extend the PPP-over-ATM Architecture across Local LANs," *ADSL Forum Contri bution 98-187* (November 1998).

6. Mamakos, L. et al., "A Method for Transmitting PPP Over Ethernet (PPPoE)," RFC 2516 (February 1999).

7. Tai, C. et al., "BMAP: Extending PPP/ATM Services across Ethernet/USB/1394," *ADSL Forum Contribution 98-018R2* (September 1998).

8. Wang, R. et al., "The Architecture of PPP Extensions for Customer Premises LANs," *ADSL Forum Contribution 98-216* (November 1998).

9. Information is available at the official Website at **www.osgi.org**.

10. Balcer, R., Eaves, J., Legras, J., McLintock, R., and Wright, T. "An Overview of Emerging CCITT Recommendations for the Synchronous Digital Hierarchy: Multiplexers, Line Systems, Management, and Network Aspects," *IEEE Communications Magazine* (August 1990).

11. Kasai, H., Murase, T., and Ueda, H. "Synchronous Digital Transmission Systems Based on CCITT SDH Standard," *IEEE Communications Magazine* (August 1990).

12. Asatani, K., Harrison, K. R., and Ballart, R. "CCITT Standardization of Network Node Interface of Synchronous Digital Hierarchy," *IEEE Communications Magazine* (August 1990).

13. Brown, T. S. "A Functional Description of SDH Transmission Equipment," *British Telecommunications Technology Journal*, Vol. 14, No. 2 (April 1996).

14. Vogel, M. O., and Menendez, R. C. "Fibre to the Curb Systems: Architecture Evolution," *Global Telecommunications Conference and Exhibition: Communications—Connecting the Future*, Globecom (1990).

15. Oakley, K. A., Guyon, R., and Stern, J. "Fibre in the Access Network," *British Telecommunications Engineering*, Vol. 10 (April 1991).

16. Farina, A., Rosi, R., and Trondoli, A. "Fibre Loop Architectures: A Comparative Evaluation," *Supercomm/ICC '92, Discovering a New World of Communications*, Vol. 4 (1992).

17. Cook, A., and Stern, J. "Optical Fibre Access—Prospectives Toward the 21st Century," *IEEE Communications Magazine* (February 1994).

18. Barber, J., and Guyon, R. "Evaluating the Potential of Passive Optical Networks," *IRR Conference "Hybrid Fibre Coax in the Access Network"* (April 1996).

Service Architecture

Overview

It is the services reference architecture that determines the functions performed by the various network elements in the access environment. This chapter addresses how various services, voice, data, video, and integrated services, are transported over a XDSL-based access infrastructure. We will also discuss some of the control, signaling, and operating support issues involved in transporting these services over a common access infrastructure. For service providers the operations support is important because it has become a showstopper for DSL deployment. Traditional organizational structures and processes tend to be designed around assigned services and access facilities. However, DSL can support multiple services. So, this becomes an ownership of access issue for management and the organization as a whole. It will take some bold leadership to sort out the organizational issues and address the process for deploying DSL services in the network.

Reference Architecture[1,2,5,7,12]

Reference architecture requires that the interfaces between the access network and other elements of the telecommunication network be more precisely defined. Figure 9.1 summarizes the main components in the overall reference architecture. To comply with the needs of service providers, this overall architecture has to meet some basic requirements:

■ The network must be compatible with FSAN (full service access network)-like systems and generic access systems required or already installed.
■ The network should be designed to support multiple service (voice, data, video) over a common infrastructure.

The service node is connected through the core network of service providers who have servers to provide various services (e.g., information for file access, video for video servers, etc.) The connection to these servers may be vary among service providers. One may use ATM while other may use IP connections for narrowband and broadband services.

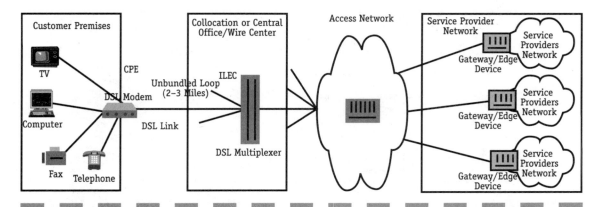

Figure 9.1   *Reference service architecture*

The effective implementation of an end-to-end broadband platform requires carefully defined, preferably standardized, access-network building blocks. These building blocks are (1) the customer premises network (including network termination and home wiring; (2) DSLAM and collocation sites; and (3) the interface to the service providers and their services. The final element of the implementation is the network/service management aspect. This is not shown in the figure, but is critical for a service provider's relationship to its customers.

Figure 9.2 depicts a functional DSL-based reference architecture over which voice, data, and video services are transported. Here the function performed by each element in the access network is identified. Usually the devices closest to the customer are software based and require limited hardware modification, which enables service providers to download upgrades to customer equipment as needed. The CPE is an important element in enabling a wide range of services to the home or office. In fact, there are a variety of configurations for CPE in the home, and DSL-based CPE is an integral part of a burgeoning area called home networking. Figure 9.3 shows just one potential configuration for home networking using DSL that will enable voice, data, and video services. In this case, the CPE is capable of performing functions such as voice adaptation (i.e., packetization of voice), signaling, ringing, data support, and video protocols. The interworking details of how this is done are discussed later in this chapter.

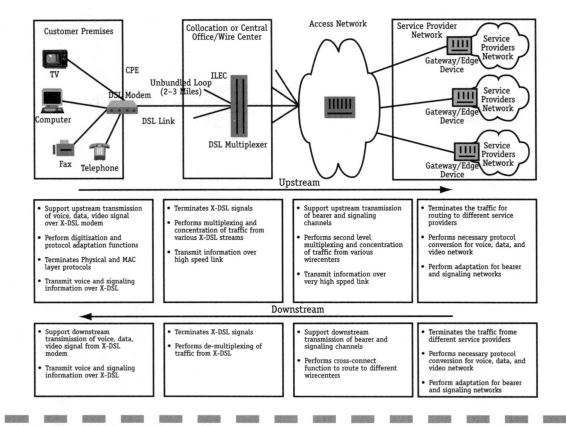

Figure 9.2 *Functional DSL-based reference architecture*

Figure 9.3
*Example of
DSL-based home*

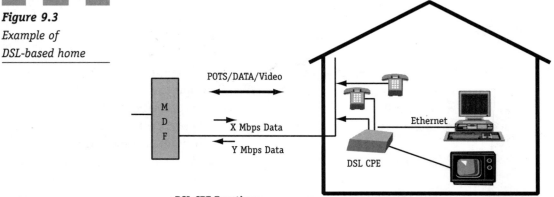

DSL CPE Functions
- No in-building rewiring
- No POTS splitter (in case of ADSL G.lite)
- No phone filters/isolators
- Phones connected to DSL CPE could be packetized voice or baseband voice
- Multiple TV channel

The copper loop transports a standard formatted signal where layer 1 is the XDSL frequency spectrum and layer 2 technologies are the ATM AAL format, IP, or MAC. The device that terminates the loop in the central office or colocation provides the concentration, multiplexing, security, and adaptation functions. The traffic from the DSL line is transported over a high-speed interface to the service/switching point in the network. At this point, various services are separated and sent to the appropriate devices to perform service-specific functions and inter-workings to various networks such as the PSTN, IP, and ATM.

The model is the same for business and residential customers with respect to the functions performed. However, if it is a business customer, some of the functions are colocated at the customer site with a device that is capable of performing the CPE and colocation functions. In this case, the traffic from the business customer is backhauled directly to the switching or service sites.

Service and Higher-layer Architecture

The enormous bandwidth enabled by DSL technology over copper when compared with traditional technology has opened the floodgates to a wide range of services to the mass market in a cost-effective way. Currently each service is transported over an infrastructure that is not shared by multiple services, which means it has a predefined, standardized protocol. Now, the services have to coexist over a common infrastructure, and, at the same time, maintain backward compatibility so that existing CPE devices—analog telephones, fax machines, computers with 10Base-T interfaces, etc.—can operate in the new environment.

Layers 2 and 3 are the service layers of the access network while layers above the OSI* protocol model are part of the CPE devices. Figure 9.4 shows the three layers (1–3) that reside in the customer site or device.

The following sections discuss the various services and their interworking over the DSL infrastructure. Each of these services is designed to maintain backward compatibility so that it operates seamlessly over the existing infrastructure at the customer end and in the public net-

---

* OSI: Seven-layer protocol stack standard.

work. The section on full integrated services discusses plans to provide these services in a homogeneous way. Major service providers are now involved in an FSAN project to standardize multiple services over a common infrastructure such as DSL.

Figure 9.4

CPE protocol stack

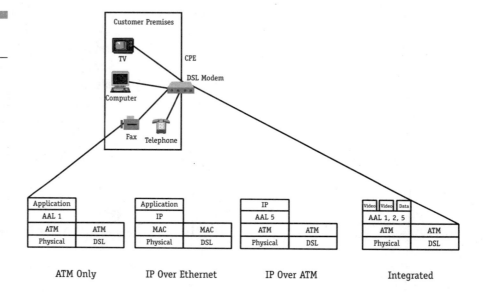

<table>
<tr><td>Application</td><td></td></tr>
<tr><td>AAL 1</td><td></td></tr>
<tr><td>ATM</td><td>ATM</td></tr>
<tr><td>Physical</td><td>DSL</td></tr>
</table>

ATM Only

<table>
<tr><td>Application</td><td></td></tr>
<tr><td>IP</td><td></td></tr>
<tr><td>MAC</td><td>MAC</td></tr>
<tr><td>Physical</td><td>DSL</td></tr>
</table>

IP Over Ethernet

<table>
<tr><td>IP</td><td></td></tr>
<tr><td>AAL 5</td><td></td></tr>
<tr><td>ATM</td><td>ATM</td></tr>
<tr><td>Physical</td><td>DSL</td></tr>
</table>

IP Over ATM

<table>
<tr><td>Video</td><td>Video</td><td>Data</td></tr>
<tr><td colspan="3">AAL 1, 2, 5</td></tr>
<tr><td>ATM</td><td colspan="2">ATM</td></tr>
<tr><td>Physical</td><td colspan="2">DSL</td></tr>
</table>

Integrated

Voice Services and Architecture[35,37,39,40,41]

Although DSL is designed for high-speed data, transporting voice along with data makes economic sense for service providers. Especially when voice accounts for 90 percent of the proits for service providers. Any new architecture for advanced services must accommodate voice services, as well as the new high-speed data and video services.

Protocol Stack

Voice services can be provided in a couple of ways over XDSL architecture, depending on the type of DSL technology. The options are (1) baseband voice (<4 kHz band) or (2) derived or packetized voice. Irrespective of the type of voice service, it is imperative that the user be unable to perceive a difference in the quality of service.

Baseband voice services use the same bandwidth (64 Kbps) and frequency (0 to −4 kHz) as POTS over copper wire pairs. ISDN uses the 4–10 kHz frequencies, and new DSL services are provided above the baseband frequency (10 Hz+). At the CPE, the voice traffic is multiplexed onto a common copper pair and transmitted to the central office (CO). From there, the reverse function (demultiplexing) is performed—voice is separated and routed to a traditional voice switch in the central office, and the data traffic is routed to the data network via a router or an ATM switch. Figure 9.5 illustrates this concept, where the XDSL technology, such as ADSL or ADSL lite, is applicable in this environment.

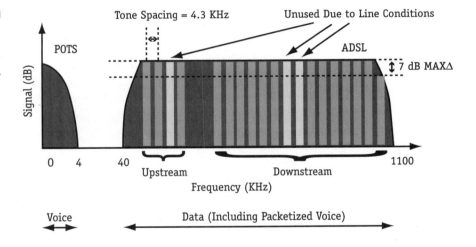

Figure 9.5

Separation of voice and data

One of the advantages of this type of voice service is that the user hears no difference in service. It has the same quality, lifeline capability, and 64-Kbps bidirectional TDM channel as traditional POTS. The disadvantage is that only one POTS line can be provided, i.e., if the user needs a second voice line, another copper pair is required.

However, not all DSL technologies are designed to support this. ADSL, ADSL lite, and the new HDSL2 are the only variations of DSL that support this type of voice service. Chapters 4 and 5 describe these technologies in detail. These DSL variations are designed to maintain backward compatibility with the existing legacy environment.

The other type of voice service is called *derived voice* or *packetized voice*. Engineers have been trying to put voice into packets ever since the first packet network (ARPANET) was built. One of the big hurdles, however, was maintaining the real-time criteria for voice. This is critical in a voice conversation.

Today, technology has advanced to the point that voice can be packetized and transported across a network without any degradation in quality of service. In fact, various packetized versions of voice are available today from different vendors. Some of the packetized solutions are voice over frame relay, voice over ATM, and voice over IP. Each has its own merits. It is this author's opinion, however, that, from the outset, voice over ATM was designed by the standards community to be the best option for voice with respect to managing the delay budget. This, along with the ability to provide multiple voice lines over a single copper pair makes it an attractive alternative. This has become the driving force, especially for the new competitive local exchange carriers who must lease copper pairs from the ILECs. Now, with the ability to provide multiple voice lines over a single pair, along with other services, CLECs have an opportunity to target medium and small business customers not currently addressed by traditional solutions.

There are many driving factors behind the voice-over-packet network solutions. They include:

∎ Voice-over-ATM standards in the ATM Forum and AAL1, AAL2, and AAL5 voice specifications in the ITU.

∎ End-user demand for low-cost solution technologies, such as DSL, echo cancellation, and silence suppression and detection.

Figure 9.6 shows the protocol stack used for the voice-over-ATM-over-DSL scenario.

The advantage of this architecture is that, since it is a derived service, it shares the bandwidth used by the data or other services. Every piece of information is packetized, thus enabling the service provider to integrate voice and data. In addition, the bandwidth used by voice services is only reserved during a voice conversation. Once the conversation is completed, the bandwidth is released for other services. Therefore, sufficient bandwidth is available on an as-needed basis whenever the user needs the telephone service. Here, the CPE device provides dialtone, adapts to fax traffic, and performs other user-selected options while supporting the traditional functions provided by the telephone service.

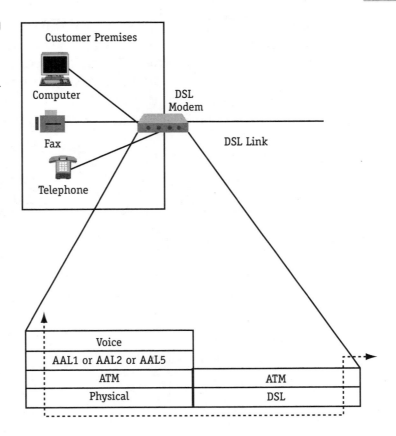

Figure 9.6
Protocol stack for voice over ATM

Key Architecture Element

To enable voice over ATM, customer premises, colocation, and switching equipment are needed to perform specific functions.

DSL-based CPE

The CPE device at the user end is connected to traditional telephones. If one of the telephones goes off hook, the CPE will provide a dialtone for the user to dial the digits. The process of how a CPE call goes through is illustrated in Figure 9.7.

First-generation voice over ATM on DSL was either in a baseband or AAL1 format if it was over ATM. This enabled easy mapping of the 64-Kbps voice channel. It had a significant disadvantage in terms of bandwidth and benefits to end-users such as small businesses and home-office users. The second generation of customer premises equipment takes advantage of the development in standards with respect to voice over ATM as variable bit-rate (VBR) traffic. This, along with advances in DSP (digital signal processing) technology, enabled has vendors to develop cost-effective voice encoding along with echo cancellation.

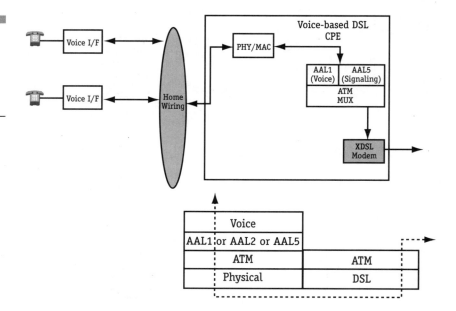

Figure 9.7
Voice-capable DSL CPE (Source: Coppercom Website)

VBR techniques (silence detection and suppression) for voice exploit the inherently bursty nature (brief periods of silence) of voice communications, which can result in increased efficiency. These silence periods (in decreasing levels of importance) occur:

- When no call is up on a particular trunk; that is, the trunk is idle during off-peak hours (trunks are typically engineered for a certain call-blocking probability: at night, all trunks could be idle).
- When the call is up, but only one person is talking at a given time
- When the call is up, and no one is talking.

The addition of more bandwidth-effective voice coding (e.g., standard voice is coded using 64 Kbps PCM) is economically attractive, particularly over long-haul circuits and T-1 ATM interfaces. Various compression schemes have been standardized in the industry (e.g., G.720 series of standards). Making these coding schemes dynamic provides the network operator with the opportunity to free up bandwidth under network-congestion conditions. For example, with the onset of congestion, increased levels of voice compression could be dynamically invoked, thus freeing up bandwidth and potentially alleviating the congestion without diminishing the quality of the voice.

A further enhancement to the support of voice over ATM is voice switching over virtual connections. This entails interpreting voice signals and routing the calls to the appropriate destination, as shown in Figure 9.8. The advantage from a traffic management perspective is that connection admission controls can be applied to new voice calls. Under network congestion conditions, these calls could be rerouted over the public networks, and therefore not cause additional levels of congestion.

Figure 9.8

SVC-based voice call via DSL infrastructure

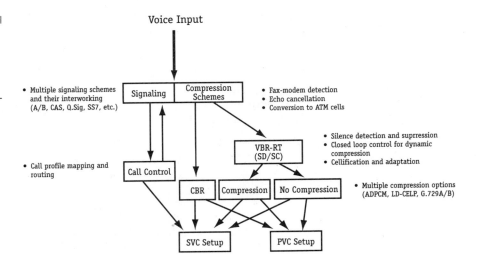

Once the decision-making process is completed, the CPE requests bandwidth in the DSL transport by sending the appropriate in-band signaling message to the network via the AAL 5 protocol. The message returns with the signal regarding the availability of switch ports and other network resources. Once the signal is obtained, the CPE provides the necessary dial tone for the user and waits for the digits based on the configuration (predefined user profiles), i.e., whether compression is activated; if so, at what speed, etc. This is done either by the service provider or by the user. In addition to this type of compression, the CPE device performs other options such as silence detection, silence suppression, and echo cancellation. Once a voice sample goes through this process, it is mapped to an ATM cell.

These cells are transported via a DSL-based copper pair per ADSL Forum or ITU standards. In transport, each voice conversation is assigned a specific VP/VC channel. The ATM layer provides the necessary requirements for low-error performance and low latency to meet the various QoS requirements set by the users. The minimum standard set by the ANSI T1.413 for BER* is 10^{-7} with a 6 dB margin.

ATM mapping incurs some delay (with ITU recommendation) for the voice sample. The two most popular voice compression algorithms are ADPCM and LD-CELP (see Chapter 7). The delays for these algorithms are 0.15 ms for ADPCM and 0.6 ms for LD-CELP. This is just one-way compression, but for voice, one must take into consideration the end-to-end delay in selecting the appropriate algorithm.

Thus, the following are the functions performed by CPE for supporting voice services:

- Fax/modem detection
- Echo cancellation
- Multiple compression algorithms (PCM, ADPCM, LD-CELP, and other ITU standards)
- Silence suppression and detection
- Closed-loop control for dynamic compression
- Packetization and adaptation of ATM and DSL
- Allocation of bandwidth on demand
- Multiplexing
- Signaling either in SVC, SPVC, or PVC using Q.2931 protocol.

Central Office or Colocation Function

In the downstream (i.e., toward the customer) direction, the cells are routed on a VPI and/or VCI to the appropriate DSL modem. In the upstream direction (i.e., toward the network), the cell streams are com-

---

* BER: Bit error rate.

bined/concentrated at the transmission convergence layer to form a single ATM cell stream in a link to the network for switching. Here, the basic function of mapping the ATM cells from the DSL line to a high-speed transmission, such as DS3 or OC3, is performed. The cells are then multiplexed and transported to a switching center. Figure 9.9 shows the basic functions and its associated protocol stack.

Figure 9.9
Typical colocation functions

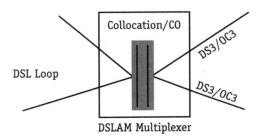

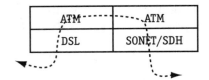

- Multiplexing and concentration
- Grooming
- Adaptation and protocol internetworking

In some cases, the voice switch is placed at the location where the DSL line terminates. Otherwise, the voice samples from the colocation are carried over high-speed SONET or DWDM to a switch site where the voice cells are terminated either at a gateway or an ATM switch.

The gateway functions as a mediation device between the public switched telephone network and the DSL access network. Here, the incoming cells from the DSLAM access concentrator are mapped to a TDM-based trunk connected to a class 5 switch using a standard GR-303 or V.5.2 interface. This path is predefined during the signaling process. Thus, the process of converting cells to TDM occurs here with some delay at the originating and terminating sides. This is shown in Figure 9.10.

Figure 9.10

TDM-to-ATM mapping

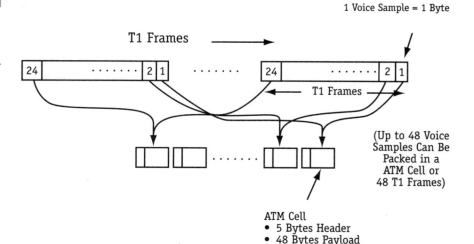

The other option is where the ATM cells are carried directly to the ATM switch, which is connected to the call control system performing the switching. This system processes the cell destination address information and routes it to the appropriate ATM trunk using a unique VP/VC combination, while maintaining the ATM encoding done at the CPE. The reverse process occurs at the customer's CPE or device where the handoff to the PSTN network is done.

Architecture Requirements

One of the basic objectives of the service provider is to provide traditional POTS service that is transparent to the end-user. To do this, however, there are numerous implementation issues that must be addressed. In general, the overall system should meet the following requirements:

- High availability
- Transparency to end-users using the traditional analog phones
- Easy access to existing value-added features
- Ease of installation and use (plug-and-play)
- Lower cost than current service offering (need to add value to end-users)
- Backward compatibility with existing telephones, faxes, modems, and other traditional devices (and transparent to the end-user).

These system requirements can be mapped to each element in the access network to identify the specific technical and marketing requirements to meet the end-user voice service expectation.

To meet the market requirements, the DSL-based network for telephone service should:

- Support toll-quality service for up to 16 POTS lines on a single DSL line
- Have the ability to dial domestic, international, and emergency numbers
- Have the ability to select local and long distance providers for telephone services
- Support fax- and modem-based services
- Support existing wiring and DTMF equipment
- Support basic, class, and custom calling features.

Some of the basic features are 411 directory assistance, 911 emergency service, 311 non-emergency service, 611 repair service, 800 and 888 toll free service, 900 numbers, AIN service, and access to calling card features. Custom features include call waiting, caller ID, call forwarding, three-way forwarding, and voice mail.

Architecture Issues

There is a wide range of architecture issues that need to be addressed to support telephone services. Some of the major issues are:

Network-level Security

In the traditional PSTN network, each line or circuit is reserved for the duration of the call. This is still valid in ADSL/ADSL lite-based DSL access. But in the case of a POTS line derived from DSL bandwidth, the voice line is a logical entity with its bandwidth shared by multiple users. In this case, care must be taken to guarantee security.

Traffic Modeling

With its bandwidth availability, DSL in the access network presents opportunities to provide multiple lines (for data and voice of varying bandwidth) on a single line. This has implications for the rest of the network. Consequently, detailed traffic analysis needs to be performed to address such issues as having low blocking probability, low latency, etc.

Network Reliability

Since multiple lines can be supported on a single copper pair, a cut or failure in the line affects many services. Therefore, care must be taken to ensure end-to-end reliability via alternate sources at a reasonable cost.

Interoperability and Interworking Issues

With the advent of new technologies and protocols, the need to support existing services and infrastructure becomes greater. With the development of standards, interworking has also become a critical issue. Interworking means that the network is not totally dependent upon any one vendor's products or solution; different vendors' equipment can communicate with each other.

Data Services and Architecture[43,44]

Driven by the growth in Internet traffic, data services have been primary drivers in expediting DSL deployment in the local access network. A favorable regulatory environment and affordable technology has fueled deployment as a means to enable new broadband integrated services.

Internet access has forced vendors to develop DSL-based solutions that can be deployed worldwide. Also, the regulatory environment favored a network where any new service provider could lease copper pairs and provide DSL-based services. This has enabled the deployment of DSL by new competitors (i.e., CLECs), which, in turn, has forced the incumbent LECs to begin offering DSL-based services.

Protocol Stack

There are many protocol options for providing data services in the DSL-based access network. (Refer to Chapter 7 for a discussion of some of the key technologies.) We will discuss some of the more important data services and their associated protocol stack. Other data services can be derived using a combination of these protocols. The protocols for data services to be discussed are:

- Channelized T1/E1-over-DSL
- Frame-relay-over-DSL
- IP-over-DSL
- ATM-over-DSL.

In general, any of these protocols can be mapped onto any of the DSL technologies discussed earlier in this chapter. However, due to the nature of certain applications in the real world, a particular combination of DSL technologies along with layer 2 and layer 3 technologies is used. For example, channelized T1/E1 is used on HDSL and is planned for HDSL2, whereas frame relay is used with IDSL (ISDN DSL) and IP with ADSL. As more applications emerge, vendors will develop more combinations, which will drive DSL technology choices.

Channelized T1/E1 Service

Traditional leased T1 in the United States and E1 in Europe is the most widespread and expensive high-speed option available today. It is provided during the circuit provisioning process by connecting various network elements manually. At the beginning of the 1990s, service providers started using HDSL (see Chapter 5) in the place of T1, which has proven to be the most cost-effective way of providing high-speed services because it enables automated provisioning. This channelized HDSL acts as the transport for the T1 service, thus making the HDSL transparent to the users.

In Figure 9.11, the customer premises has an HDSL modem with DSX-1 interface on the customer side connected to a PBX or a router. On the network side are two RJ-45 interfaces for terminating two pairs of copper. On the CO side, the HDSL modem has two RJ-45 terminating interfaces that are connected to a 1/0 cross-connect. This configuration is based on these ITU standards: V.35/G.703/G.704 for E1.

As shown in Figure 9.12, HDSL uses two copper pairs to provide the T1 bandwidth. Recently, however, a non-standard version of HDSL, called SDSL, has emerged; it is capable of carrying the equivalent of T1 traffic using a single copper pair. This is what has enabled new service providers, such as the CLECs, to compete effectively with incumbent providers—they lease the one copper pair and then provide T1 service.

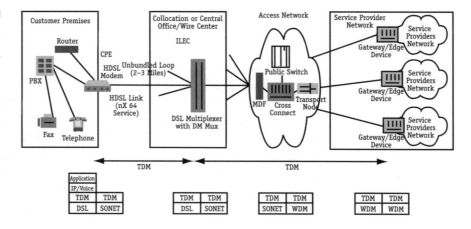

Figure 9.11

Channelized T1/E1

Frame Relay (FR) over DSL

One of the first services offered over DSL technology was based on frame relay for Internet access. Figure 9.12 shows layer 1, layer 2, and layer 3 protocols in the access network. As an example, UUNet, using the CopperMountain platform, currently offers this DSL-based solution.

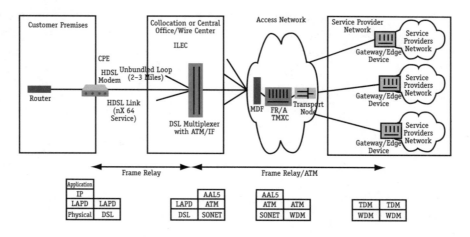

Figure 9.12

Protocol stack for FR over DSL

Frame relay is a packet-based service that offers significant savings over traditional private leased lines. It allows service providers to offer more speed for less money when compared to a leased-line network because they can take advantage of shared bandwidth, user traffic patterns, and over subscription (addition of more subscribers than usual). The reason users prefer frame relay over DSL is because of the current service offering. By mapping FR to DSL, the user can take advantage of the DSL cost without affecting service.

Users traditionally access the frame relay network via dedicated leased lines from the customer location to the service provider's POP. The cost of a traditional FR access line (56 Kbps or T1/E1) is computed in multiples of 56 Kbps and is distance sensitive (i.e., distance-based pricing). In other words, the higher the bandwidth and the longer the distance, the higher the cost. This is where DSL can play a critical role.

By using DSL instead of a traditional leased line, the user can have the same 56 Kbps bandwidth as usual, even though the underlying transport is capable of carrying much higher bandwidth. Then, when the user needs additional bandwidth, it can easily be activated on demand without delay and without reengineering the access network. This makes service provisioning easier and more cost effective because FR starts at the customer site. The service provider can more easily manage a single service. In general, DSL offers two significant benefits for frame relay:

- It can reduce the local access portion of the service cost for end-users. This is a particularly good deal for providers who lease traditional facilities from local service providers.
- It brings higher bandwidth access closer to the customer, so that a service provider can offer other value-added services.

IP over DSL

Over the last few years, the Internet has had a profound effect on the way information flows. Two trends have emerged that have increased the bandwidth requirement by at least an order of magnitude:

1. The addition of audio and video media to Internet-based information exchange.
2. The integration of context presentation, a push paradigm, into the existing pull paradigm.

In addition, the first thing that comes to mind about the Internet is IP (Internet protocol). This packet-based architecture is currently the backbone of the core Internet. Therefore, when it comes to accessing the Internet, the default protocol everyone prefers is IP. In addition, it is the protocol of choice for Internet service providers. As a result, with DSL technology and native IP, it has become the default choice for service providers. Figure 9.13 shows the protocol for IP-over-DSL-based access architecture.

Figure 9.13

IP-over-DSL protocol

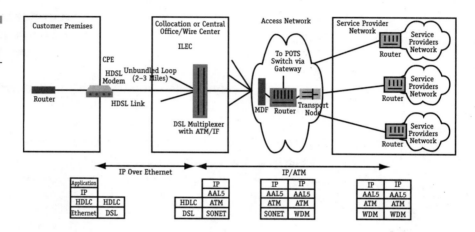

ATM over DSL

Although ATM is similar to frame relay, there are some basic differences:

- Frame relay has variable frames; ATM has fixed cells.
- ATM external signaling and networking parameters result in more control over QoS and performance for a broader range of traffic types.

For economic reasons, ATM currently resides in the Internet backbone where it provides the benefit of QoS and sharing of bandwidth at very high speeds. Since ATM was designed for a high-speed environment, it never made any inroads in the access environment. However, with DSL enabling a low-cost solution and the push for network convergence, ATM has gained some long-overdue credibility. Of all the technologies discussed, ATM was, from the beginning, designed for a converging network. To support data traffic, ATM maps IP traffic over AAL5 and transports across the access network. Figure 9.14 illustrates the protocols from the CPE to the switch in the access network.

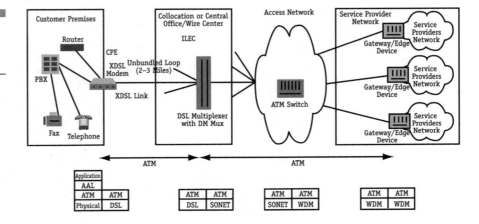

Figure 9.14

ATM-over-DSL protocol

By statistical multiplexing and logical connection, ATM provides the necessary QoS needed for various types of data services from the same customer. In addition, one could add voice and video services without any modification to the infrastructure, thus increasing revenue without incurring higher costs.

Architecture Requirements

Some of the requirements for providing data services from an end-user point of view include:

1. Easy migration from an existing dial-up mode
2. Connectivity to the Internet and corporate intranets
3. Multi-protocol support at layers 2 and layer 3
4. Security
5. QoS.

Migration from Dial-up Mode

Since most Internet access today is via dial-up PPP (see Chapter 7), any new DSL-based solution should support this mode. In the PPP model, most of the higher-layer functions such as networking, management, and administration (such as IP address, domain name, etc.) are transparent to the user. The only exception is better response time when accessing the Internet and/or Intranet.

One of the major market segments for high-speed connectivity is the telecommuter working from home. A basic requirement is Internet access while users are connected to the corporate network. This means the telecommuter accesses the corporate network through a firewall and then passes through the firewall to the public network. Another way is to dial a public ISP directly and enter a secure Website with appropriate authentication. In all cases, with the exception of faster speed and additional value-added services at a lower cost, there should be no differences.

Multi-protocol Support

One of the biggest hurdles for data access is the number of protocol options available, especially at layers 2 and 3. Therefore, any new system should be designed to support a wide range of protocols that are capable of interworking smoothly.

Security

One of the critical requirements for all users, e-commerce applications, and other public access services is guaranteed security. Any system should be able to provide authentication, authorization, and privacy.

Quality of Service (QoS)

Depending on the type of application, a user may need various levels of quality of service within the data application. If the telecommuter is a customer service agent, then access to the corporate database for customer information should have as high a priority as downloading e-mail or accessing a Web page. QoS also implies control, therefore bandwidth measurement becomes necessary. This also implies that QoS should be implemented in such a way that one user's service or performance is never degraded to the frustration level.

Architecture Issues

1. Like any other application, there are issues related to carrying data services in the DSL environment. The overall DSL architecture

should support simultaneous access to multiple network service providers.

2. Since multiple service providers will be offering service, the system design must support the ability to provision discontinuous IP domains. This will facilitate better management of the IP address space.

3. The architecture must be scalable.

4. The services should be easy to provision and must support dynamic connections between service providers and users.

5. Dynamic IP address assignment, as well as static IP address assignment, is required.

6. Security issues should be addressed.

7. Service providers must be able to authenticate users before they access the network.

8. Both business customers and service providers have significant investment in the infrastructure today. Hence, any new DSL-based solution should not require significant changes in existing customer requirements.

9. Private IP addresses must also support the unregistered addresses that many businesses use.

10. The architecture must support non-IP protocols.

11. The architecture should provide QoS capability.

12. The architecture should meet regulatory requirements.

Video Services and Architecture[14–21,23–26]

As service providers modernize their networks, interest is increasing in planning for the services that require higher speed in real-time services such as Internet video and high-speed interactive communication. The potential for revenue from business and residential services in the future is huge. Some of the applications that fall under the heading of video services are videoconferencing, video on demand, and interactive video games. These applications have high-quality, real-time requirements, thus making it more expensive and complex to deliver them over a DSL-based access network.

In order to reduce the cost and optimize the capacity, the system must have load-balancing capability under normal conditions. To achieve this and meet performance and reliability criteria for a large-scale interactive video system, network control functions may need to be distributed throughout the network. In this section, we will discuss the issues and architectures pertinent to the transport of video over a DSL-based access network.

Protocol Stack

The basic protocol stack for video services over a DSL-based access network is shown in Figure 9.15. This shows the generic mapping of MPEG-2 into the network.

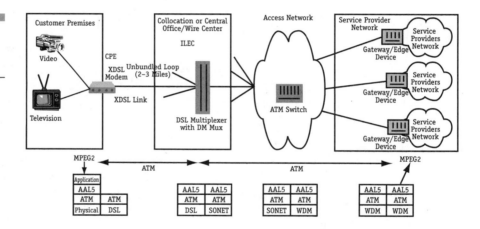

Figure 9.15

Protocol stack for video

The network adaptation uses AAL 5 by mapping MPEG-2 TS (transmission sublayer) packets into AAL5 cells. The specification allows for the following mapping:

1. Every AAL5-SDU shall contain n MPEG-2 SPTS packets unless there are fewer than n packets left in the SPTS. (Remaining packets are placed in the final CPCS-SDU.)
2. The value of n is established via ATM signaling using n = AAL5 CPCS-SDU size divided by 188. The default AAL5 CPCS-SDU size is 376 octets, which is equal to two TS packets (i.e., n = 2).
3. In order to ensure a base level of interoperability, all equipment should support the value n = 2 (AAL5 CPCS-SDU size of 376 octets).

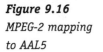

Figure 9.16

MPEG-2 mapping
to AAL5

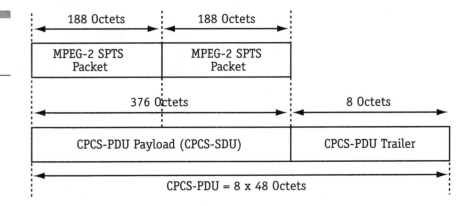

A NULL service-specific convergence sublayer is indicated by specification.

Mapping MPEG-2 Bitstreams into ATM Cells

There are two main options for mapping MPEG-2 bitstreams into ATM cells. One is constant bit-rate (CBR) transmission over ATM adaptation layer 1 (AAL1), and the other one is transmission over AAL5. Originally, AAL2 was envisioned as the adaptation layer that would provide the necessary support for video services over ATM, but currently, AAL2 for video is undefined. Therefore, we will discuss the merits of AAL5 and AAL1.

Figure 9.16 shows the mapping of transmission sublayer packets into AAL5 protocol data units (PDUs). Two TS packets will map exactly into eight ATM cells. The ATM Forum has adopted this mapping in the video-on-demand specification 1.1. A major drawback of using ATM AAL5 is that it lacks a built-in mechanism for timing recovery. Also, AAL5 does not have a built-in capacity for supporting forward error correction (FEC). A major advantage of using AAL5, however, may be financial. Since video applications require a signaling capability, AAL5 is already being implemented in ATM equipment. Another advantage of using AAL5 is that, if a NULL convergence sublayer (CS) is adopted, no additional network functionality will need to be defined. There are two major categories of video that would likely be transmitted over ATM using AAL5. Video which is being sent over heterogeneous networks would likely be sent via AAL5. This video would probably be carried as IP packets over ATM and encoded in proprietary formats such as QuickTime or AVI. The AAL5 would provide no quality of service guarantees from the network for this class of video. The second class of video would be variable bit-rate traffic native to the ATM network. This video would be able to benefit from the AAL5 quality of service guarantees.

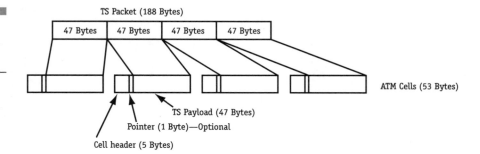

Figure 9.17
MPEG-2 mapping to AAL1

As shown in Figure 9.17, a TS packet will map neatly into four ATM AAL1 cells. One major advantage of AAL1 over AAL5 is that it is designed for real-time applications. The major disadvantage of AAL1 is that it only supports constant bit-rate applications, and future video applications will probably want to take advantage of variable bit-rate transmission options. AAL1 will also need to be supported in end equipment as will the AAL5 functionality. AAL1 does provide forward error correction, which is important for some video applications, especially over media prone to errors such as wireless ATM. AAL1 is expected to be the medium of choice to support video from H.261 or H.263 encoders. H.261 video has traditionally been transported over lines that are multiples of 64-Kbps or ISDN lines.

Architecture Requirements

The following are some of the requirements for DSL architecture to support video transport:

1. A wide range of QoS needs to be supported with the baseline being traditional VCR.
2. The BER (bit error rate) to support video should be 10^{-6} or greater.

3. CDV (cell delay variation) or "jitter" in the case of mapping MPEG to ATM cells should be limited to one or two cells to avoid any degradation in QoS.

4. CPE devices should be designed to support a wide range of interface options at a low cost.

Architecture Issues

Quality of Service Issues

To deliver acceptable-quality video to the user, the network must provide a certain level of service. Cell delay variation, bit errors, and cell loss can have severe effects on the quality of the video stream received. The bit error rate should be 10^{-6} or better, otherwise serious degradation in the quality of the received video is visible to the end-user. Similarly, cell delay, cell loss, and rate control issues have a significant impact on the quality of video received. This section examines some of these issues.

Cell-delay Variation (CDV)

Cell-delay variation or jitter can have a significant impact on the quality of a videostream. MPEG-2 video systems use a 27 MHz system clock in the encoder and the decoder to synchronize the operations at the decoder with those at the encoder. This enables video and audio streams to be correctly synchronized and also regulates the retrieval of frames from the decoder buffer to prevent overflow or underflow. To keep the encoder and decoder in synchronization with one another, the encoder places program clock references (PCRs) periodically in the transmission sublayer. These are used to adjust the system clock at the decoder as necessary. If there is jitter in the ATM cells, the PCRs will also experience jitter, and jitter in the PCRs will propagate to the system clock, which is used to synchronize the other timing functions of the decoder. This will result in picture quality degradation.

One proposed solution for traffic over AAL1 is to use SRTS (synchronous residual time stamps). In this method, both ends of the transmission would need to have access to the same standard network clock. This reference clock could then be used to determine and counter the effects of the cell-delay variation. Whether this clock would be readily available is unknown. Also, there is some question about whether AAL1 would provide enough bits for SRTS to be effective.

A lengthy discussion of sources of jitter and ways to estimate jitter in ATM networks is provided in Appendix A of the *ATM Forum Video on Demand Specification.*[16]

Bit Error Rate (BER)

Encoded video streams are highly susceptible to loss of quality as a result of bit errors. Bit error rates are media dependent, with the least error rates expected from optical fiber. Bit error rates for a 5 Mbps video sequence are given in Table 9.1 below.

Table 9.1

Bit error rate in ms

Bit Error Rate	Average Interval between Errors
10^{-5}	20 ms
10^{-6}	200 ms
10^{-7}	2 sec
10^{-8}	20 sec
10^{-9}	3 min, 20 sec

The encoding method of MPEG-2 video makes it susceptible to picture-quality loss due to bit errors. When a bit error occurs, it can propagate both spatially and temporally throughout the video sequence. A spatial error occurs because the variable length codes (VLC) that make up the blocks and slices are coded differentially and utilize motion vectors from the previous VLC. If a VLC is lost, then that error will propagate to the next point of absolute coding. In an MPEG-2 stream, this point is at the start of the next video slice. Therefore, a bit error can degrade the picture quality of a larger strip in the frame. Temporal error propagation occurs as a result of the forward and bidirectional prediction in P and B frames. An error that occurs in an "I" frame will propagate through a previous B frame and subsequent P and B frames until the next I frame occurs. The original frame is the strip in error due to the loss of VLC synchronization. The error is propagated temporally through a group of pictures. In a typical video sequence, a GOP (Group of Pictures) can last for 12 to 15 frames. At 25 to 30 frames per second, the error could persist for 0.5 seconds. This would be long enough to make the video quality objectionable in many cases. Bit errors that occur in P frames will be propagated in a similar manner to surrounding B frames, generating a similar, but more limited, effect. Bit errors in B frames would only affect that frame.

Cell-loss Rate (CLR)

For the reasons described in the previous section, the cell-loss rate also plays a critical role in the quality of the decoded videostream. The CLR can depend on a number of factors including the physical media in use, the switching technique, the switch buffer size, the number of switches traversed in a connection, the QoS class used for the service, and whether the videostream is constant or variable bit-rate. Cell losses in ATM networks are often the result of congestion in the switches. Providing appropriate rate control is one way to limit the loss.

Error Correction/Concealment

Bit errors and cell losses in video transmissions tend to cause noticeable picture quality degradation. Error correction and concealment techniques provide methods for the decoder to deal with these errors in a way that minimizes the quality loss. Error correction techniques remove the errors and restore the original information. Error concealment techniques do not remove the errors, but manage them in a way that makes them less noticeable to the viewer. Encoding parameter adjustments can also be made to reduce the effects of errors and cell loss.

Error Correction

Error correction is more difficult for real-time data than it is for non-real-time data. The real-time nature of video streams means that they cannot tolerate the delay that would be associated with a traditional retransmission-based error correction technique such as automatic repeat request (ARQ). Delay is introduced in the acknowledgment of the receipt of frames, as well as in waiting for the timeout to expire before a frame is retransmitted. For this reason, ARQ is not useful for error correction of videostreams.

Forward error correction is supported in ATM by AAL1. FEC takes a set of input symbols representing data and adds redundancy, which produces a different and larger set of output symbols. FEC methods that can be used are Hamming, Bose Chaudhuri Hocquenghen (BCH), and Reed-Solomon. Forward error correction presents a tradeoff to the user, however. On the positive side, it allows lost information to be recovered. On the negative side, this ability is paid for in the form of a higher bandwidth requirement for transmission. The added traffic can introduce additional congestion to the network, leading to a greater number of lost cells. These additional lost cells may or may not be recoverable with FEC. Consequently, the role of FEC in video is still a topic of discussion.

Error Concealment

Error concealment is a method of reducing the magnitude of errors and cell loss in the videostream. These methods include temporal concealment, spatial concealment, and motion-compensated concealment. With temporal concealment, the bad data in the current frame are replaced by the error-free data from the previous frame. In video sequences where there is little motion in the scene, this method will be quite effective. Spatial concealment involves interpolating the data that surround a frame block in error. This method is, however, most useful if the data do not contain a high level of detail. Motion compensated concealment involves estimating the motion vectors from neighboring error-free blocks. This method could be used to enhance spatial or temporal concealment techniques. "I" frames cannot be used with this technique since they have no motion vectors.

Encoding-parameter Adjustment

The encoding parameters for a videostream can be adjusted to make a stream more resistant to bit errors and cell loss. MPEG-2 (as well as Motion JPEG) support scalable coding, which allows multiple qualities of service to be encoded in the same videostream. When congestion is not present in the network, all the cells will arrive at the decoder, and the quality will be optimal. When congestion is present, the coding can be performed so that the cells that provide a base layer of quality will reach the decoder, while the enhancement cells will be lost. Temporal localization involves adding additional I frames to the videostream. These additional I frames prevent long error propagation strings when a cell is lost, since errors are rarely propagated beyond the next I frame encountered. The additional I frames are larger than the P or B frames they replace, however, so compression efficiency will be reduced. In addition, the greater bit rate required for these added I frames can contribute to network congestion. A third technique that can be performed at the encoder is to decrease the slice size. Since re-synchronization after an error occurs at the start of the next slice, decreasing the slice size will allow this resynchronization to occur sooner.

Full Integrated Services (Integrated Voice, Data, Video)

The previous sections in this chapter assume that only one service is provided and the underlying DSL access network is designed to meet its requirements. In a full-service access network,* the architecture is designed to support integrated services and share the common infrastructure. FSAN-based access assumes an optical fiber section based on highly flexible, bandwidth-efficient technology. The remaining network leverages the existing copper infrastructure to the customer. Chapter 8 discusses all the various combinations of fiber- and copper-based architectures and addresses how various versions of DSL fit into the architectures.

In the FSAN initiative, the group has to agree on a common broadband architecture for the provisioning of both broadband and narrowband services using ATM as the technology of choice. It relies on a Passive Optical Network (PON) (i.e., shared downstream distribution based on fiber splitters) whose border nodes are optical network units interfacing with VDSL or ADSL. On the user side of these optional XDSL links, the network termination (NT) provides an ATM user-network interface (UNI) at 25 Mbps, Ethernet, or I.430/I.431 interfaces. The FSAN specification covers everything from the UNI to the service node interface (SNI), which ends in an optical line terminator (OLT) interfacing VB5 to the service provider. The user-server signaling should be carried transparently from the terminal equipment to a service node that can reside anywhere in the ATM network. At the user access (UA) device, the network termination equipment can be an XDSL modem, an ONU, or a full ATM-over-fiber site such as the service node. Attached to the UNI is the residential gateway that would carry out control, multiplexing, and adaptation functions for the different user devices and terminal equipment.

---

* FSAN Initiative: an initiative by a consortium of telephone operators around the world to define a broadband architecture capable of supporting integrated services (voice, data, and video).

An important problem when trying to provide the highest XDSL rates (25 Mbps to 50 Mbps) is that the required loop lengths imply a very high density of network nodes interfacing with XDSL and the optical network. A number of experiments trying to extend the applicability scope of XDSL in an effort to avoid the massive deployment of fiber are also under way. Advanced Communications Technologies and Systems (ACTS) project AC309, ITUNET, is evaluating a full-copper solution employing inverse multiplexing of ATM-over-VDSL, together with statistical multiplexing. Again, ATM is the chosen data transport technology.

The key to enabling full service begins at the customer premises. Traditional CPE currently available with a DSL interface is designed to support a specific type of service. However, the CPE for a full-service network should be capable of supporting multiple services and a wide range of protocol options within each service. The details of the CPE are discussed later in this section.

Protocol Stack

For a full-service access network to support a wide range of services requires the right combination of underlying infrastructure. Figure 9.18 shows the protocol stacks for a FSAN-based access network. Here, the copper-based segment is enhanced by DSL with the layer 2 protocol being ATM. The service specific protocols are in layer 3 and above. For instance, IP-based services are in layer 3 and above.

Figure 9.18
FSAN protocol stack

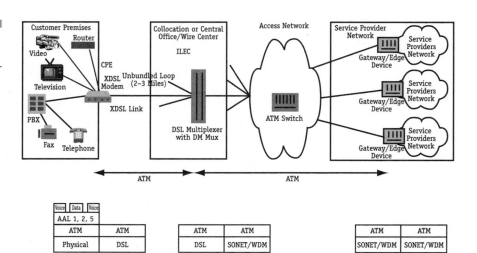

Architecture Requirements

In order to achieve the objective of an integrated access network, several requirements involving various access network elements must be met. These can be grouped into CPE-, transport-, and OSS-related requirements. Other elements will have a combination of these requirements for an integrated architecture.

CPE Requirements

In a rich multitechnology context, it is quite useful to have a common network architecture that enables the coexistence of new and legacy access techniques. The proposed model calls for dividing the UA device into two units: the network termination (NT) unit and the residential gateway (RG) unit, as shown in Figure 9.19. The NT provides homogeneous access to any broadband technology by means of an ATM UNI (e.g., a XDSL modem with an ATM interface).

Figure 9.19

Home network architecture elements and the protocol

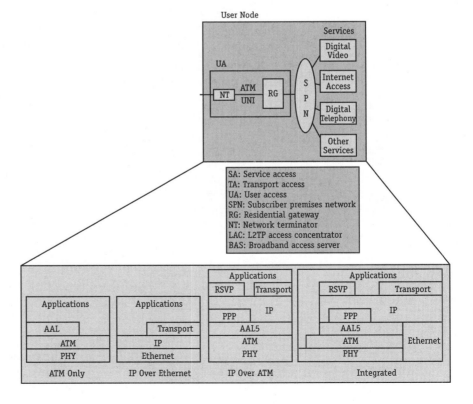

The residential gateway bridges from the ATM interface to any of the service interfaces (e.g., analog TV, POTS, IEEE 802.3, USB, etc.). This gateway unit multiplexes the traffic from interactive and non-interactive applications into a single data stream and performs the necessary control plane functions (not tackled here) carried transparently from/to the service node by the integrated transport access network.

The key issue is the incorporation of an ATM interface into newly deployed UA devices for data integrated services, while keeping the required legacy interface(s) (analog telephone, ISDN, analog TV, etc.) in the device. This means that network termination modems should include an ATM interface to connect to any QoS universal multiplexer residential gateway. At the other end of the access network, SA and TA devices provide interfaces to the service provider's ATM LAN and the network operator's ATM transport WAN infrastructure, respectively. In this way, ATM becomes the common layer for all protocols across all the platforms providing uniform end-to-end QoS management.

So far, several ATM-based architectures have been cited, although not much rationale has been given for the introduction of ATM throughout the access network. There are some technical factors that make it convenient to use ATM as the standard interface technology.

First, ATM as the universal access interface would enable seamless integration of access and backbone networks. Existing broadband networks are based on ATM technology. In addition, many headend-to-central office (or multiservice center) networks use synchronous digital hierarchy (SDH) or ATM multiplexing capabilities. If ATM were used across all access media, statistical multiplexing would be possible over the whole path, and no protocol adaptation would be needed for each and every access technology.

Second, ATM appears to integrate a wide range of applications with different QoS requirements within a given bandwidth capacity, regardless of the underlying physical media. It enables efficient bandwidth management to carry heterogeneous traffic types. ATM's small and fixed packet length makes it possible to guarantee bounded delays, makes channelized statistical multiplexing feasible, and enables finer implementation of scheduling policies for higher-level data units.

From a network management point of view, bringing ATM to the user's access units has other implicit advantages thanks to its connection-oriented nature. Here, most of the billing, security, maintenance, and operation support infrastructure used by current switched service networks can be supported, thereby making it possible to charge on the basis of usage with an end-to-end managed QoS.

Access Transport

ADSL transport of data in synchronous transfer mode (STM) provides a serial bit interface with up to seven simplex/duplex subchannels with synchronous multiplexing. However, ADSL transport of data in an ATM format supports up to two frame-based data transport "paths" with asynchronous multiplexing. The *fast* data path is intended to provide a low data-transfer delay, up to 2 ms, as appropriate for real-time interactive applications. The *interleaved* path is intended to provide a very low error rate and greater latency (tens of milliseconds).

Several other ways of implementing ATM over copper technologies are already commercially available. ITUNET splits a geographical area into cells with a radius of ~1 km with its nodes interconnected by a shared ATM gigabit backbone. This backbone is physically built on top of existing telephone pairs using VDSL bundling in transit before connecting to an optical network node (i.e., using XDSL not only in the last mile). This could be a migration strategy with respect to fiber deployment, which requires a huge investment in comparison to the cost of extra modems and multiplexer equipment. Again, ATM is the data-transport technology chosen.

The advantages of the different architectures, plus the provision of an Ethernet interface for data SPN, may be integrated into the IP-over-ATM architecture. The only requirement to integrate the ATM-only architecture is to allow the applications to access the ATM service directly. This may be very convenient in some cases, such as video distribution applications that require low delay, jitter, and low transmission overheads. Be aware that applications wanting to use ATM service directly will need end-to-end ATM service. The provision of an Ethernet interface may be accomplished by an IP routing function in the RG. Actually, the router approach allows any networking interface supported by IP to be provided, with the obvious restriction that the internetworking protocol must be IP.

The main drawback of the integrated architecture is the complexity and cost of the NT and RG devices.

OSS Requirements

The cost of owning a product over its life is linked to its capital or replacement cost, and in many instances can be of similar magnitude or greater. Operating, administrative, and management costs are therefore of significant importance to service providers. The main requirements for minimizing OAM costs are:

- Use of standard interfaces and information models where possible.
- Minimizing the need for special operational skills and maintenance tools.
- Provision of accurate maintenance information by the system.
- Promoting reuse by locating complex functions at common points to enable platform sharing.
- Provision of functions for locating faults and detecting network degradation.
- The need to build in the ability to evolve and upgrade the network and its management capability.
- Adopting a single flexible multiservice delivery platform with a common management system avoids the old vertical management structure, which contains separate processes and tools on a per-service basis.

Just after Table 9.2 we discuss the major impact on the operating processes after introducing XDSL technologies to existing service providers.

ATM-based Access Network Performance

One of the important requirements for an integrated network is maintaining the performance of various services in spite of the fact they all share the same resources. This requirement spans all the elements in the access network. Key performance parameters for an ATM-based access network are bandwidth, access transmission delay, access delay, response time, and cell loss/discard ratio. The values of these parameters depend on the services being transported. Table 9.2 shows the performance requirements of an ATM-based access network.

Table 9.2

*Performance
requirements*

Service	Traffic type	Band-width (peak)	Access trans-mission delay	Access delay	Response time	Cell loss
Data (IP- based)	UBR (best effort, minimum throughput guaranteed)	10 Mbps (typical)	1.5 ms[1]	< 1s	—	Cell discard on IP packet basis (one cell loss results in one IP packet loss)
ATM SVC (voice, video, data, etc)	CBR/VBR/ABR	<150 Mbps	1.5 ms[1]	—	—	<10–5
VoD	CBR	6 Mbps (typical)	1.5 ms [1]	<3 ms[3]	<500 ms	< 10-8
POTS/ISDN	CBR	<2 Mbps	1.5 ms[2]	—	—	< 10-5

[1] Provisional value (under assumption G.982 can be applied to these services)

[2] 1.5 ms is specified in ITU-T G.982 for POTS service

[3] ITU-T I.356

Architecture Issues

Providing an integrated network is not a technological problem, since much of it is already available. Larger businesses are already using broadband networks over point-to-point fiber. The key costs are fiber installation and the multiplexing technology at the ends. Using HDSL can save a little, but this is only suitable for rates up to 2 Mbps; even then, it is a good solution only if two or more metallic pairs are available.

Studies have shown that the dominant costs—digging and terminal equipment—have been addressed in two ways. A fiber-based transport that allows a common feeder, transceiver, and multiplexing system at the central office or co-location site reduces the overall equipment requirement, but not the costs at the customer end where resource sharing is not possible. Reusing the existing metallic pair (e.g., with VDSL technology) can save digging costs.

The major issue with integrated networks is security, since multiple services and users are shared. This leads to a traffic problem in terms of QoS, congestion, and management. Although these are not a top priority, they can differentiate services between two providers in a competitive environment.

Impact of DSL-based Services on Operational Processes

Introduction of XDSL-based technologies and supporting services has a huge impact on the operational process, and it is even larger for a legacy service provider than for a new entrant. Network operations and maintenance can be expressed using business processes that can be broken down based on the service functions they provide. One critical step is to ensure that operational processes and actions can deal effectively with the changes introduced by the new technology. This involves some or all aspects of operational processes such as:

■ Order negotiation
■ Service configuration
■ Service activation—plan and build network
■ Service assurance—maintain and repair service
■ Service assurance—performance monitoring
■ Billing and collection.

In addition, the data requirements of each process must be understood to ensure consistency across the operational support systems.

Figure 9.20 depicts a simplified example of possible impacts as a result of adding requirements onto existing processes. In this example, new requirements are identified which allow for the precertification of an unbundled loop. The shaded area shows the impact on the service configuration process and the corresponding OSS functions. These requirements will raise some of the questions listed below:

■ How can the remote and local line testing be done for owned and leased loops?

■ What is the process for rejecting loops that do not meet specifications?

■ What access privileges should be given to other service providers' operational systems?

Figure 9.20
Service process
overview

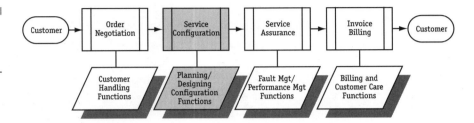

New processes and functions will have to be developed, or existing functions will need to be modified to satisfy these requirements. There may also be new data that indicate the service configuration process needs to deal with no access to qualified loops.

Each operational process will comprise a number of OAM&P functions. Some of these functions relate to people, others relate to the systems and technology being used. For example, the introduction of XDSL technology may involve new installation skills, which will entail additional training for employees. The network operator may need to re-engineer the service delivery process because of a new technology and point of interconnection with another service provider.

New OAM&P functions can be derived from re-engineered processes, which will need to be implemented in an XDSL system. The repair process may indicate a new fault isolation and repair process. This can be translated into a requirement for remote monitoring and electronic exchange of information between service providers.

Cost-effective ways of operating and maintaining an XDSL system must be considered if this type of network is to be viable. The network should be designed to require minimal human intervention with respect to installation, service configuration, performance monitoring, fault management, trouble repair, etc. Also, it should be kept very simple, using remote management systems to achieve optimal results. Automatic configuration should result in inventory information (e.g., serial number, equipment type, next maintenance date, and software/hardware versions being supplied by the equipment), which is vital for network maintenance. It can be used to determine XDSL compatibility, locate versions of equipment that have been identified as being prone to errors, or locate equipment that is due for maintenance.

Customer expectations of service availability will result in the network operator's needing to have accurate information so fault isolation and correction can be performed quickly. Fault corrections should not require special fault-finding techniques or tools. For example, with MCI's proactive network monitoring, network operators do not wait until a customer reports a fault. The operators' expectation is that faults will be corrected before a customer is even aware there is a problem. To support this proactive fault-handling feature, the network will need to provide accurate event reports. To simplify the task, the reports should include the location, type, time of the event, and possible repair action in an easy-to-understand format. Another consideration is how to prevent a large number of events from being reported. This requires that filtering be performed, which is of direct significance to network equipment and OSS design. Fault correction can be further simplified by the use of correlation functions in upper-level operational systems.

Some test functions are necessary, and these too affect the design of the XDSL system. An example is the requirement for a copper-line test function, which detects stated conditions that must be identified at the design stage. It is important to have a cost-effective implementation of this functionality in XDSL hardware/software, connecting network elements, and operational systems. If this is not done, then suppliers will provide their best estimate of what is needed. At best, this may provide unnecessary functionality, and at worst, may completely miss functionality, which is considered essential to network operations.

Summary

In this chapter, we described the various voice, data, video, and integrated services that can be provided over a DSL-based access infrastructure along with any of their specific requirements. We identified the supporting architecture, protocols, requirements, and issues that are pertinent to the various service offerings.

Broadband Access Network Design

Overview

In this chapter, the design of the DSL-based access network is discussed. Figure 10.1 shows the reference access network in the United States; it is used for DSL-based access to the public network. However, this is applicable to rest of the world where the loop length is smaller than in the U.S. The access network is where all traffic is originated and terminated because without it, no traffic could exist in the backbone. This network is so critical to backbone network providers such as IXCs that they are willing to pay up to 40 percent of their revenues to the providers of local access networks* for voice services. We can only assume that this proportion will be applicable to new broadband services, as well.[†]

Figure 10.1
Reference access network

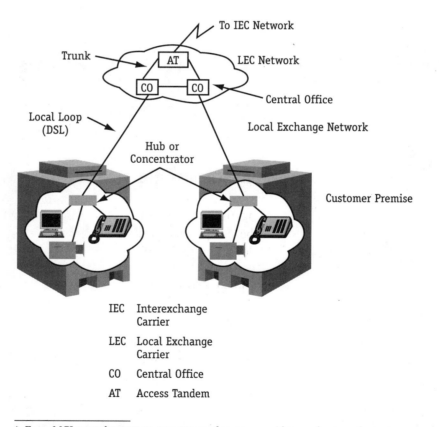

IEC   Interexchange Carrier

LEC   Local Exchange Carrier

CO   Central Office

AT   Access Tandem

---

* From MCI annual report on percentage of revenues paid toward access charges.
† With the changing regulatory environment, this amount is decreasing. In spite of that, the overall access cost is still a major expense for all long-distance service providers.

This chapter addresses the design issues of estimating traffic, link (transmission system) sizing, and node (switching/multiplexing system) sizing in the access network. The different components in a network are:

- **DSL CPE.** Owned by the public carrier or the customer, but located at the customer premises. This is part of the premise domain mentioned in Chapter 1 (Figure 1.7).
- **DSL-based local loop.** The link connecting the CPE and the central office (CO), which is the first interface to the public network. This is part of the access domain identified in Figure 1.7.
- **Backhaul trunk.** The links connecting different COs/co-location sites to switching centers. This is part of the transport domain identified in Figure 1.7.
- **Switching sites.** The first level of switching providing traffic switching or routing and access to advanced services and to other networks. This is part of the services domain identified in Figure 1.7.

Each of these components must be considered in designing the DSL-based access network.

Broadband Access Network Definition

Before listing the requirements, we should define an access network, which can be categorized into two possible access levels: the user or application level and the public network level. Figure 10.2 depicts these two levels.

The user level defines the access required to the public network resources. The link between the customer's hub and the central office is the access interface to the public network. This level provides the greatest diversity in interfaces, protocols, architecture, technologies, and standards compared with any other level in the network hierarchy. User network design is based on the amount of traffic generated from different sources, such as voice from PBX, data from LANs, and video from the desktop or a videoconferencing center on the user's premises. All this traffic must be identified during the design at the user/application level. This information is usually available within the company.

Figure 10.2

Two levels of access network

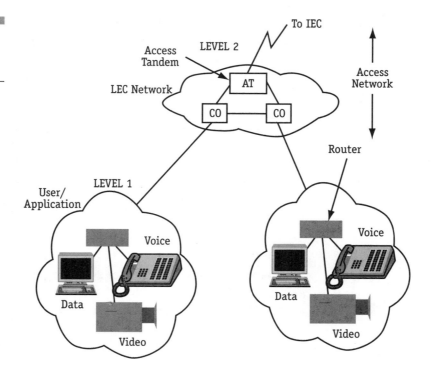

The next level is the access interface to the switching sites. This is usually a connection to a router or switch. In the broadband environment, this device should be able to handle all types of traffic (voice, video, and data). Based on the service requested, the switch/router gains access to advanced services with the help of a gateway either within the same site or from a remote site. This location also enables the interconnection point for other carrier and network systems.

Network Design Requirements for Broadband Access

In a DSL environment, design of the access network is primarily based on existing traffic projections for each customer or user. The most difficult part here is identifying the traffic requirements of a customer. Customers can design their own private networks, because they already have their traffic and usage pattern information. For a public network provider, however, it is difficult to project the exact traffic that could be generated from the many user applications, especially if the traffic is nonexistent, such as with broadband traffic. Network requirements are designed to meet these uncertainties and allow for expansion if the projections don't match the actual traffic. In this section, we divide the public access network requirements into the following general categories:

- Traffic
- Protocol
- Architecture/technology
- Services, features, and functions
- Access route diversity (redundancy).

Traffic

One of the most important aspects of the network design is estimating traffic. In a broadband network, user traffic comes from various applications, each with different characteristics and usage patterns. It is difficult to estimate broadband traffic now, because no such traffic currently exists. Below we look into the details of this engineering problem.

Protocol

To interface with the public network, the interface protocol should meet the public-network specifications set by the standards committees, such as Bellcore, ANSI, IEEE, etc. For a DSL network, the protocol (layers 1 to 3) should be capable of handling integrated voice, video, and data traffic in addition to what the published standards call for. Some of the important protocol and related technologies are discussed in Chapter 7. For economic reasons and the business objective a service provider might initially focus on either data or voice services. Thus, service providers pursue products that meet their price point for the current services. Usually this is without the consideration of long-term implication of such a decision. For example, the UUNET primary service offering is data. Their first protocol for their solution is IP-over-frame relay. In case of COVAD Communication, which plans to offer data and voice, the choice is for ATM with packetized voice and IP-over-ATM. Both service providers' decisions depend on their long-term service objectives.

Architecture

Having decided on the protocol, the service provider determines what architecture should interact with the selected protocol. Usually, architecture and technology go together. For DSL architectures, the key is selecting the appropriate one; this largely depends on the end-user applications. In the design of DSL access networks, the selection of architecture/technology should provide both short-term and long-term benefits in terms of handling current and future traffic and user growth. Figure 10.3 shows a typical XDSL-based architecture.

Figure 10.3
XDSL-based architecture

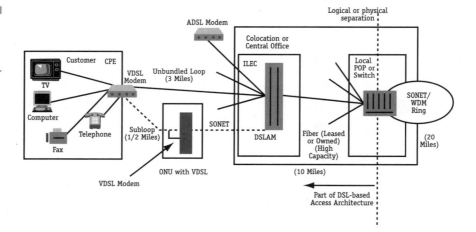

Services, Features, and Functions

When a user interfaces to the public network through a DSL device, certain features, functions, and services are required for standard interface and protocol support. These vary depending on the protocol. For example, the virtual path identifier (VPI) and virtual channel identifier (VCI) are based on the ATM protocol. All service to be provided on ATM is mapped to this VPI and VCI. The services, features, and functions provided with the protocol must be considered in the design of the DSL-based access network. Today, providers can offer service transparency through mapping, i.e., the user's service characteristics are preserved regardless of the access protocol. The disadvantage of service mapping is that the overhead involved in each protocol consumes additional bandwidth, which inefficiently uses system resources. The user should therefore consider all services offered when deciding on a particular protocol.

Access Route Diversity

Integrated broadband networks are unique in that all traffic going outside the premises exits via a single very high-speed interface. Thus, the customer's only communication to the outside world is through one interface. Any damage to or failure of this interface makes the customer vulnerable to service interruption. It is therefore necessary to provide a completely diverse, redundant access route to each interface point in the network, as shown in Figure 10.4.

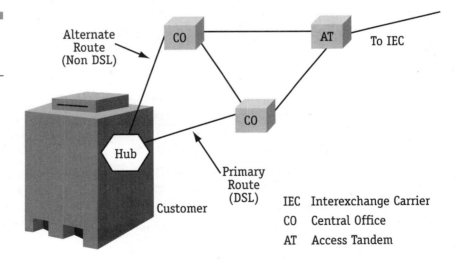

Figure 10.4
Access route diversity

In this section, we look at a high-level process for designing a broadband access network and explain the steps involved in the design of such a network. There are many ways to approach the design of a broadband network. Let us assume a bottom-up approach, as illustrated in Figure 10.5.

DSL-based Access Network Design

In this section, we look at a high-level process for designing a broadband access network and explain the steps involved in the design of such a network. There are many ways to approach the design of a broadband network. Let us assume a bottom-up approach, as illustrated in Figure 10.5.

Figure 10.5
*Bottom-up approach
for design*

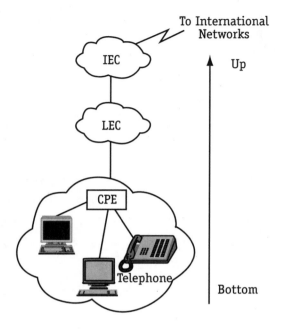

As mentioned earlier, designing a DSL network is a lot more compli-
cated than designing a conventional circuit-switched telephone network
or a packet-switched computer network because the latter two networks are designed for specific applications. For example, the telephone
network is designed to carry voice and is optimized for voice traffic.
Thus, although data can be carried over the network, they are not transmitted efficiently. Likewise, a computer network is designed to transfer
data, whose requirements are quite different, and full-duplex communication is necessary. Voice transmission is not possible at all over a computer network. In broadband networks, integrating these two types of
traffic onto a single network so that its resources are used efficiently is
a formidable task.

In Chapter 1, we noted that a typical network consists of nodes
(switching centers) and links (transmission lines). Broadband networks
are no exception. For the design of a broadband access network, we
need a reference network architecture. Figure 10.6 shows the reference
access architecture for broadband.

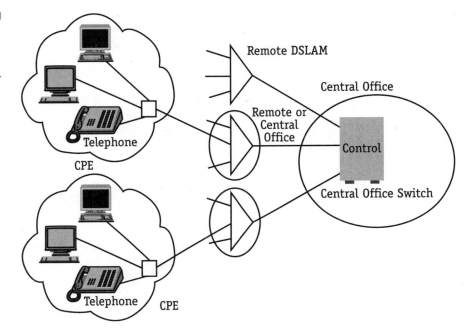

Figure 10.6
*Reference access
network architecture*

To design a broadband access network, one must estimate the traffic from each customer (business or residential) and understand the traffic patterns and usage characteristics in the busy hours. For example, residential customers' telephone usage peaks between 4:00 P.M. and 6:00 P.M. In broadband networks, video traffic, which peaks between 7:00 P.M. and 10:00 P.M. (prime time), is included. Figure 10.7 shows approximate traffic patterns for different traffic types.

We can see from the figure that each type of traffic has a different usage pattern. In designing a network, the busy hours for each pattern should be considered. For example, if traffic peaks around 11:00 A.M., the network needs to be able to accommodate it.

The following are design issues that must be addressed:

- Traffic engineering
- DSL technology selections
- Broadband network link design
- Broadband network node design

Each is described below.

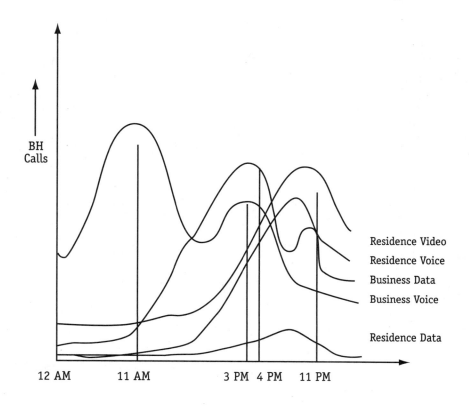

Traffic Engineering

The most important component when designing a network is estimating
the traffic, which becomes difficult to estimate when potential sources,
such as switched video, voice, personal communication service (PCS),
etc., do not yet exist. While some data are available, these new traffic
sources have made projection more guesswork than concrete engineering.
Here, we propose a simplified way to estimate different types of traffic.

In traffic analysis, the basic assumption is that all traffic can be categorized into voice continuous bit rate (CBR) or data variable bit rate
(VBR). If voice traffic is packetized, then it is treated as data with real-
time criteria.

Assume that video traffic belongs to one of the two categories. If
video traffic is packetized, it is VBR; if not, it is CBR with high band-

width and large holding time (compared with voice traffic holding time). Voice traffic can also be packetized to become VBR traffic. Usually, each traffic type has its own units. Voice traffic is measured in CCS (hundred call seconds, where the first C represents the Roman numeral symbol for 100) or *Erlangs*, depending on the country of origin. Data traffic is usually measured in packets, messages, frames, or cells. For simplicity, assume all types of traffic are converted into bits per second. Data traffic units are usually expressed in packets per second or bits per second, and video traffic units are usually bits per second.

The first step is to convert all traffic into bits per second. Usually, the available information on traffic is in bytes per day, but network designs are based on busy hour (BH) traffic. It is customary to assume BH traffic to be 25% of the day's traffic, depending on the country and the service type. Figure 10.8 shows a typical traffic source for a business customer (including home-office users). Figure 10.9 shows typical traffic sources for residential customers.

Figure 10.8

Business access traffic

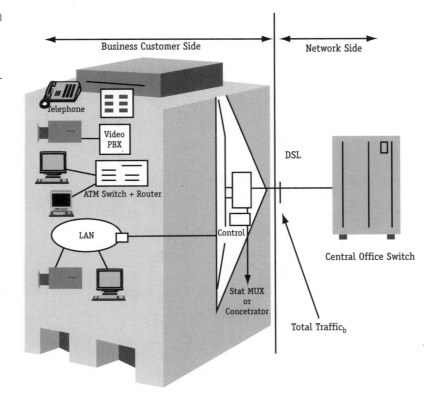

Figure 10.9
*Residential access
traffic*

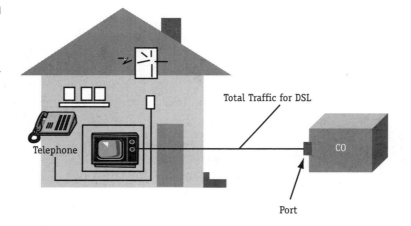

From Figures 10.9 and 10.10, the following can be derived. Note that the busy hours for voice, video, and data might be different.

Total traffic offered to the network (at the first point of interface) =

$$\sum_{i=1}^{n} (\text{Total residential traffic})_i \, |_{BH} + \sum_{i=1}^{m} (\text{Total business traffic})_i \, |_{BH}$$

where
n = Total number of residential customers
m = Total number of business customers
BH = Busy hour

Residential traffic $|_{BH}$ = Σ voice traffic $|_{\text{res BH}}$ + Σ data traffic $|_{\text{res BH}}$ + Σ video traffic $|_{\text{res BH}}$

where

res.BH = residential busy hour.

Business traffic $|_{BH}$ = Σ voice traffic $|_{\text{bus BH}}$ + Σ data traffic $|_{\text{bus BH}}$ + Σ video traffic $|_{\text{bus BH}}$

where

bus BH = business busy hour

Once traffic is estimated, all the traffic in BH is summed to estimate total offered traffic to the network. This estimate is based on extremes, such as peak traffic for data. The offered traffic varies with the usage pattern. Because the effective traffic changes with time, it is important to keep the user traffic information up to date. The network should be engineered toward the worst-case BH traffic.

Typical voice/data traffic distribution is as follows. Note that these are approximate distributions; distribution varies from application to application and customer to customer.

■ Offered traffic = 100%
■ Intra-premises = 70%
■ Extra-premises = 30%

Of the extra-premises traffic, 80% is intra-LATA and 20% is inter-LATA. For instance, switched video traffic for residential customers is usually intra-LATA (93%), where the traffic originates from the local video store server. These traffic distribution patterns can be used while designing the links and nodes in the public access network.

Access Technology Selection

Having estimated the traffic, the customer must decide on the best technology for access to the public network. Many factors exist in selecting the right technology. For the sake of simplicity, assume that the selection is based on technology capability. Table 10.1 shows the appropriate technology based on customer traffic.

Link or Transmission Network Design

One of the components of the network is the physical connection between the different switches or nodes in the network. In a broadband network, these links operate at very high speeds and carry huge amounts of information. These links are part of the transmission system of a network, which carries the traffic from point A to point B error-free, as shown in Figure 10.10.

Table 10.1

Access technology for different traffic rates

Traffic type	Effective BH traffic in b/s in the access line*	XDSL variation	Layer 2 and 3 technology
Voice only	< 64 Kbps	IDSL,ISDN	2B1Q;ISDN;Circuit
Voice + data	< 1.544 Mbps	HDSL,ADSL, SDSL, ADSL Lite	ATM, ADPCM,PCM, LD-CELP
Data	<1.544 Mbps	DSL,SDSL, ADSL	IP, ATM, FR
Data	>1.544Mbps– <45 Mbps	ADSL	IP, ATM
Voice + data	>1.544 Mbps	ADSL,VDSL	ATM, IP, ADPCM
Voice + data + video	>1.544 Mbps	VDSL	ATM, MPEG2, IP, ADPCM, LD-CELP

* Including allocation for uncertainties.

Figure 10.10

Typical transmission network

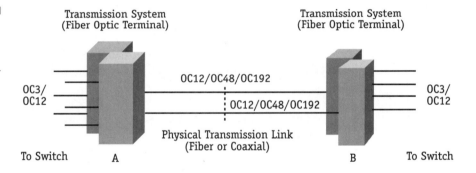

It is easy to estimate the traffic on the link if it originates at A and terminates at B. Because points A and B are usually several hundred miles apart in an access network, care must be taken that the link is reliable and error free. In a real-world network, not all nodes are fully connected as direct links. Usually, each node is connected to two or three adjacent nodes. To reach other nodes, the traffic must traverse the intermediate nodes that connect the source and destination node. Thus, to size the intermediate link, all traffic passing through the link in the BH must be taken into consideration. The most critical portion of the access network is the link between the customer and the network.

Because only one link exists from the customer premises to carry the traffic to the network, the capacity of the link should be designed based on the effective sum of *all* the traffic. Care must be taken to consider the utilization of the link caused by each type of traffic. Figures 10.8 and 10.9 showed typical sources of traffic from a business and residential customer, respectively, in a DSL network environment. Table 10.2 shows the typical traffic characterization required in designing the access network.

Assume each device at the customer premises generates x_1–x_n Mbps 1_{BH} traffic, which is calculated based on the traffic characteristics in Table 10.2.

Table 10.2

Broadband traffic characteristics

Service	Peak Rate (b/s)	Duration (hrs)	Burstiness	BH call attempts
NTSC video	45 Mbps	0.25	1.70	4
HDTV	150 Mbps	0.25	1.35	4
Voice	8–64 Kbps	0.05	1	2
Videoconferencing	1.5 Mbps	0.67	1	2
Imaging	1 Mbps	0.25	15.60	3
File transfer	1.5 Mbps	0.0056	15	6
Transaction data	100 Kbps	0.0083	200	20

Reference: BISDN resource management by J.Burgin, North Holland, 1990

Therefore, total offered traffic is

$$\sum_{i=1}^{n} \chi i \text{ Mbps } 1_{BH} = K$$

where
n = total number of traffic sources
i = traffic source
K = total offered traffic

A certain percentage of the traffic usually remains within the premises. The total traffic leaving the premises is calculated as shown below.

Let A be the percentage of traffic leaving the premises.

Total traffic leaving the premises = K * A = Z Mbps 1_{BH}

Using the traffic values from Table 10.1, one can find the appropriate technology suitable for access to the offered traffic. It is necessary to predict the customer's traffic over the next five to 10 years. Then, using the traffic numbers, the required access technology can be determined from the table. In the real world, economic, regulatory, and actual applications define the selection of DSL technologies. These calculations enable the carrier/customer to select the appropriate access technology and speed required by the traffic. The speed is the link-access speed required from the customer premises to the network node.

To design the rest of the links in the public access network, the effective utilization of each DSL link (customer link) must be calculated. Summing across the busy hour traffic of all customers provides the effective trunk-side link size. The trunk-side links are typically connected to more than one node, usually two or three for redundancy purposes. In that case, the traffic must be distributed to reach its destination via the shortest path, which is based on either distance or link utilization. Figure 10.11 shows the relationship between the access side and trunk side, its traffic distribution, and how the trunk traffic is calculated in the network.

Figure 10.11
Trunk-side traffic on the access network

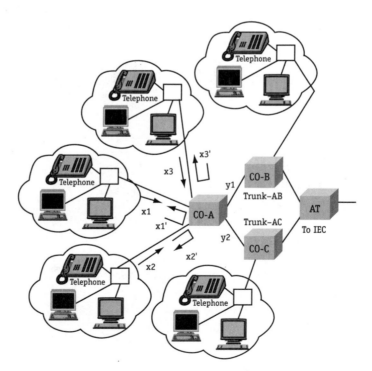

Total traffic = x1 + x2 + x3 = K Mbps

Of these, $x_1' + x_2' + x_3'$ Mbps = M terminate within the CO-A switch

Total trunk traffic on node

CO-A = $y_1 + y_2$ = N traffic units or CO-A = K − M = N traffic units

Once the traffic for the node is known, especially the traffic between each node pair, one can estimate the traffic on the link between the pair. This calculation is simple if a direct link exists between the two nodes, because the traffic that passes through the link is the same as the traffic between the node pair. In the real world, however, no network is fully connected, even in the broadband backbone network. Thus, the traffic must pass through many nodes via links to reach its destination. Figure 10.12 shows a typical four-node network. If the traffic from A to D is given, then the network is connected as shown in the figure.

Figure 10.12

Example network

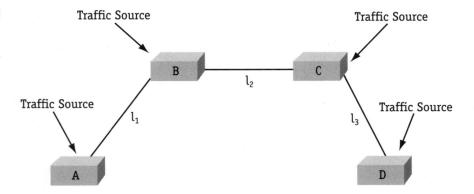

The network consists of four nodes: A, B, C, and D, and three links l_1, l_2, and l_3. For traffic from node A to reach node D and vice versa, it must pass through nodes B and C via links l_1, l_2, and l_3 before reaching destination node D. The traffic of link l_1 is

Traffic in one direction = traffic (AD)
Traffic in other direction = traffic (DA) + traffic (BA) + traffic (CA)

Thus, the size of link l_1 should be based on the maximum of two directions. Similarly, for links l_2 and l_3, the total traffic in both directions of each link must be estimated, and the link size should be the maximum of the traffic in two directions. When the link is designed, all the intermediate traffic must be taken into consideration. There should also be sufficient room for uncertainties—some extra capacity must be allocated for each link. This variable is simple if the network is small, but in today's complex broadband network, link sizes must be optimized without incurring additional cost and while taking every possible uncertainty into consideration. Link size depends on how much traffic is routed via that link. Routing depends on the type of algorithm used. Many algorithms are based on distance, cost, or a combination of both. For broadband networks, the routing algorithm must be dynamic and should be adjustable, depending on the situation of the traffic at the time of routing.

In most cases, carriers use existing routing methods unless they are convinced that a new algorithm is easy to use and cost effective. Broadband networks require dynamic routing algorithms because of the bandwidth-on-demand type of traffic. In dynamic routing algorithms, routing decisions reflect changes in traffic patterns. Because we do not know the exact broadband traffic between each pair of nodes, it is crucial to estimate the traffic correctly. Without knowing the traffic between the nodes, it is difficult to perform routing. In the case of broadband networks, because of uncertain traffic patterns, the best one to use is the distributed routing algorithm, which uses dynamic routing. In distributed routing, each node periodically exchanges explicit routing information with each of its neighbors.

Once the traffic between the node pair is available, we can find the route taken by the traffic under various conditions using any routing algorithm. It is possible to estimate the link size from the results of the routing algorithm. The link is sized considering the loading factors during the busy hour. Usually, it is assumed that the busy-hour traffic is 80% of the link capacity. In other words, almost 80% of the link is utilized at any given time. Thus, the link bandwidth in one direction (we can assume the same bandwidth in the opposite direction) can be given as:

Link bandwidth $(1_{ij}) = \sum\limits_{j=1;\ i=1}^{j=m;\ i=n}$ point to point traffic from i to j + transit traffic between i and j = 20% loading factor

where

i, j = Node on the network
l_{ij} = link between nodes i and j

Total link capacities are obtained by summing all the individual link capacities. This does not represent the total traffic traversing the network. Sufficient bandwidth is reserved in the link to handle the uncertainties so that almost all traffic can be rerouted without bringing down the network. The link capacity should be designed to existing traffic, and handle future traffic whose characteristics are not very predictable. All the traffic calculations are currently based on trials and theoretical research work. These theories and extrapolation of existing traffic by the public carriers can help estimate traffic.

Thus, to summarize the steps in designing the broadband backbone link:

1. Calculate the total access traffic on a per-node basis.
2. If node-to-node traffic is not available, use the gravity method to estimate node-to-node traffic.
3. Use a routing algorithm to route the traffic in the given network topology.
4. Sum all the traffic on a link-by-link basis to calculate link traffic.

Add "fudge factors" for uncertainties, such as 20% of additional bandwidth to provide an estimated link capacity.

On the trunk side, the effective utilization of the link is usually 70% to 80% of link capacity, including overhead. On the access side, the utilization is less than 10%, on average.

Broadband DSLAM Node Design

Figure 10.13 shows a typical broadband access node, including the access side, access ports, trunk side, and trunk ports.

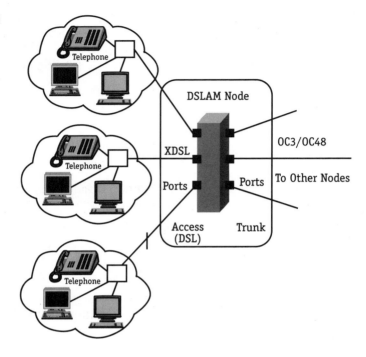

The broadband DSLAM access node design is based on three main factors:

- Total DSL ports required (access links)
- Total trunk ports required (trunk links)
- Total traffic offered to the switch, i.e., the sum of all the port speeds (total ports).

Thus, the DSLAM size is determined by its traffic-handling capability and number of ports (line side + trunk side). These parameters can be calculated in the following way:

Total broadband access node size (in Mbps) = 'S' traffic per port laccess and trunk) l_{BH}

Broadband access node size in number of ports = 'S' DSL links + 'T' trunk links

The number of DSL ports is usually higher than the number of trunk ports because of the lower utilization of the link (<5%) on the DSL side compared with the trunk side. Trunk-side links are more effectively utilized (70% to 80%), and the link speed on the access side is usually less than that of the trunk side. Typically, the ratio of the number of access ports to the number of trunk ports is 10:1. In design of the node, plans for future expansion in terms of switch capacity and additional ports must be considered.

Broadband Access Network Topologies

The typical topology for a telecommunications or computer network is a star or double star, as shown in Figure 10.14.

Figure 10.14
Star and double star topologies

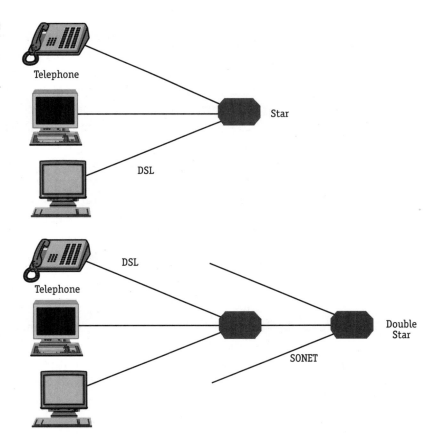

Because of the requirements set forth in the design of broadband communications, as well as the availability of new technologies such as fiber for high capacity, numerous topologies are under consideration for access networks. Some of these topologies are:

- Physical star/logical star
- Physical ring topology
- Logical star/physical ring
- Physical ring/logical ring
- Bus/star and bus/bus.

We describe the most popular architectures in the following subsections.

Physical Star/Logical Star Topology

Figure 10.15 illustrates physical star/logical star topology. This topology offers maximum potential bandwidth to each customer and hence provides maximum flexibility in service provisioning and future upgrading. This architecture simplifies the administration of bandwidth requirements for individual customers and the diagnosis and partitioning of faults, so life-cycle operational costs are low. The dedicated facilities of this architecture provide a high degree of security and limit unauthorized access, but can result in higher initial costs. This topology should be considered when potential demand exists for a full range of services, when high reliability/security is required, or when the distances are short and initial cost is not a sensitive parameter.

Physical Ring/Logical Star

This topology is physically interconnected as with a ring, but a wavelength or time slot is allocated for each node. The initial cost for this architecture might be lower than that for a physical star because of the sharing of fiber, but the electronics cost can be higher because of the higher bandwidth requirements. This architecture takes advantage of the logical-star topology by providing security and privacy.

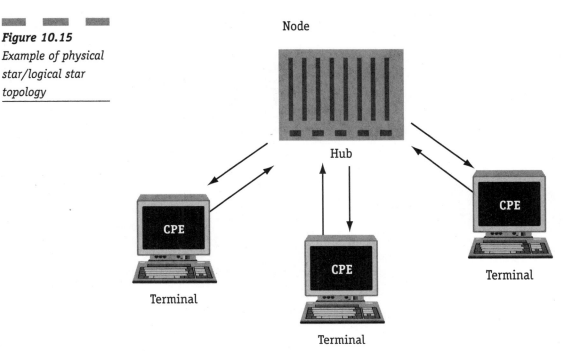

Figure 10.15

Example of physical star/logical star topology

Physical Star/Logical Ring

This topology has the same physical layout as the physical star/logical star, as shown in Figure 10.16. At the central node, however, the receive fiber from each node is connected to the transmit fiber of another node. Because information passes through all nodes and links on the ring, bandwidth is more difficult to administer, higher-speed interfaces are required, and fault partitioning is more difficult.

In this topology, the number of switch ports can be reduced. For example, the nodes at either end are connected to the switch directly, and the other nodes are connected to a patch panel (cheap manual connector board) and looped back. In the case of broadcast video, the signal can be broadcast to one CPE and can flow to other CPEs without utilizing switch resources.

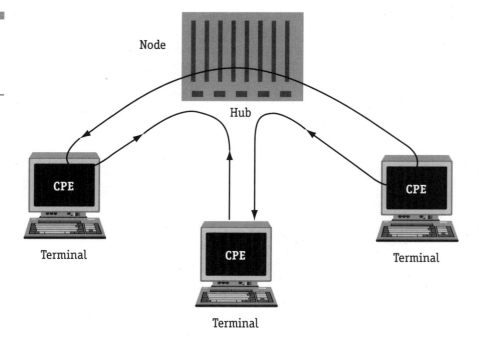

Figure 10.16
*Example of physical
star/logical ring
topology*

Summary

Design of the broadband DSL-based access network poses interesting design issues mainly because of the types of traffic carried via common facilities. Because each type of traffic has its own characteristics with respect to the parameters such as QoS, end-to-end delay, and security, it must be treated accordingly. To meet these requirements, traffic, link, and node must also be considered separately and designed with care. In addition, the appropriate access topology must be selected to meet both short- and long-term requirements based on external environments such as market requirements, regulations, and product availability.

Telecom Operating Companies Pursuing DSL

Company Name	Internet Address	Major Service Offering
ACC Corp.	http://www.acccorp.com/	Long distance, international, and local service telephony provider
AGT	http://www.agt.net/agt1/agt.home.html	Telecom operating in Alberta, Canada
AirTouch Communications	http://www.airtouch.com/	Wireless service provide
Ameritech	http://www.ameritech.com/	News, people, speeches, school, current commercials
AT&T	http://www.att.com/	Prototype with history and '93 annual report
BC Telecom	http://www.bctel.com/	Provider of voice, data, and enhanced telecomm. to British Columbia
Belgacom	http://www.belgacom.be/	Belgium
Bell Atlantic	http://www.ba.com/	Legislation, archives, history, exec profiles and speeches, media info
Bell Canada	http://www.bell.ca/	Ontario and Quebec; English and French language
BellSouth	http://www.bellsouth.com/	US RBOC serving southeastern states
Bezeq	http://bezeq.macom.co.il/	Israel
British Telecom (BT)	http://www.bt.com/	Largest United Kingdom operator
Cable & Wireless	http://www.mercury.co.uk/othersit.html	Competive long distance service provider in United Kingdom and international
Cincinnati Bell	http://www.cinbelltel.com/	Regional Bell operating company in Cincinnati
DDI	http://www.ddi.co.jp/	Japanese long distance company
Deutsche Telekom	http://www.dtag.de/english/index.htm	Germany
ED TEL	http://www.edtel.com/	Telco for City of Edmonton, Alberta, Canada
Electric Lightwave, Inc.	http://www.eli.net/	Vancouver, Washington-based full service business telecommunicacions provider
EnerTel	http://www.enertel.nl/	Telecom operator in the Netherlands
Finnet Group	http://www.finnet.fi/	Finland
France Telecom	http://www.francetelecom.fr/	France
Frontier Corporation	http://www.frontiercorp.com/	Formerly known as Rochester Telephone; provides telecom service in United States with new name.

continued on next page

Company Name	Internet Address	Major Service Offering
GTE Corporation	http://www.gte.com/	Largest independent local operating company. Now provides local, long distance, and other telecom services
Helsinki Telephone Company	http://www.hpy.fi/	Finnish telecom company
Hongkong Telecom	http://www.hkt.net/index.shtml	Hong Kong telecom provider
ITC Long Distance	http://www.itctelecom.com/	Preferred provider of long distance for non-profit trade associations
KDD	http://www.kdd.co.jp/	Long distance provider in Japan
Korea Telecom	http://melon.kotel.co.kr/ktrl/koreatelecom.html	Korea
MATAV	http://www.matav.hu/index_e.html	Hungary
MAXITEL GROUP	http://www.maxitel.pt/	Portuguese private telecom operating company
MCIWorldcom	http://www.wcom.com/	Long distance carrier
McLeod USA	http://www.mcleod.net/	Local phone service, long distance, voice mail, paging, Internet access, and e-mail
Mercury Communications	http://www.mercury.co.uk/	Second largest long distance provider in United Kingdom
MTS NETCOM	http://www.mts.mb.ca/	Manitoba Telephone System, Canada
NBTel	http://www.nbnet.nb.ca/	New Brunswick, Canada; English and French language
NewTel Communications	http://enterprise.newcomm.net/ntc/	Newfoundland, Canada
Nippon Telephone and Telegraph (NTT, Japan)	http://www.ntt.co.jp/	Japan's telephone operator
NorthWestel Inc.	http://www.yukonweb.wis.net/business/nwtel/	Northern Canada
NYNEX	http://www.nynex.com/	Regional operating telephone company in the United States. Now merged with Bell Atlantic
NYNEX CableComms	http://www.nynex.co.uk/nynex/	UK telephony and cable provider
OTE	http://www.gsc.net/business/ote/ts.htm	Greece
Pac-West Telecomm, Inc.	http://www.pacwest.com/	Full service telephone company serving California with special services for ISPs

continued on next page

Company Name	Internet Address	Major Service Offering
PlusNet	http://www.plusnet.ch/	Germany
Portugal Telecom	http://www.telecom.pt/	Portugal
PTT Telecom	http://www.ptt-telecom.nl/	Netherlands
Quebec Telephone	http://www.quebectel.qc.ca/ qtel/qt0000ag.htm	Canada
Radiotel	http://www.radiotel.ro/	Romania
Rogers Communications, Inc.	http://www.rogers.com/	Wireless, long distance, cable systems, and multimedia in Canada
SaskTel	http://www.sasktel.com/	Sasketchewan, Canada
SFR	http://www.sfr.fr/	Société Française du Radiotéléphone; in French
Singapore Telecom	http://www.singtel.com/	Singapore national telephone operator
Southern New England Telephone	http://www.snet.com/	SNET
Southwestern Bell	http://www.swbell.com/	Texas, Kansas, Arkansas, Oklahoma, and Missouri
Sovam Teleport	http://www.sovam.com/	Russia
Sprint/United Telephone—Florida	http://www.utelfla.com/	
Sprint Communications Company	http://www.sprint.com/	
SPT Telecom	http://www.spt.cz/html/ indexa.htm	Czech Republic
Stentor	http://www.stentor.ca/	Alliance of Canadian operating companies
Swiss PTT	http://www.vptt.ch/	Swiss national telephone carrier
Tampere Telephone	http://www.tpo.fi/	Finland
Tele Danmark	http://www.teledanmark.dk/	In Danish and English
Tele2	http://www.tele2.se/	Sweden
Telebec	http://www.telebec.qc.ca/	Quebec, Canada
Teleboss (Pty Ltd.)	http://www.teleboss.co.za/	Telephony solution provider in South Africa
Telecom Argentina	http://www.telecom.com.ar/	Argentina
Telecom Eireann	http://www.telecom.ie/	
Telecom Eireann (Ireland)	http://www.broadcom.ie/ telecom/dupjmc/teprofile.html	Ireland
Telecom Finland	http://www.tele.fi/	Finland

continued on next page

Company Name	Internet Address	Major Service Offering
Telecom Italia	http://www.telecomitalia.interbusiness.it/	Italy
Telecom Malaysia	http://www.telekom.com.my/	Malaysia
Telecom New Zealand	http://www.telecom.co.nz/index.html	New Zealand
Telecom Poland	http://www.tpsa.pl/	Poland
Telecom Portugal	http://www.telecom.pt/uk_pages/uk_index_pt.htm	Portugal
Telecom UK	http://telecom.co.uk/	UK telecommunications and phone card company
Telefonica de Argentina	http://www.telefonica.com.ar/	Argentina
Telefonica de Espana	http://www.telefonica.es/	Spanish PTT
Telefonica del Peru	http://www.telefonica.com.pe/	Peru
Teleglobe Canada (English language)	http://www.teleglobe.ca/	International telecommunication carrier
Telenor	http://www.telenor.no/	Norway: local, long distance, satellite and cellular; Norwegian only
Telenordia	http://www.telenordia.se/	Sweden
Teleport Communications Group	http://www.tcg.com/	Competitive local exchange carrier
Telepost (Norway)	http://web.telepost.no/	Norwegian only
Telesat Canada	http://www.telesat.ca/	
Televerket Research Institute (Norway)	http://www.nta.no/xtf/xtf.html	Norwegian only
Telfort	http://www.telfort.com/	Dutch long distance operator
Telia	http://www.telia.se/	Swedish "national" telephony company
Telkom Indonesia	http://www.telkom.co.id/	Indonesia Telecom Company
Telkom SA	http://www.telkom.co.za/	South Africa
TELMEX	http://www.telmex.com.mx/	Mexico
Telstra Corporation, Ltd.	http://www.telstra.com.au/	Australia
Turku Telephone	http://www.ttl.fi/	Finland
US WEST	http://www.uswest.com/	US regional operating company serving midwestern states

Regulators

Company Name	Internet Address	Major Service Offering
Australian Telecommunications Authority	http://www.austel.gov.au/	The Australian telecommunications regulator
BAKOM - OFCOM	http://www.admin.ch/eved/m/bakom/main.html	The Swiss telecoms regulator
BAPT	http://www.bapt.de/English/default.htm	Bundesamt für Post und Telekommunikation
Bundesministerium für öffentliche Wirtschaft und Verkehr	http://iis.joanneum.ac.at/BMWV/Telekom/	Austrian Ministry of Public Works and Communications
Bundesministerium für Post und Telekommunikation	http://www.government.de/inland/ministerien/post.html	German regulator for telecoms
Canadian Radio-television and Telecommunications Commission	http://www.crtc.gc.ca/	CRTC–Canadian regulatory body for telecommunications
Deparpostel	http://www.telkom.go.id/postel.htm	Departement of Tourism, Post, and Telecommunications in Indonesia
The Department of Trade and Industry	http://www.dti.gov.uk/	UK regulatory responsibility in conjunction with OFTEL
Federal Communications Commission	http://www.fcc.gov/	US telecom regulatory commission
Hoofddirectie Telecommunicatie en Post	http://www.minvenw.nl/hdtp/home.html	Dutch telecommunications and post regulator
Industry Canada–Telecommunications/Spectrum Management	http://info.ic.gc.ca/ic-data/telecom/telecom-e.html	Canada
Ministére francais poste, telecommunication, espace	http://www.telecom.gouv.fr/	French telecommunications regulator
Ministry of Information n and Communicatio	http://www.mic.go.kr/	Republic of Korea
Ministry of Posts and Telecommunications	http://www.mpt.go.jp/	Japan
ISPO	http://www.ispo.cec.be/	The Information Society Project Office of the European Commission and general information from EC DGXIII
I'M Europe	http://www.echo.lu/	Telecoms and Information Industries
New Zealand Ministry of Commerce	http://www.govt.nz/ps/min/com	Includes information on communications policy, spectrum management, and IT usage statistics in New Zealand

continued on next page

Company Name	Internet Address	Major Service Offering
OFTA	http://www.ofta.gov.hk/	Office of the Telecommunications Authority—the Hong Kong regulator
OFTEL-	http://www.oftel.gov.uk/	UK telecom regulator
Post & Telestyrelsen	http://www.pts.se/	Swedish telecoms regulator
Secretaria de Communicaciones y Transportes	http://www.sct.gob.mx/	Mexico's telecoms regulator
Telehallintokeskus– Telecommunications Administration Centre	http://www.thk.fi/	Finnish regulatory agency under the Ministry of Transport and Communications
Telestyrelsen– National Telecom Agency Denmark	http://www.tst.dk/	Denmark
WTO Protocol on World Trade in Telecoms	http://www.mft.govt.nz/ Business/File/vol3no1.htm	New Zealand government summary of WTO agreed protocol on world trade in telecommunications

Telecommunications
Journals, Magazines and Other Electronic Media

Journals and Magazines

Access (monthly)
Telecommunications Research
P.O. Box 12038
Washington, DC 20005

ACM—Transactions on Information Systems (quarterly)
Association for Computing Machinery
11 West 42nd St.
New York, NY 10036

AT&T Technical Journal (bimonthly)
AT&T
550 Madison Ave.
New York, NY 10022

AT&T Technology (quarterly)
Richard A. O'Donnell
550 Madison Ave.
New York, NY 10022

British Telecom Journal (quarterly)
British Telecom
81 Newgate St. Fl.
A2 London EC1A 7AJ
England

Business Communications Review (monthly)
BCR Enterprises, Inc.
950 York Rd.
Hinsdale, IL 60521

COMSAT Technical Review (semiannual)
COMSAT Corp.
22300 COMSAT Dr.
Clarksburg, MD 20871

Canadian Communications (semimonthly)
MACLEAN-Hunter Ltd.
Business Publication Div.
MACLEAN-Hunter Bldg.
777 Bay St.
M5W 1A7 Toronto, Ontario
Canada

Communication (weekly)
NTIS
5285 Port Royal Rd.
Springfield, VA 22161
Congressional Report on

Communications (semimonthly)
New Media Publishing
1117 N. 19th, #200
Arlington, VA 22209

Data Communications (monthly)
McGraw-Hill
1221 Ave. of the Americas
New York, NY 10017

Data Communications Management (bimonthly)
Auerbach Publishers
1 Penn Plaza
New York, NY 10119

Datamation (monthly)
Reed Publishing Co.
44 Cook St.
Denver, CO 80206

Datapro Reports on Telecommunications (monthly)
Datapro Research Corp.
600 Delran Pkwy.
Delran, NJ 08075

European Telecommunications (semimonthly)
Probe Research Inc.
3 Wing, #240
Cedar Knolls, NJ 07927

FCC Rulemaking Reports (biweekly)
Commerce Clearing House
4025 W. Peterson Ave.
Chicago, IL 60601

FCC Week (weekly)
Capital Publications
1101 King St., #444
Alexandria, VA 22314

Fiber to Home (biweekly)
Information Gatekeepers Inc.
214 Harvard Ave.
Boston, MA 02134

Fiber and Integrated Optics (quarterly)
Taylor and Francis, Inc.
79 Madison Ave. #1110
New York, NY 10016

Focus on Communications (monthly)
Business Communications
3190 Miraloma Ave.
Anaheim, CA 92806

Global Communications (monthly)
Cardiff Publishing Co.
6300 South Syracuse Way
Englewood, CO 80111

Globe Communications IEEE (annually)
IEEE
345 East 47th St.
New York, NY 10017

GTE Telenet Packet (monthly)
GTE Telenet
Communication Corp.
12490 Sunrise Valley
Reston, VA 22096

IEEE Communications Magazine (monthly)
Institute of Electrical and Electronics Engineers
345 E. 47th St.
New York, NY 10017

IEEE Journal on Selected Areas in Communications (bimonthly)
Institute of Electrical and Electronics Engineers
345 E. 47th St.
New York, NY 10017

IEEE Network (bimonthly)
Institute of Electrical and Electronics Engineers
345 E. 47th St.
New York, NY 10017

IEEE Spectrum (monthly)
Institute of Electrical and Electronics Engineers
345 E. 47th St.
New York, NY 10017

IEEE Transactions on Communications (monthly)
Institute of Electrical and Electronics Engineers
345 E. 47th St.
New York, NY 10017

International Telecommunications Union Operational Bulletin (monthly)
International Telecommunications Union
Place de Nation, CH1211
Geneva 20, Switzerland

ISDN Report (semimonthly)
Probe Research Inc.
3 Wing Dr. #240
Cedar Knolls, NJ 07927

ISDN User (bimonthly)
Information Gatekeepers Inc.
214 Harvard Ave.
Boston, MA 02134

Lightwave (monthly)
Penn Well Publishing Co.
P.O. Box 987
1 Technology Park Dr.
Westford, MA 01886

List of International Telephone Routes (annually)
International Telecommunications
Union
Place des Nation, CH1211
Geneva 20, Switzerland

Military Fiber Optics Communications (biweekly)
Information Gatekeepers Inc.
214 Harvard Ave.
Boston, MA 02134

Mobile Communications Business (monthly)
Phillips Publishing Inc.
7811 Montrose Rd.
Potomac, MD 20854

Network World (weekly)
IDG Communications
161 Worcester Rd.
Framingham, MA 01701

Networking Management (monthly)
Penn Well Publishing Co.
P.O. Box 987
1 Technology Park Dr.
Westford, MA 01886

NTT Topics (quarterly)
Ruder, Finn, and Rotman
110 E. 59th St.
New York, NY 10022

Perspective on AT&T and BCR Products and Marketing (monthly)
BCR Enterprises, Inc.
950 York Rd.
Hinsdale, IL 60521

Planning Guide 1 Inter-LATA Telecommunications Rates and Services
(monthly)
McGraw-Hill
50 S. Franklin Tpk.
Ramsey, NJ 07446

Planning Guide 2 Inter-LATA Telecommunications Rates and Services
(monthly)
McGraw-Hill
50 S. Franklin Tpk.
Ramsey, NJ 07446

*Planning Guide 3 Value Added Networks and Data Private Line
Telecommunications Rates and Services* (monthly)
McGraw-Hill
50 S. Franklin Tpk.
Ramsey, NJ 07446

Soviet Journal of Communications Technology and Electronics
(16 per year)
John Wiley/Scripta Technica
7961 Eastern Ave.
Silver Spring, MD 20910

Telecom Bulletin (semiannual)
Fleural Management Co.
Ottawa, Ontario
Canada

Telecom Insider (monthly)
International Data Corp.
Framingham, MA 01701

Telecom Today (monthly)
British Telecommunications
London, England

Telecommunication Journal (monthly)
International Telecommunications Union
Place de Nation, CH1211
Geneva 20, Switzerland

Telecommunications (monthly)
Horizon House—Microwave Inc.
685 Canton St.
Norwood, MA 02062

Telecommunications Abstracts (10 per year)
R.R. Bowker, EIC
New York, NY

Telecommunications Alert (monthly)
Management Telecommunications Pub
New York, NY

Telecommunications Authority of Singapore Telecoms Annual Report
(annual)
Telecommunications Authority of Singapore
Singapore

Telecommunications Journal of Australia (3 per year)
Telecommunications Society of Australia
Melbourne, Australia

Telecommunications (Norwood) (monthly)
Horizon House
Norwood, MA 01105

Telecommunications (Potomac) (quarterly)
Phillips Publishing Inc.
7811 Montrose Rd.
Potomac, MD 20854

Telecommunications Product plus Technology (monthly)
Penn Well Publishing Co.
P.O. Box 987 1 Technology Park Dr.
Westford, MA 01886

Telecommunications Product Review (monthly)
Marketing Programs and Services Group
1350 Piccard Dr.
Rockville, MD 20850

Telecommunications Reports (weekly)
Telecommunications Reports
1036 National Press Bldg.
Washington, DC 20045

Telecommunications Sourcebook (annual)
North American Telecommunications Association
1036 National Press Bldg.
Washington, DC 20045

Telecommunications Systems and Services Directory (irregular)
Gale Research Inc.
835 Penobscott Bldg.
Detroit, MI 48226

Telecommunications Week (weekly)
Telecommunications Reports
1036 National Press Bldg.
Washington, DC 20045

Teleconnect (monthly)
Telecom Library Inc.
12 West 21 St.
New York, NY 10010

Telematics and Informatics (6 per year)
Pergamon Press Inc.
Journal Division
Maxwell House
Fairview Park
Elmsford, NY 10523

Telephony (weekly)
Intertec Publishing Co.
55 East Jackson Blvd.
Chicago, IL 60604

U.S. Telecom Digest (23 per year)
Capitol Publishers, Inc.
1101 King St. #444
Alexandria, VA 22314

Journals and Other Electronic Media

ADSL Resource Guide Resource of information about DSL and cable modems.

ATM Component Review (http://www.tiac.net/users/kma) A quarterly publication which contains detailed technical information on components available for use in ATM products.

Bellcore's DIGEST of Technical Information (http://www.bellcore.com/DIGEST/)

Business Communications Review (http://www.bcr.com/)

Cable World (http://www.mediacentral.com/index/CableWorld)

Communications Standards Review (http://www.csrstds.com/): publishes journals reporting on formal telecommunications standards in the ITU, ETSI and TIA.

Communications Standards Review (http://www.csrstds.com) ITU, TIA wireline, wireless and fiber optic work-in-progress.

Communications Week Interactive (http://www.emap.com/cwi/)

Communications Week Interactive (http://tw2.cmp.farm.barrnet.net/cw/current/default.htm) The page for corporate network managers.

Computer Telephony (http://www.computertelephony.com/ct/ct_home.html) The magazine for computer and telephone integration.

d.Comm (http://www.d-comm.com/) Daily news and features for information and communications workers, published by The Economist.

Data Communications Magazine (http://www.data.com/)

Dataquest (http://www.indiaworld.com:1500/subscribe/pubs/dq/current.html) Covers developments in infotech industry, trade, company information of relevance to the Indian and Asia/Pacific infotech industry.

David Fannin ADSL How To for Linux systems

Dan Kegel's ADSL Page ADSL Information page

DBS Online (http://www.dbs.digifix.com/DBS/index.html)

ICB Toll Free News (http://www.icbtollfree.com/) Regulatory and telecom industry reporting.

IEEE Communications Magazine (http://www.ieee.org/comsoc/commag.html)

IEEE Internet Computing (http://www.computer.org/pubs/internet/) Bimonthly magazine covering Internet-based computer applications and technologies.

Indepth Magazine (http://www.eclipse.net/~indepth/index.html) Coverage of the global telecommunications market.

Indian Techonomist (http://www.c2.org/~rishab/techonomist/) Newsletter on India's information industry: computers, communications and broadcasting.

Interactive Age (http://www.wais.com/techweb/ia/current/off-the-pages.html)

Internet Australasia (http://www.interaus.net/magazine)

InternetTelephony (http://www.internettelephony.com/) Online home of *Telephony* magazine.

Network World Magazine (gopher://ftp.std.com/11/periodicals)

NII Scan (http://www.ncb.gov.sg/nii/scan.html) Journal from Singapore devoted to "Tracking NII Policies Worldwide."

NSF Network News (http://www.internic.net/newsletter/) Official newsletter on the evolution of the NSFNET and the Internet.

Satellite Journal International (http://itre.uncecs.edu/misc/sj/sj.html)

Technology In Government (http://www.plesman.com/tg/) Information systems and telecommunications.

Technology Online (http://www.tol.mmb.com) Technology Cyber-Magazine.

Technology Review (http://web.mit.edu/afs/athena/org/t/techreview/www/tr.html) MIT's magazine.

TechWeb (http://techweb.cmp.com/) Entry point to all CMP publications (e.g., "Communications Week", "Information Week").

tele.com (http://www.teledotcom.com/) Focus and analyze the new common business and technology issues.

tele.com (http://www.teledotcom.com/) Publication for telecom service providers.

Telecom A.M. (http://www.telecommunications.com/am/) Daily telecommunications information resource covering all aspects of the telecommunications industry—mobile data, local competition, regulation, online activities, interactive markets.

Telecom Digest (WWW interface) One of the oldest mailing lists covering telecom topics. Also appears as newsgroup comp.dcom.telecom.

Telecom Information Clearinghouse

Telecom Market Monitor Daily telecom newsletter featuring market column and company news.

Telecom Publishing Group (http://www.cappubs.com/tpg) Produces a range of titles in telecommunications.

Telecom Regulation (WWW interface) Mailing list on regulation of telecom (includes searchable archives).

Telecom Reseller Opportunities Magazine Information for companies that buy, sell, resell, and bundle telecom products and services.

Telecom Tribune English-language publication dealing with Japan's telecommunications industry.

Telecom Tribune (http://teltrib.techjapan.co.jp/yearmap.html) News, information, and statistics on Japanese telecoms.

Telecom Update Weekly summary of telecom news, focus on Canada.

TELECOM-CITIES Discussion List Practical and theoretical aspects of the changes advanced telecommunications and telematics are rendering on our urban centers.

TeleacomEuropa (http://www.telecoms.com/text/te-pubs.htm) Produces newsletters for the telecommunications industry.

Telecommunications Magazine (http://www.telecoms-mag.com/tcs.html)

Telecommunications Policy Journal on economics, politics regulation of telecoms, and information systems.

Telecommunications Reports International, Inc. (TRI) (http://www.tr.com/) News and information services for the communications, multimedia and electric utility industries.

Tele-Consumer Hotline Information to help consumers better understand the broad new array of communications products and services.

TeleGeography Home Page (http://www.telegeography.co./)

Teletronikk magazine (Norway)

Telemanagement Monthly review of telecommunications management issues in Canada.

Telephony (http://www.internettelephony.com/)

TELE-Satellit News Continuous newsfeed.

Televak Uitgeverij Trade magazine for cable, telecoms, and media.

The Technology Reporter Resource for those interested in computers, technology, the Internet, and telecommunications.

The Telecom Virtual Storefront Computer telephony, voice and fax processing, Internet telephony, and speech recognition.

Total Telecom (http://www.totaltele.com/) A daily news and information service for telecommunications professionals.

Total Telecom News and information about the global telecoms business.

Voice & Data (http://www.voicendata.com/) Indian communication magazine.

Washington Telecom Newswire (http://wtn.com/wtn/wtn.html)

World Telephone Numbering Guide Links and news regarding telephone numbering in various countries and regions.

World Wide Web Journal

World Wide Web, WWW An international, archival, peer-reviewed journal which covers all aspects of the world wide web.

ZD Internet Life Web-based magazine from Ziff-Davis Publishing.

ZD Internet Magazine MegaSite Internet and intranet computing.

Ziff-Davis Interactive Gateway to all Ziff-Davis online publications, including. *PC Magazine, PC Week, Inter@ctive Week*, etc.

Znews.com E-zine covering the latest Net and Tech issues.

Index

Note: Boldface numbers indicate illustrations.

ABOUT THE AUTHORS

PADMANAND WARRIER is the Marketing Program manager for Texas Instruments (TI). He is responsible for product planning and development with TI's Broadband Access Group on the convergence of broadband access and home networking. Previously, he was President and Principal Consultant of Thuriya Consulting. He is a recognized expert on power management issues, protocol stacks, and network architectures.

BALAJI KUMAR is Vice President of Systems Planning at Telehub Communication. Previously, he was a Senior Manager at MCIWorldcom, responsible for local network strategy and architecture. He has vast experience within the tecommunications industry in the areas of network planning, process reengineering, and strategic development in the broadband environment. He is the author of *Broadband Communications, Signature Edition* (McGraw-Hill, 1998).